Springer Tracts in Mechanical Engineering



For further volumes:
http://www.springer.com/series/11693

Jingshan Zhao • Zhijing Feng • Ning Ma • Fulei Chu

# Design of Special Planar Linkages

Springer

Jingshan Zhao
Zhijing Feng
Ning Ma
Fulei Chu
Tsinghua University
Beijing
China, People's Republic

ISSN 2195-9862 ISSN 2195-9870 (electronic)
ISBN 978-3-662-51353-8 ISBN 978-3-642-38448-6 (eBook)
DOI 10.1007/978-3-642-38448-6
Springer Heidelberg New York Dordrecht London


Softcover reprint of the hardcover 1st edition 2014


Printed on acid-free paper

Springer is part of Springer Science+Business Media (www.springer.com)

# Preface

Linkages are widely used in our life and work. Almost all planar linkages can be decomposed with a number of equivalent four-bar linkages which are considered to be one of the fundamental mechanism forms. As the simplest closed chain mechanism, planar four-bar linkages are extensively used from mechanical engineering to aerospace engineering and even civil engineering.

This book focuses on the design of a kind of special planar linkages which consist of a number of pivot jointed scissorlike elements (SLEs) in one or more different planes. These planar linkages are overconstrained in structure but allow to stretch or compact in some directions; therefore, they have the advantages of both truss and mechanism. However, the kinetostatics and dynamics of these linkages are different from their existing counterparts of trusses or mechanisms. The statics, kinematics, and dynamics of such planar linkages are important topics in applications. We sincerely hope the book will be beneficial for mechanical engineers, robot design engineers, and researchers in other related fields.

Chapter 1 presents an overview of the pivot jointed planar linkages that will be investigated in this book. Some definitions, such as planar kinematic chain, mobility, planar linkages, workspace and singularity, design problems including statics design, kinematics design, and structural dynamics of planar linkages, are discussed in this chapter.

Chapter 2 develops a uniform position design equation for planar four-bar linkages. According to Burmester theory, a general planar four-bar linkage could trace five arbitrary accurate positions at most. However, modern design often requires a mechanism to precisely or approximately trace more positions. The text starts from proposing a uniform equation for position design and then provides a proof for the sufficiency and necessity. The algorithm can search the accurate results when there are such solutions for the specified $n$ precise positions; otherwise, the best approximate solutions will be found. This permits the engineers to directly obtain the planar four-bar linkages when the specified positions are prescribed.

Chapter 3 proposes a foldable stair which remains the walking conversions of human being and all advantages of concrete stair in civil engineering. The stair is discussed as a mechanism synthesis example with screw theory in accordance to the common concrete stairs between two floors. It consists of a number of identical deployable SLEs that form the staircases when expanded. The instantaneous screw of every stair represents the relative rotation and the rotation center of the stair. The actuator is a planar four-bar linkage which is synthesized in line with the two extreme positions of the stair, folded and unfolded.

Chapter 4 starts from the discussion of the number of points that a planar four-bar linkage could precisely trace and then proposes a noncircular gear-coupled five-bar linkage as a steering mechanism. It strictly follows the Ackermann steering rule compared with the existing four-bar steering linkages. Therefore, the method can be used to design any kind of planar five-bar linkage to satisfy a specified function of curves.

Chapter 5 introduces the theoretical foundation for the workspace of planar parallel mechanism under rotational actuations and the algorithms to search the reachable and the dexterous workspace. Then singularity workspaces of such mechanisms are presented with an application example of a planar 3-RRR linkage. The method and algorithm discussed in this chapter can be used to analyze other planar linkages.

Chapter 6 addresses the statics of some special planar linkages with the example of foldable stair proposed in Chap. 3. The statics of rigid-body system of the stair mechanism is first discussed. Because of the redundant constraints, the internal forces are examined by using the second Castigliano's theorem. And then the inner forces of every link are analyzed so that the engineers can execute an identical strength design.

Chapter 7 investigates the kinetostatics of overconstraint mechanisms which are made of double planar linkages. The mechanism is redundantly constrained in structure and therefore has both merits of high structural stiffness and strength of a truss structure and motion flexibility of a mechanism. The kinetostatics is discussed with the synthesis of a deployable wing frame and the lift mechanism.

Chapter 8 discusses a structural dynamics algorithm for foldable linkages based on transfer matrix. The foldable stair and deployable wing are all typical planar linkages which are made up of a number of identical units. Therefore, the dynamics of each link between every two adjacent revolute joints is precisely expressed by the transfer matrix of Euler-Bernoulli beam with the variables of boundary conditions of the joints. In this way, the structural dynamics of the whole structure can be developed by using the least variables compared with the traditional methods. In addition, this algorithm avoids the problem of the traditional transfer-matrix method that the number of variables greatly increases when there are a huge number of cross joints within a structure.

Chapter 9 proposes a foldable tower for supporting wind turbine whose height is changeable. The advantage of the flexibility to fold and unfold can be utilized for protecting the wind turbine when destructive weather comes. The dynamic equivalent stiffness of the foldable tower is investigated as the function of deployment angle of the actuator. To expand the applications of the foldable structures, multi-planar linkages and the topology changeable linkages are discussed at the end of this chapter.

Chapter 10 introduces the design of foldable mounting equipment for solar cell panels. The equipment consists of a number of identical brackets to fix the solar cell panels. Therefore, this chapter first investigates the geometry of each foldable unit and then focuses on statics of the structure and simulation experiments. When the bracket is deployed, the frame of each unit forms the space for fastening the solar cell panels through fixtures. The merit of the structure is that the whole structure occupies less space when completely folded. This is most convenient for transporting and manufacturing because of its less occupied space when folded. Meanwhile, it is also available to form arc-shaped or any spatial curved space panels by changing the positions of the middle joints of scissor units in assembly.

I am pleased to express my gratitude for the contribution of many teachers and colleagues whose work over the years has developed and clarified the design and analysis theory for overconstraint planar linkages. This book consists of results of the insight, commitment, and hard work by Jianyi Wang, Houlin Fang, Xiang Liu, Li Ye, Zhengfang Yan, Zheng Cai, and Huichan Zhao. I have also benefitted from the insight of my coauthors Zhijing Feng, Ning Ma, and Fulei Chu. In addition, I am grateful for the inspiration of Wenxiu Lu, Ketao Zhang, Guowu Wei, Ligang Yao, Daniel Martins, Jian S. Dai, Yuefa Fang, and J. Michael McCarthy.

Finally, I gratefully acknowledge the support of the Natural Science Foundation of China, the Natural Science Foundation of Beijing, the Foundation for the Author of National Excellent Doctoral Dissertation of China, and the Program for New Century Excellent Talents in University of Education Ministry of China and the support of State Key Laboratory of Tribology in Tsinghua University.

Tsinghua University, Beijing, China Jingshan Zhao

# Contents

# Chapter 1
# Design Requirements of Planar Four-Bar Linkages

Generally, a mechanism is a device that transforms motion and force, while a machine typically contains one or more mechanisms that are designed to provide significant forces and transmit significant power [1]. A mechanical system, or machine, generally consists of a power source and a mechanism for the controlled use of this power [2]. Mechanical design is a complex process, requiring many skills. It is an iterative process with many interactive phases [3]. In general, mechanical design often starts from mechanism design. In the design process, there is no clear-cut dividing line between mechanisms and machines. A number of links connected in sequence via joints form a kinematic chain. It is called a mechanism when the kinematic chain has definite motion under one or more actuations. Among the mechanisms, planar linkages play a very important role in mechanical engineering. Simply the best has become the nowadays design principle. Planar four-bar linkages as the simplest closed-chain mechanism are widely used from mechanical engineering to civil engineering and aerospace engineering. This book therefore focuses on some special planar mechanisms which typically consist of a number of four-bar linkages and discusses the position design, kinematic synthesis, workspace and singularity of planar parallel linkages, stiffness of the linkage structure, kinetostatics, and structural dynamics. Possible applications of these special planar linkages in mechanical engineering and civil engineering are illustrated via examples.

## 1.1 Analysis of Planar Linkages

A linkage is a collection of interconnected links; the physical connection between two of which is called a joint [2]. The analysis of planar linkages mainly focuses on the motion generation, position, velocity, and acceleration. Because planar linkages are the simplest but widely used mechanism form, the objective of motion generation of a planar linkage is to calculate the mechanism parameters required to

J. Zhao et al., *Design of Special Planar Linkages*, Springer Tracts in Mechanical Engineering, DOI 10.1007/978-3-642-38448-6_1,

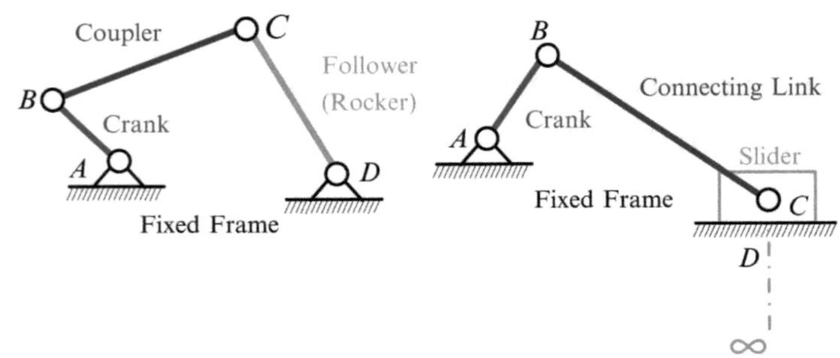


**Fig. 1.1** Two typical planar kinematic chains

achieve a set of specified poses which include the positions and orientations of a link [2]; the velocity and acceleration analysis can be accomplished with geometry as well as analytical methods. This section will briefly introduce some terms that will be used in what follows.

### *1.1.1 Planar Kinematic Chain*

A system of rigid bodies connected together in sequence by joints is called a kinematic chain. A kinematic chain is called closed if it forms a loop; otherwise it is called open if it does not form any loop. A kinematic chain is called serial kinematic chain if each link within the chain, except the first one and the last one, is only connected to two other links. A kinematic chain is a planar kinematic chain if all links of the chain are constrained to move in or parallel to a same plane. In planar kinematic chain, the mainly used joints are revolute joint (*R* joint) and prismatic joint (*P* joint). The revolute joint is often called rotary or pin joint and the prismatic joint is also called sliding joint. An *R* joint allows the two links connected to have a relative rotation. The joint used to attach a door to the frame is one of such examples. A *P* joint permits the two links connected to make a relative translation. The best example of this is the connection between a piston and a cylinder in an internal combustion engine. Figure 1.1 shows the two typical planar kinematic chains.

As shown in Fig. 1.1, a side link which can fully revolve relative to the frame is called a crank; correspondingly any link which does not fully revolve is called a rocker.

### *1.1.2 Planar Linkages*

When a rotary actuator is assigned to any joint of the left kinematic chain shown in Fig. 1.1, it forms a four-bar linkage as there is a determined motion for the link

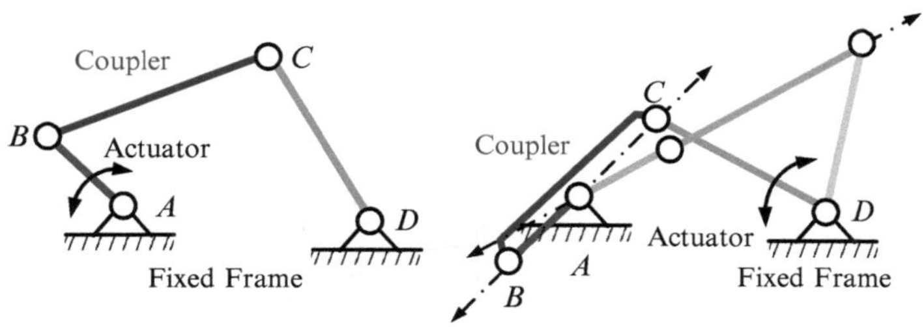


**Fig. 1.2** A same planar four-bar kinematic chain but with different actuators

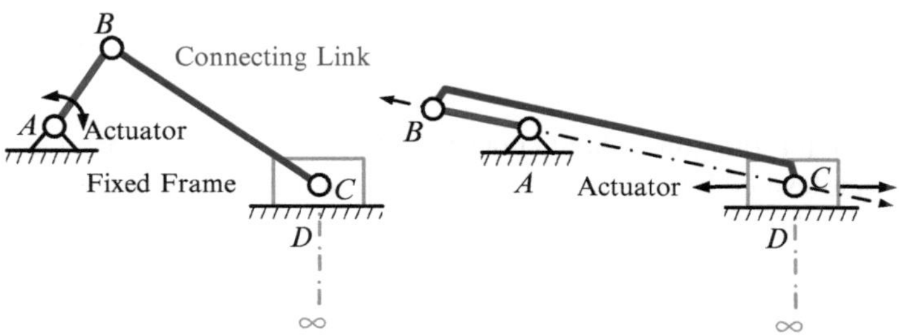


**Fig. 1.3** Identical slider-crank kinematic chains with different actuators

system. In a four-bar linkage, if the shortest side link revolves and the other side link oscillates, it is called a crank-rocker mechanism. In a four-bar linkage, if both of the side links revolve, it is called a double-crank mechanism. In a four-bar linkage, if neither of the side links can make a full rotation, it is called a double-rocker mechanism.

The linkage has continuous and smooth motion when the actuator is set on revolute joint $A$, shown in Fig. 1.2 (left). However, it has two positions where the force or motion from rocker $BC$ could not be transformed to the crank $AB$ as $AB$ and $BC$ are collinear when the actuator is set on revolute joint $D$ which are shown in Fig. 1.2 (right). These two positions are called dead points. Although the actuator could not transfer any force or motion to the follower at any dead point, it could be overcome through inertia of the linkages in applications.

The same is true for the second slider-crank kinematic chain. The linkage has continuous smooth motion when the actuator is set on revolute joint $A$, shown in Fig. 1.3 (left). However, it also has two dead points when the actuator is set on slider $D$, one of which is shown in Fig. 1.3 (right). The dead points can only be passed through inertia of the linkages in applications.

Therefore, avoiding or taking measures to conquer the effect of the dead points is one of the design tasks. There are some techniques introduced in many textbooks [1, 4, 5]; this book does not repeat these problems.

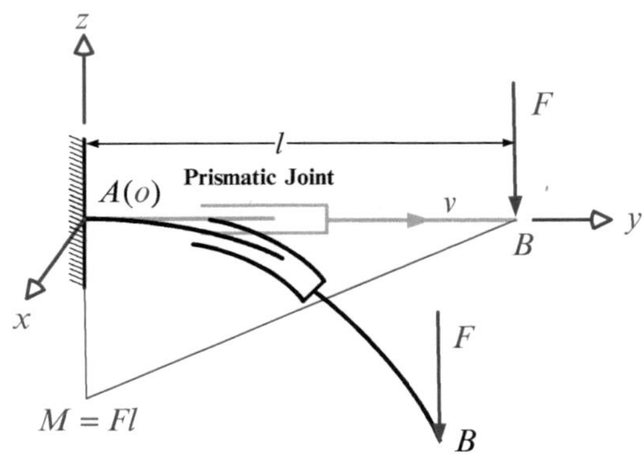


**Fig. 1.4** Vertically loaded cantilever with one prismatic joint

Prismatic joint sometimes should not be used as passive joint. A passive prismatic joint in a cantilevered beam whose free end is being exerted one transversal load is discussed. As shown in Fig. 1.4, the free end part connected with the fixed wall through a prismatic joint is sliding rightward at a speed of $v$ under a vertical load $F$.

The deflection of every point on the cantilever can be obtained in line with energy method:

$$z(y) = -\int_0^y \frac{1}{EI_x(y)} M_x(y) \frac{\partial M_x(y)}{\partial F} \, \mathrm{d}y \tag{1.1}$$

where the sign "–" indicates that the deflection is along the negative direction of $z$-axis, $E$ represents the Young's modulus of elasticity of the cantilevered beam, $I_x(y)$ stands for the second inertia moment of the beam about the $x$-axis at $y$, $M_x(y) = F(l - y)$ which denotes the bending moment of the beam at position of $y$, and $l$ is the whole distance of the cantilevered beam at the instant.

Equation (1.1) can be expanded as

$$z(y) = -\frac{Fl^3}{3EI_x} + \frac{F(l-y)^3}{3EI_x} \tag{1.2}$$

where $I_x = I_x(y)$ is assumed to be a constant for the beam.

Equation (1.2) shows that the deflection of every position on the beam depends on the parameter of $I_x$ if all of the other parameters are the same for both segments of the beam. Therefore, the deflections of the prismatic joint on the two segments are often different in most cases. And this will surely induce a great resistance for the sliding motion. Therefore, the prismatic joint shown in Fig. 1.4 is not fit for passive connection under the transversally loaded conditions. However, there are a lot of missions in engineering that need such passive prismatic joints. So, it is urgently necessary to synthesize a simple but robust structure to fulfill this requirement. Chapter 9 will discuss this topic.

### 1.1.3 Workspace and Singularity

The reachable workspace of a planar linkage is the set of positions, consisting of both a reference point and the orientation about this point, that are reachable by its end effector. An ideal design is that the end effector of the linkage has unconstrained free motion within its reachable workspace. However, the workspace does have boundaries, defined in part via the extreme reach allowed by the chain, and sometimes contain the singularities. The singularity distribution and the shape and size of the workspace for a planar linkage are primary considerations in the design.

For the analysis of position, velocity, and acceleration of a linkage, there are a lot of methods, such as geometry method, complex number method, and vector method which all suit the planar linkages.

## 1.2 Design of Planar Linkages

Linkage design is often divided into three major categories of tasks, which are called motion generation, function generation, and point-path generation [1, 2, 5]. A good planar linkage should keep a continuous smooth movement while transmitting forces or torques. Therefore, another problem for the design of planar linkages is that the actuator could remain in continuous motion and the transmitting angle should keep within a reasonable range.

If a linkage is to be used in a continuous operation, the input crank should be able to fully rotate so that it can be driven by a rotating power source. A study of the configurations of a 4R linkage led Grashof to conclude that, for the shortest link of length $s$ and the longest link of length $l$, the shortest link will fully rotate if

$$s + l \leq a + b \tag{1.3}$$

where $a$ and $b$ are the lengths of the other two links.

Equation (1.3) is known as Grashof's criterion, and linkages that have a rotatable crank are called Grashof linkages. There are four linkage types that satisfy Grashof's criterion shown in Fig. 1.5. If the input link is the shortest, then the crank-rocker is obtained (subfigure (a) in Fig. 1.5). If the ground link is the shortest, then both the input and output links will fully rotate relative to the ground; this is the double-crank linkage (subfigure (b) in Fig. 1.5). Finally, if the floating link is the shortest link, then the input and output links are rockers (subfigure (c) in Fig. 1.5); this is the Grashof double-rocker [2]. The planar four-bar linkage is a non-Grashof double-rocker (subfigure (d) in Fig. 1.5) if the lengths of the links do not satisfy Eq. (1.3).

For a Grashof linkage, it is also a very important requirement to keep a good transmission angle. This is a check problem after geometry design.

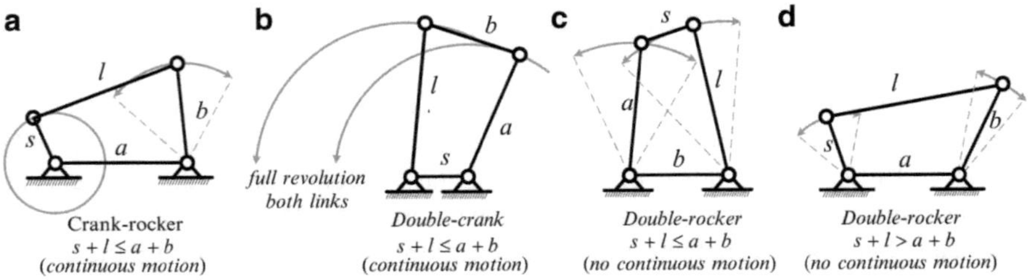


**Fig. 1.5** Grashof's criterion for a planar four-bar linkage

### 1.2.1 Position Design

The design of any planar linkages is always available to be transformed to the design of a planar four-bar linkage. Position design, which is also called motion generation, only focuses on planar four-bar linkages. The objective of planar four-bar motion generation is to calculate the mechanism parameters required to achieve a set of prescribed coupler poses which include the positions and orientations of a link [6]. According to Burmester theory, a general planar four-bar linkage could trace five arbitrary accurate positions at most. However, the modern design often requires tracing more positions. This book proposes a uniform design theory for planar four-bar linkages. It can search the accurate results when there are such solutions for the specified $n$ precise positions; otherwise, the best approximate solutions will be found. The merit of the method proposed in this book is that the engineers could directly obtain the planar four-bar linkages when the specified positions are known.

### 1.2.2 Kinetostatic Design

Kinetostatic design mainly focuses on the synthesis of kinematics and load capacity of a mechanism. For example, an ideal mechanical steering system used in the automobiles should satisfy the pure rotation criterion when turning, which is called Ackermann turning geometry. Figure 1.6 indicates that the four wheels must rotate about a pivot on the rear axle to keep a pure rotation so that the slip could be eliminated.

Figure 1.7 presents a foldable stair. This stair remains the walking conversions of human being and all the advantages of a concrete stair in civil engineering but allows the users to completely fold it after use. Therefore, both the kinematics and structural dynamics of the stair in application should be analyzed and properly designed. For the kinetostatics, the foldable stair consists of a number of identical planar four-bar linkages; therefore, it should first meet the motion requirement in folding and unfolding, and then, the whole structure should have enough strength and stiffness, but the self-weight should not be too high so that it can be easily folded.

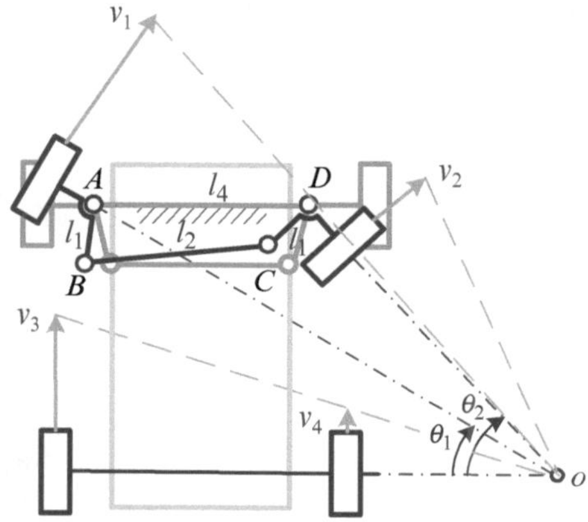


**Fig. 1.6** A planar linkage satisfies the kinematic geometry of Ackermann criterion

**Fig. 1.7** Foldable stair. (**a**) Deployed stair. (**b**) Folded stair

Kinetostatic analysis could provide the theoretical evaluation for the structure design on kinematics and statics aspects. To reduce the self-weight of a mechanism, the theoretical strength and stiffness of the structure should be considered in structure and geometry design. After manufacturing and assembly, experiment is another means to appraise the design. Figure 1.8 shows the load test for the foldable stair. Identical strength theory was adopted in the design of the foldable stair. But some approximations were also used for the sake of manufacturing. So the kinetostatic experiment must be made after design.

What is worth pointing out is that only the kinetostatic analysis and design is not enough for the operation of a linkage.

### *1.2.3 Dynamics Design*

For some complex planar linkages, they often form a truss structure under the working conditions. The dynamics stability and vibration mode of the structure have great effects on a mechanism. To research the dynamic effect on complex

**Fig. 1.8** Load capacity of the foldable stair

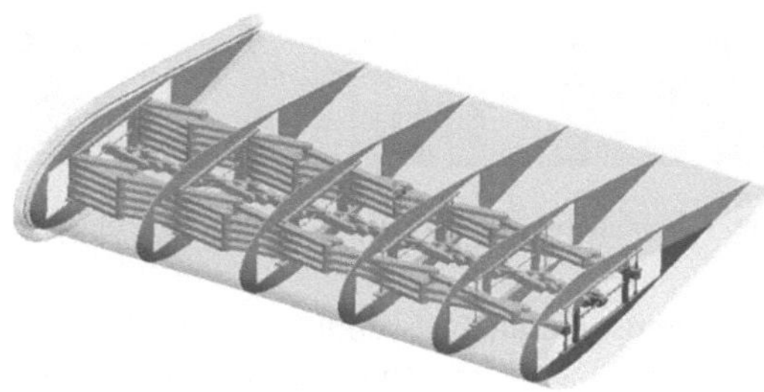

**Fig. 1.9** Concept design of a deployable wing

planar linkages, this book focuses on a foldable stair shown in Fig. 1.6 and proposes a structural dynamics method based on transfer matrix. The stairs are made up of a number of identical scissor-like elements (SLEs) which are complex planar linkages. The linkage has both characteristics of a mechanism and those of a truss structure. The structural dynamics is more complicated than both. Therefore, how to build the structural dynamics of the whole stairs and the similar complex linkages is the core of dynamics design.

The most widely used elements in planar linkages are long and thin links which can be accurately modeled as Euler-Bernoulli beams. The dynamics of each segment beam between every two adjacent revolute joints could be precisely expressed via the transfer matrix of the segment with the variables of boundary conditions of the joints. Therefore, the structural dynamics of the whole stair structure is built using the least number of variables compared with the traditional methods. Figure 1.9

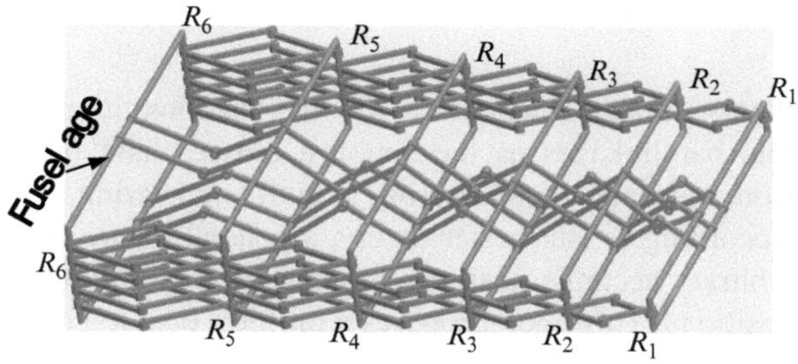


**Fig. 1.10** Deployable wing structure

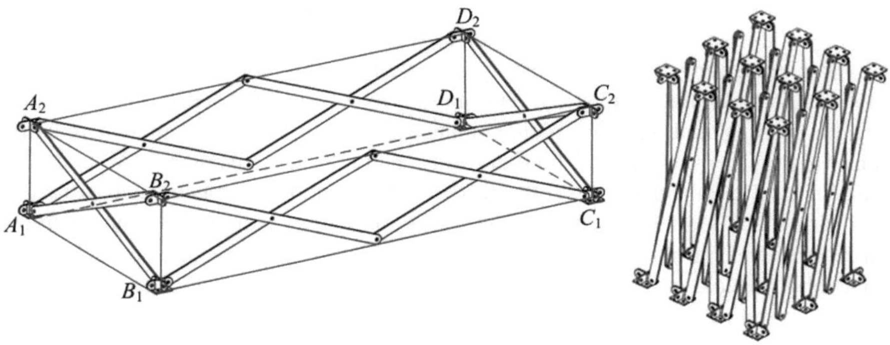


**Fig. 1.11** Foldable brackets for solar panels

shows a concept design of a metamorphic wing structure. The structure is made up of planar linkages in two perpendicular planes. Figure 1.10 presents the linkage of the metamorphic wing. The terraced linkage is obtained in line with the identical strength which will be discussed in Chap. 7.

Figure 1.11 shows the foldable brackets for the requirements of photovoltaic industry. Every side of a fixing unit is a planar linkage. The working conditions require the foldable structure be deployed to form a cell bracket after assembly. When the structure is completely folded, the space that the whole structure occupied is little, which is very convenient for fabricating and transportation.

All these examples only consist of planar linkages. Of course, there are overconstraints in the double and multi-planar linkages. The overconstraints offer the mechanism higher stiffness and better stability. The advantage of the flexibility to fold and unfold for complex linkages can be utilized for many applications in mechanical engineering, structural engineering, and space structures. To realize the design target, the structural dynamics analysis must be investigated in the concept design stage so that the dynamic stability, strength, and stiffness and even the durability which measure the whole mechanical features can be taken into the geometry design of the mechanism and structure.

## 1.3 Contents of the Book

Planar linkages witness wide applications in mechanical engineering. Planar four-bar linkages as the simplest closed-chain mechanism are extensively used from mechanical engineering to civil engineering and even aerospace engineering. According to Burmester theory, a general planar four-bar linkage could trace five arbitrary accurate positions at most. However, modern design often requires more positions. This book proposes a uniform position design theory for planar four-bar linkages. It can search the accurate results when there are such solutions for the prescribed $n$ precise positions; otherwise, the best approximate solutions will be found. The merit of the method proposed in this book is that the engineers could directly obtain the planar four-bar linkages when the specified positions are known.

After presenting the design theory and method, this book discusses the kinematics and reachable workspace and singularity of planar 3-RRR linkage which can be used to analyze other planar linkages. Then, a foldable stair that remains the walking conversions of human being and all the advantages of concrete stair in civil engineering is synthesized; a lifting guidance mechanism structure has virtues of high strength, high stiffness, and light weight of overconstraint structures, and the motion flexibility of mechanisms is designed. The deployable solar photovoltaic power generation support equipment is developed to provide a flexible system in both manufacturing and utilization. The statics and kinetostatics of the double planar overconstraint linkages are discussed to measure the stiffness of the complex structure. Because of the high stiffness of the planar linkages, the structural dynamics based on transfer-matrix method is investigated for the engineering applications.

## 1.4 Conclusions

This chapter has introduced the primary design requirements of planar linkages which are the focus of the whole book. Planar linkages are famous for their simple structures and wide applications in engineering. They are made up of a number of links and joints which can be constructed from hinges and sliders to form planar motion, and therefore, they are called planar linkages. The size and shape of the workspace of planar linkages characterizes the performance of the device with full planar degrees of freedom. The goal of the design theory presented in the following chapters is to equations and solve the design problems of special planar linkages that ensure the necessary strength, stiffness, and dynamics requirements in engineering applications.

## References

1. Norton RL (2004) Design of machinery: an introduction to the synthesis and analysis of mechanisms and machines, 3rd edn. Tata McGraw-Hill, Boston
2. McCarthy JM, Soh GS (2011) Geometric design of linkages. Springer, New York
3. Budynas RG, Nisbett JK (2010) Mechanical engineering design. McGraw-Hill Science, New York
4. Mahie HH, Reiholtz CF (1987) Mechanisms and dynamics of machinery. Wiley, New York
5. Shigley JE, Ucker JJ (1980) Theory of machines and mechanisms. McGraw-Hill Book Company, New York
6. Shen Q, Al-Smadi YM, Martin PJ, Russell K, Sodhi RS (2009) An extension of mechanism design optimization for motion generation. Mech Mach Theory 44(9):1759–1767

# Chapter 2
# Design of Planar Linkages with Specified Positions

This chapter proposes a uniform design equation for planar four-bar linkages with $n$ prescribed positions. The coordinates of circle point at the first position are used as the design variables to build the distance constraint equations for the following successive positions through matrix transformation. Expanding the quadratic equations and canceling the quadratic items of the coordinates of the center point, a set of linear equations are obtained. The augmented coefficient matrix consisting of the coordinates of circle point at the first position is obtained to construct a uniform expression for the design of planar four-bar linkages.

## 2.1 Analysis of Planar Four-Bar Linkages with Multispecified Positions

For planar four-bar linkage synthesis, Burmester problem as a classic issue has been extensively discussed in the previous literatures [1–5]. Hunt [1], Suh and Radcliffe [2], Bottema and Roth [3], Angeles [4], and McCarthy [5] investigated the planar Burmester problem with a focus on five-pose synthesis by intersecting two center/circle-point curves of four-position combinations for the specified five positions.

The objective of planar four-bar motion generation is to calculate the mechanism parameters required to achieve a set of specified coupler poses which includes the positions and orientations of a link [6]. To satisfy this requirement, the crank center must be located within certain area of the fixed plane which can be designed in advance [7]. This mechanism design target is particularly useful when the coupler must achieve a specific displacement sequence for effective operation [6]. Burmester circle-point and center-point curves could give the driving cranks of crank-rocker or slider-crank linkages when combined with a prescribed driven crank [8–12]. For the

J. Zhao et al., *Design of Special Planar Linkages*, Springer Tracts in Mechanical Engineering, DOI 10.1007/978-3-642-38448-6_2,

centenary of Burmester's classical textbook, Koetsier [13] gave some biographical data on Burmester and briefly commented on the importance of the book. Historical surveys [14–16] trace the early events relating to mechanical inventions and their applications that were instrumental in, but preceded, the eventual development of the science of linkage mechanism as originally expounded by Reuleaux. Analytical studies of the properties of coupler curves commenced in 1875 [15], when Samuel Roberts published the first account of the algebraic properties of the planar four-bar curve. Planar synthesis for orderly progression through four specified positions can be solved with Burmester theory [17, 18]. As to the comment on the Burmester theory [19], Koetsier addressed the importance of Burmester's classical textbook to the development of planar linkage synthesis [13].

There are three problems that have to be considered in the synthesis: branch, order, and crank problem [17]. Different kinds of methods have been proposed to solve the Burmester problem. To provide parameters of value in selecting linkage dimensions to meet design requirements, Davies et al. [20] described transformations which map the dimensions of planar 4-bar linkages onto finite regions of three-dimensional space. Waldron [8] developed a method to locate regions of the Burmester circle-point curve, which gives the driving cranks of crank-rocker or slider-crank linkages when combined with a specified driven crank. Waldron and Strong [9] presented a method to resolve the ambiguity in Waldron's solution to the order problem of planar Burmester synthesis by inspection of an image pole circle. They also proposed another method to allow the mapping of permissible segments of the circle-point curve in Waldron's solution to the branch problem. Prentis [21] discussed the classical problem of choosing a four-bar linkage to move its coupler through four prescribed positions. To satisfy this requirement, the crank center must be located within certain areas of a fixed plane. The result shows that these areas can be determined in advance. Simple extensions of the analysis offer a detailed picture of the movements of both driving crank and followers. Dyads method modeled by complex numbers [12] is presented in several different forms for three prescribed positions of either motion, path generation with the prescribed timing, or function generation. Complex number method [22] is applied to the dimensional synthesis of planar triads, for six exact positions, and movement generation with the prescribed timing. For rigid-body guidance mechanism synthesis reaching four specified task positions, Burmester curve equations are fit for being presented by a displacement matrix method [23].

Kim and Kwak [24] proposed a selective accurate synthesis technique by adopting the displacement matrix approach. For more than five prescribed poses, a four-bar linkage can be generally synthesized to access the poses approximately [18]. Although adjustable driving crank for 4-bar linkages could allow different motions [11], the requirement of an additional control of the driving crank increases the complexity and expense of the system. Therefore, this chapter focuses on the planar four-bar linkage and discusses the uniform expression of multiposition design.

## 2.2 Uniform Equation of Planar Four-Bar Linkages with *n* Specified Positions

For a planar four-bar linkage, any point on the perpendicular bisector of the two corresponding position points can be selected as the fixed joints on the base when two positions are specified. Therefore, there are numerous solutions for two-position synthesis of a planar four-bar linkage. The fixed joints can also be obtained by finding the crosspoint of perpendicular bisector of every two corresponding position points when three positions shown in Fig. 2.1a are prescribed. Figure 2.1b shows the process to find the fixed joint corresponding to the 3-position points $P_1$, $P_2$, and $P_3$. Connecting $P_1$, $P_2$, and $P_3$ in turn and sketching the perpendicular bisectors of $P_1P_2$ and $P_2P_3$, one can find a crosspoint, denoted by $A$, which is the pivot corresponding to position points $P_1$, $P_2$, and $P_3$. Similarly, the pivot $D$, corresponding to position points $Q_1$, $Q_2$, and $Q_3$, could be obtained, the process of which is shown in Fig. 2.1c. Figure 2.1d shows the case that the planar four-bar linkage is in the first position.

However, the solution obtained through the above process is based on the assumption that the joints are just the position points on the link. There would be other solutions if the joints were chosen differently. As a result, the number of solutions for 3-position synthesis of a planar four-bar linkage is numerous. What is discussed above is obtained through a geometric method. Because there are too many geometry lines to obtain the solution of four or more position problems, what follows will concentrate on the analytical means to get the solutions of multiposition synthesis of planar four-bar linkages.

### 2.2.1 *General Requirements for Multiposition Design of a Planar Four-Bar Linkage*

Suppose that the link of a planar four-bar linkage shown in Fig. 2.2 is required to pass a number of specified positions, $P_1Q_1, P_2Q_2, \ldots, P_nQ_n (n \geq 3)$. The target of the synthesis is to find the pivots on the fixed base and the link as well.

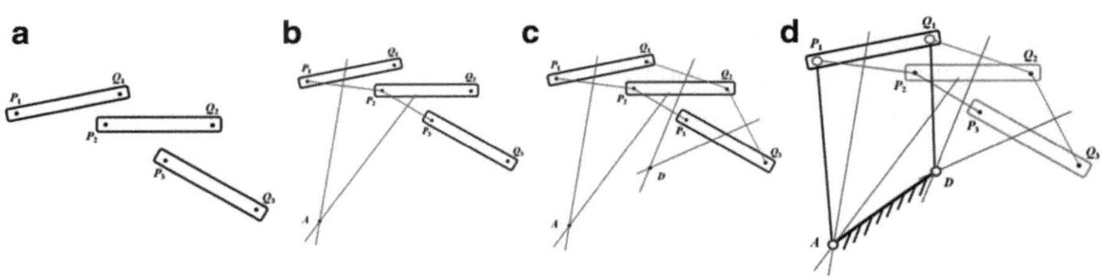


**Fig. 2.1** 3-position synthesis for a planar four-bar linkage. (**a**) Prescribed 3 positions (**b**) Left center point (**c**) Right center point and (**d**) Four-bar linkage

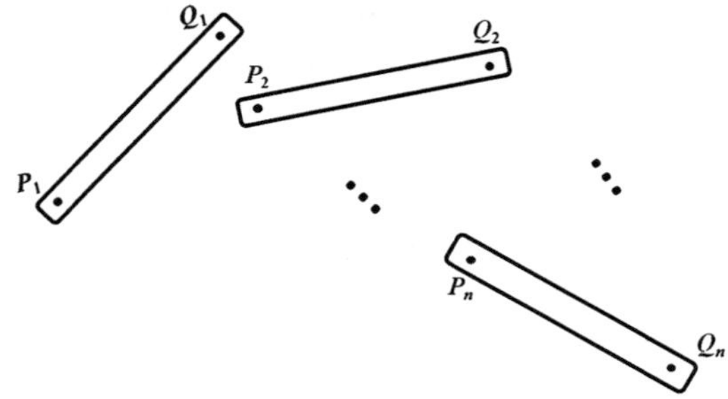


**Fig. 2.2** General multiposition synthesis of planar four-bar linkage

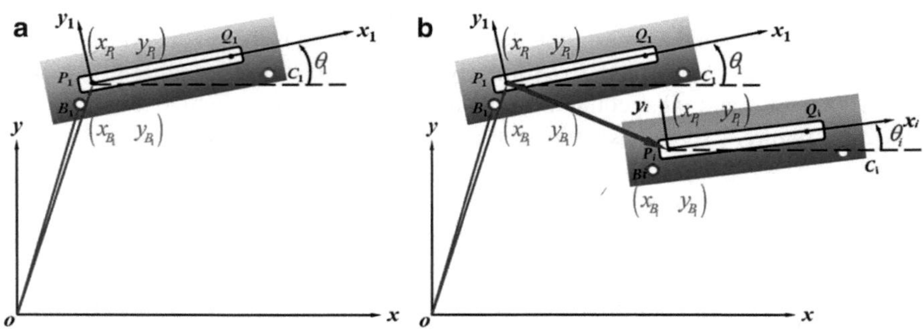

**Fig. 2.3** Transformation between two positions. (**a**) The first prescribed position and (**b**) The second prescribed position

Figure 2.3a shows the case that link $P_1 Q_1$ is in the first position. In the absolute coordinate system, the position is represented by the absolute coordinates of joint $P_1$ and the rotational angle $\theta_1$ with respect to the absolute coordinate system about the $z$-axis which is perpendicular to the $xoy$-plane and therefore is omitted here. Assume that the pivots on the link corresponding to the first position are denoted by $B_1$ and $C_1$. When the $i$th ($i \leq n$) position, $P_i Q_i$, shown in Fig. 2.3b, is specified, the transformation from the first coordinate system, $x_1 B_1 y_1$, to the $i$th one, $x_i B_i y_i$, can be uniquely determined:

$$\boldsymbol{r}_{B_i} = \boldsymbol{r}_{P_1} + \boldsymbol{r}_{P_1 P_i} + \mathbf{R}(\theta_{1i})\, \boldsymbol{r}_{P_1 B_1} \tag{2.1}$$

where $\boldsymbol{r}_{B_i}$ denotes the absolute vector coordinates of joint $B_i$, $\boldsymbol{r}_{P_1}$ denotes the absolute vector coordinates of joint $P_1$, $\boldsymbol{r}_{P_1 P_i}$ ($i = 1, 2, \ldots, n$) denotes the absolute vector coordinates of $P_1 P_i$, and

$$\mathbf{R}(\theta_{1i}) = \begin{bmatrix} \cos\theta_{1i} & -\sin\theta_{1i} \\ \sin\theta_{1i} & \cos\theta_{1i} \end{bmatrix} \tag{2.2}$$

where $\theta_{1i} = \theta_i - \theta_1$.

Because

$$\boldsymbol{r}_{P_i} = \boldsymbol{r}_{P_1} + \boldsymbol{r}_{P_1 P_i} \tag{2.3}$$

**Fig. 2.4** Circle point and center point

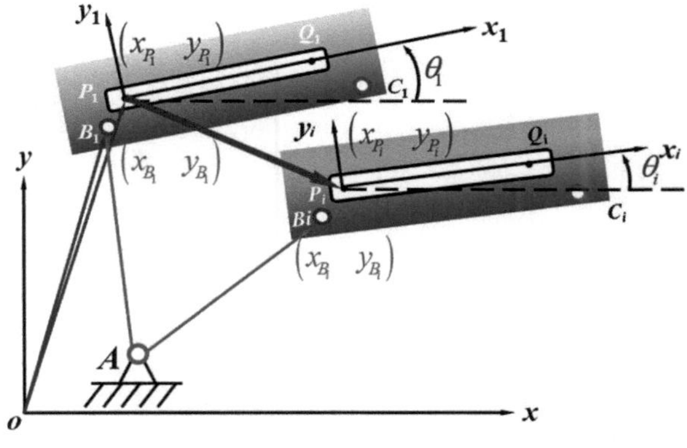


and $\boldsymbol{r}_{P_i}$ is prescribed, Eq. (2.1) can also be simplified as

$$\boldsymbol{r}_{B_i} = \boldsymbol{r}_{P_i} + \mathbf{R}(\theta_{1i})\,\boldsymbol{r}_{P_1B_1} \tag{2.4}$$

For the joints $B_1, B_2, \ldots$, and $B_i$ are all circle points, they must locate on a circle whose center is the fixed pivot on the base and which is denoted by $A$ in Fig. 2.4.

Immediately, the following equation is obtained:

$$\|\boldsymbol{r}_{AB_i}\|^2 = \|\boldsymbol{r}_{AB_1}\|^2\ (i = 2{,}3, \ldots, n) \tag{2.5}$$

where $\|\cdot\|$ represents the 2-norm of vector "·". Expanding Eq. (2.5) presents

$$c_{i1}x_A + c_{i2}y_A + c_{i3} = 0 \tag{2.6}$$

where $i = 2, 3, \ldots, n$, $x_A$, and $y_A$ denote the absolute coordinates of center point $A$ and

$$\begin{cases} c_{i1} = 2\left[x_{B_1}(1-\cos\theta_{1i}) + y_{B_1}\sin\theta_{1i} + x_{P_1}\cos\theta_{1i} - y_{P_1}\sin\theta_{1i} - x_{P_i}\right] \\ c_{i2} = 2\left[y_{B_1}(1-\cos\theta_{1i}) - x_{B_1}\sin\theta_{1i} + y_{P_1}\cos\theta_{1i} + x_{P_1}\sin\theta_{1i} - y_{P_i}\right] \\ \quad c_{i3} = 2\left(-x_{P_i}y_{B_1} - y_{P_i}x_{P_1} + x_{P_i}y_{P_1} + y_{P_i}x_{B_1}\right)\sin\theta_{1i} \\ \qquad + 2\left(x_{P_i}x_{B_1} - x_{P_i}x_{P_1} - y_{P_i}y_{P_1} + y_{P_i}y_{B_1}\right)\cos\theta_{1i} \\ \qquad - 2y_{B_1}y_{P_1} - 2x_{B_1}x_{P_1} + y_{P_i}^2 + x_{P_i}^2 + y_{P_1}^2 + x_{P_1}^2 \end{cases} \tag{2.7}$$

When three specified positions of the link are known, which are illustrated in Fig. 2.1a, the following equation set should hold:

$$\begin{cases} c_{21}x_A + c_{22}y_A + c_{23} = 0 \\ c_{31}x_A + c_{32}y_A + c_{33} = 0 \end{cases} \tag{2.8}$$

Solving Eq. (2.8) yields

$$\begin{cases} x_A = -\dfrac{\begin{vmatrix} c_{23} & c_{22} \\ c_{33} & c_{32} \end{vmatrix}}{\begin{vmatrix} c_{21} & c_{22} \\ c_{31} & c_{32} \end{vmatrix}} \\ y_A = -\dfrac{\begin{vmatrix} c_{21} & c_{23} \\ c_{31} & c_{33} \end{vmatrix}}{\begin{vmatrix} c_{21} & c_{22} \\ c_{31} & c_{32} \end{vmatrix}} \end{cases} \tag{2.9}$$

Equation (2.7) indicates that the coefficients of the linear equation set (2.8) are the function of the coordinates of joint $B_1$; therefore, there is one solution set when joint $B_1$ is prescribed as point $P_1$. There will be other solutions when joint $B_1$ is changed to another one within the plane determined by link $P_1Q_1$. When four specified positions of the link are provided, the following equation set holds:

$$\begin{cases} c_{21}x_A + c_{22}y_A + c_{23} = 0 \\ c_{31}x_A + c_{32}y_A + c_{33} = 0 \\ c_{41}x_A + c_{42}y_A + c_{43} = 0 \end{cases} \tag{2.10}$$

Equation set (2.10) has two variables, $x_A$ and $y_A$, but there are three equations. Therefore, only two independent equations in equation set (2.10) are necessary to determine the coordinates of joint $A$. In other words, the sufficient and necessary condition that Eq. (2.10) has at least one solution is

$$R\left(\mathbf{C}_{3\times3}\right) \le 2 \tag{2.11}$$

where $R(\cdot)$ denotes the rank of matrix "·" and $\mathbf{C}_{3\times3} = \begin{bmatrix} c_{21} & c_{22} & c_{23} \\ c_{31} & c_{32} & c_{33} \\ c_{41} & c_{42} & c_{43} \end{bmatrix}$.

Therefore, the conditions that there are solutions for the coordinates of center point $A$ can be alternatively expressed as

$$\left|\mathbf{C}_{3\times3}\right| = 0 \tag{2.12}$$

where $|\cdot|$ represents the determinant of matrix "·". Equation (2.7) indicates that every element of the determinant (2.12) is a linear function of the coordinates of circle point $B_1$. Equation (2.12) represents a 3-order curve of the coordinates of point $B_1$ [23].

When five specified positions of the link are known, the following equation set should hold:

$$\begin{cases} c_{21}x_A + c_{22}y_A + c_{23} = 0 \\ c_{31}x_A + c_{32}y_A + c_{33} = 0 \\ c_{41}x_A + c_{42}y_A + c_{43} = 0 \\ c_{51}x_A + c_{52}y_A + c_{53} = 0 \end{cases} \tag{2.13}$$

Equation set (2.13) also has two variables, $x_A$ and $y_A$, but there are four equations. Only two independent equations in equation set (2.13) are needed to decide the coordinates of joint $A$. In other words, the sufficient and necessary condition that Eq. (2.13) has one solution at least is

$$R\left(\mathbf{C}_{4\times 3}\right) < 3 \tag{2.14}$$

where $\mathbf{C}_{4\times 3} = \begin{bmatrix} c_{21} & c_{22} & c_{23} \\ c_{31} & c_{32} & c_{33} \\ c_{41} & c_{42} & c_{43} \\ c_{51} & c_{52} & c_{53} \end{bmatrix}$. There should be $C_4^3 = 4$ 3-order determinants if Eq. (2.14) is replaced by the zero-determinant expressions [2–5, 25], namely,

$$\begin{vmatrix} c_{21} & c_{22} & c_{23} \\ c_{31} & c_{32} & c_{33} \\ c_{41} & c_{42} & c_{43} \end{vmatrix} = 0, \quad \begin{vmatrix} c_{31} & c_{32} & c_{33} \\ c_{41} & c_{42} & c_{43} \\ c_{51} & c_{52} & c_{53} \end{vmatrix} = 0, \quad \begin{vmatrix} c_{41} & c_{42} & c_{43} \\ c_{51} & c_{52} & c_{53} \\ c_{21} & c_{22} & c_{23} \end{vmatrix} = 0, \quad \begin{vmatrix} c_{51} & c_{52} & c_{53} \\ c_{21} & c_{22} & c_{23} \\ c_{31} & c_{32} & c_{33} \end{vmatrix} = 0 \tag{2.15}$$

The circle point $B_1$ must be the intersection of these four 3-order curves represented by Eq. (2.15). Similarly, when $n$ specified positions of the link are known, the following equation set should hold when there is one solution at least:

$$\begin{cases} c_{21}x_A + c_{22}y_A + c_{23} = 0 \\ c_{31}x_A + c_{32}y_A + c_{33} = 0 \\ \quad \vdots \\ c_{n1}x_A + c_{n2}y_A + c_{n3} = 0 \end{cases} \tag{2.16}$$

where $n = 3, 4, \ldots$.

Equation set (2.16) also has two variables only, namely, $x_A$ and $y_A$, but there are $n - 1$ constraint equations. Therefore, only two independent equations in equation set (2.16) are required to determine the coordinates of joint $A$. Hence, the sufficient and necessary conditions that Eq. (2.16) has a solution could be expressed as

$$R\left(\mathbf{C}_{(n-1)\times 3}\right) < 3 \tag{2.17}$$

where

$$\mathbf{C}_{(n-1)\times 3} = \begin{bmatrix} c_{21} & c_{22} & c_{23} \\ c_{31} & c_{32} & c_{33} \\ \vdots & \vdots & \vdots \\ c_{n1} & c_{n2} & c_{n3} \end{bmatrix} \tag{2.18}$$

If Eq. (2.17) is equivalently represented as that the determinants of $3 \times 3$ matrices are zeros simultaneously, there should be $C_{n-1}^{3} = \frac{(n-1)(n-2)(n-3)}{6}$ 3-order determinants. It is a tricky job to search the intersections of so many 3-order curves.

### 2.2.2 *Uniform Equation for the Design of a Planar Four-Bar Linkage with n Specified Positions*

When $n$ specified positions of the link are known, the condition that there is one solution at least is that the rank of the augmented coefficient matrix should always be less than 3. The solution can be expressed as the determinant of the square matrix is zero if a $3 \times 3$ matrix could be conceived with the augmented coefficient matrix in Eq. (2.18). Suppose

$$\mathbf{M} = \mathbf{C}_{(n-1)\times 3}^{\mathrm{T}}\mathbf{C}_{(n-1)\times 3} \tag{2.19}$$

Therefore, $\mathbf{M}$ represented in Eq. (2.19) is a $3 \times 3$ matrix. There is the following theorem:

*The circle points satisfying the n-position requirements must meet*

$$|(\mathbf{M})_{3\times 3}| = 0 \tag{2.20}$$

*Proof.*

1. *Necessity*

*Because* $R\left(\mathbf{C}_{(n-1)\times 3}^{\mathrm{T}}\mathbf{C}_{(n-1)\times 3}\right) \leq R\left(\mathbf{C}_{(n-1)\times 3}\right)$ and $R\left(\mathbf{C}_{(n-1)\times 3}\right) < 3$, *the following equation should hold*:

$$R\left(\mathbf{C}_{(n-1)\times 3}^{\mathrm{T}}\mathbf{C}_{(n-1)\times 3}\right) < 3 \tag{2.21}$$

*As a result, there must be*

$$|(\mathbf{M})_{3\times 3}| = \left|\mathbf{C}_{(n-1)\times 3}^{\mathrm{T}}\mathbf{C}_{(n-1)\times 3}\right| = 0 \tag{2.22}$$

2. *Sufficiency*

*Because* $|(\mathbf{M})_{3\times 3}| = 0$, *the following equation should hold*:

$$R(\mathbf{M}) < 3 \tag{2.23}$$

*which means*

$$R\left(\mathbf{C}_{(n-1)3}^{\mathrm{T}}\mathbf{C}_{(n-1)3}\right) < 3 \tag{2.24}$$

*Therefore, there is at least one 3-dimensional nonzero vector* $\boldsymbol{x}_3$ *such that the following equation holds*:

$$\left(\mathbf{C}_{(n-1)3}^{\mathrm{T}}\mathbf{C}_{(n-1)3}\right)\boldsymbol{x}_3 = 0 \tag{2.25}$$

*Left multiplying* $\boldsymbol{x}_3^{\mathrm{T}}$ *at both sides of* Eq. (2.25) *yields*

$$\boldsymbol{x}_3^{\mathrm{T}}\left(\mathbf{C}_{(n-1)3}^{\mathrm{T}}\mathbf{C}_{(n-1)3}\right)\boldsymbol{x}_3 = 0 \tag{2.26}$$

*Rearranging* Eq. (2.26) *presents*

$$\left(\mathbf{C}_{(n-1)\times 3}\boldsymbol{x}_3\right)^{\mathrm{T}}\left(\mathbf{C}_{(n-1)\times 3}\boldsymbol{x}_3\right) = 0 \tag{2.27}$$

*which can be rewritten as*

$$\left\|\mathbf{C}_{(n-1)\times 3}\boldsymbol{x}_3\right\|^2 = 0 \tag{2.28}$$

*where* $\|\cdot\|$ *represents the 2-normal of vector "·". Immediately,* Eq. (2.28) *leads to*

$$\mathbf{C}_{(n-1)\times 3}\boldsymbol{x}_3 = 0 \tag{2.29}$$

*Because* $\boldsymbol{x}_3$ *is a 3-dimensional nonzero vector, there must be*

$$R\left(\mathbf{C}_{(n-1)\times 3}\right) < 3 \tag{2.30}$$

The necessary and sufficient condition expressed by Eq. (2.20) provides a theoretical uniform equation for the circle point of planar four-bar linkages with $n$ prescribed positions.

However, Eq. (2.19) indicates that each element of matrix $\mathbf{M}$ is a quadratic function of $x_{B_1}$ and $y_{B_1}$ and it is positive-semidefinite; the determinant of $\mathbf{M}$ denoted by Eq. (2.20) is larger than or equal to zero. Suppose that

$$f\left(x_{B_1}, y_{B_1}\right) = |\mathbf{M}| \tag{2.31}$$

The optimum solutions for $x_{B_1}$ and $y_{B_1}$ of the above problem must occur when $f\left(x_{B_1}, y_{B_1}\right)$ has the minimum value. Let

$$\begin{cases} \dfrac{\partial f\left(x_{B_1}, y_{B_1}\right)}{\partial x_{B_1}} = 0 \\ \dfrac{\partial f\left(x_{B_1}, y_{B_1}\right)}{\partial y_{B_1}} = 0 \end{cases} \tag{2.32}$$

where $\frac{\partial f\left(x_{B_1}, y_{B_1}\right)}{\partial x_{B_1}}$ indicates the partial derivative of $f\left(x_{B_1}, y_{B_1}\right)$ with respect to $x_{B_1}$ and $\frac{\partial f\left(x_{B_1}, y_{B_1}\right)}{\partial y_{B_1}}$ denotes the partial derivative of $f\left(x_{B_1}, y_{B_1}\right)$ with respect to $y_{B_1}$.

Solving partial derivative equation set (2.32) yields the stationary points. Assume that

$$\begin{cases} a = \dfrac{\partial^2 f(x_{B_1}, y_{B_1})}{\partial x_{B_1}^2} \\ b = \dfrac{\partial^2 f(x_{B_1}, y_{B_1})}{\partial x_{B_1} \partial y_{B_1}} \\ c = \dfrac{\partial^2 f(x_{B_1}, y_{B_1})}{\partial y_{B_1}^2} \end{cases} \tag{2.33}$$

According to calculus, the extreme values of $f(x_{B_1}, y_{B_1})$ occur at $a > 0$ and $ac - b^2 > 0$. The minimum value point can be found when these solutions are substituted into Eq. (2.31). Therefore, the circle point $(x_{B_1}, y_{B_1})$ is obtained. It will be an accurate solution when the multipositions specified meet the requirements of a circle point; otherwise it is an approximate solution. Substituting the coordinates of the circle point into Eq. (2.16), the coefficients of the equation are all known. Rearranging equation set (2.16) presents

$$\mathbf{K}\boldsymbol{x}_2 = \boldsymbol{h} \tag{2.34}$$

where $\mathbf{K} = \begin{bmatrix} c_{21} & c_{22} \\ c_{31} & c_{32} \\ \vdots & \vdots \\ c_{n1} & c_{n2} \end{bmatrix}$, $\boldsymbol{x}_2 = \begin{bmatrix} x_A \\ y_A \end{bmatrix}$, and $\boldsymbol{h} = -\begin{bmatrix} c_{23} \\ c_{33} \\ \vdots \\ c_{n3} \end{bmatrix}$. Because $R(\mathbf{K}) = 2$ always holds for different position problems, the general solution of $\boldsymbol{x}_2$ can be obtained by the least-squares method:

$$\boldsymbol{x}_2 = \left(\mathbf{K}^{\mathrm{T}}\mathbf{K}\right)^{-1}\mathbf{K}^{\mathrm{T}}\boldsymbol{h} \tag{2.35}$$

where $\left(\mathbf{K}^{\mathrm{T}}\mathbf{K}\right)^{-1}\mathbf{K}^{\mathrm{T}}$ denotes the pseudo inverse of $\mathbf{K}$ and $\left(\mathbf{K}^{\mathrm{T}}\mathbf{K}\right)^{-1}\mathbf{K}^{\mathrm{T}} = \mathbf{K}^{-1}$ when $\mathbf{K}$ is a $2 \times 2$ full rank matrix.

The results expressed by Eq. (2.35) will be the precise solution if the specified positions satisfy the circle-point requirements; otherwise, the results will be the best approximate solutions for the proscribed multiposition problem. After obtaining the solutions, Grashof conditions should be considered in the final selection and mechanism design which has been discussed in Sect. 1.2.

## 2.3 Application Examples and Discussions

Because the augmented coefficient matrices, $\mathbf{C}$, for 2- and 3-position problems are $1 \times 3$ and $2 \times 3$, respectively, Eq. (2.20) will therefore always hold. So the solutions for the circle point $B_1$ can be any point within the plane that the link locates, for

**Table 2.1** Four specified positions

| Items | 1st position | 2nd position | 3rd position | 4th position |
|---|---|---|---|---|
| $x$-coordinate (mm) | 0 | 4 | 1 | 3 |
| $y$-coordinate (mm) | −5 | 3 | 6 | 5 |
| Pose angle (°) | 90 | 45 | −60 | 30 |

2- and 3-position problems. In this section, examples will concentrate on the 4-, 5-, and multiposition designs.

### 2.3.1 *Design of Planar Four-Bar Linkages with Four Accurate Positions*

The coordinates of circle point and its center point satisfy equation set (2.10) when four specified positions of the link are known. The augmented coefficient matrix of the constraint equation set is

$$\mathbf{C}_{3\times3} = \begin{bmatrix} c_{21} & c_{22} & c_{23} \\ c_{31} & c_{32} & c_{33} \\ c_{41} & c_{42} & c_{43} \end{bmatrix} \tag{2.36}$$

Because $\mathbf{C}_{3\times3}$ is a square matrix, Eq. (2.22) renders that

$$|(\mathbf{M})_{3\times3}| = \left|\mathbf{C}_{3\times3}^{\mathrm{T}}\mathbf{C}_{3\times3}\right| = \left|\mathbf{C}_{3\times3}^{\mathrm{T}}\right| \cdot |\mathbf{C}_{3\times3}| = |\mathbf{C}_{3\times3}|^2 = 0$$

which can be simplified as

$$|\mathbf{C}_{3\times3}| = 0 \tag{2.37}$$

This proves Eq. (2.12) again. Suppose that there are four different positions of the link which are listed in Table 2.1.

Substituting these parameters into Eq. (2.37), the precise circle-point curve can be drawn. Figure 2.5 shows the circle-point loci that satisfy the four positions in Table 2.1.

Substituting the coordinates of a point from the circle-point curve in Fig. 2.5 into any two equations of Eq. (2.10) or (2.35), the exact center-point coordinates can be obtained. For example, the center point is expressed with Eq. (2.9) when the first two equations of equation set (2.10) are selected. As previous literature [12, 23, 24, 26] investigated this problem repeatedly, this chapter does not discuss it again.

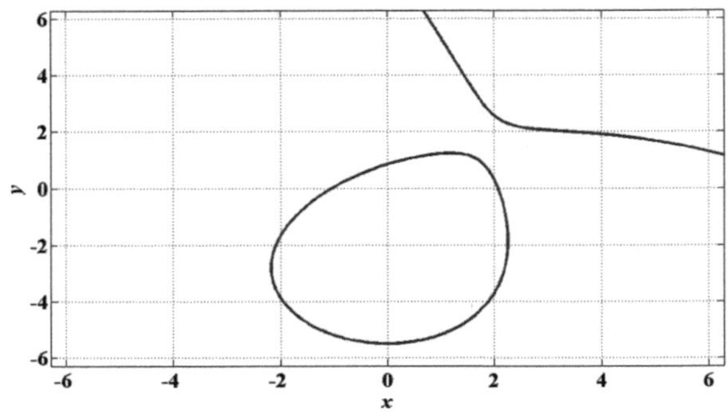


**Fig. 2.5** Circle-point curve of four-position design

**Table 2.2** Five specified positions

| Items | 1st position | 2nd position | 3rd position | 4th position | 5th position |
|---|---|---|---|---|---|
| $x$-coordinate (mm) | 10 | −6 | −13 | −17 | 6 |
| $y$-coordinate (mm) | −5 | −22 | −11 | 10 | 5 |
| Pose angle (°) | 0 | −26 | −40 | −48 | −15 |

When the parameters are prescribed differently from those listed in Table 2.1, the center-point and circle-point curves might be two-branch, thumb-shape, or knot curves [25]. For practical applications, users might prescribe a lot of design requirements, such as link position limit, dimension interval, and avoiding a certain specific defect. To solve these problems, Prentis [7], Waldron [8], Waldron and Strong [9], Chen and Fu [17], and Chen and Angeles [18] discussed different methods; Han and Qian [26] recently proposed a method of feasible mechanism solution regions where the solutions satisfy all required constraints above. This can be directly cooperated with the method proposed in this chapter to get an optimum solution within specified region. Detailed process for optimization design can be found from previous literatures [6, 14, 15, 26].

### 2.3.2 *Design of Planar Four-Bar Linkages with Five Positions*

When five positions for the link are specified, the circle points can be solved with the process discussed in Sect. 2.2. The results obtained here will be accurate if the 5-position problem has exact solutions; otherwise, the results will be the best approximate ones for the 5-position problem. Table 2.2 lists the five positions a planar four-bar linkage should trace.

Substituting the parameters of Table 2.2 into Eq. (2.19) yields

$$\mathbf{M} = \begin{bmatrix} \begin{pmatrix} 0.702454x_{B_1}^2 - 3.59562x_{B_1}y_{B_1} + 96.5332x_{B_1} \\ +4.89839y_{B_1}^2 - 258.246y_{B_1} + 3490.95 \end{pmatrix} & \begin{pmatrix} 1.79781x_{B_1}^2 - 4.19593x_{B_1}y_{B_1} + 105.939x_{B_1} \\ -1.79781y_{B_1}^2 + 97.9667y_{B_1} - 1054.19 \end{pmatrix} & \begin{pmatrix} -53.2624x_{B_1}^2 + 116.772x_{B_1}y_{B_1} - 2776.09x_{B_1} \\ +51.8288y_{B_1}^2 - 3684.60y_{B_1} + 53181.1 \end{pmatrix} \\ \begin{pmatrix} 1.79781x_{B_1}^2 - 4.19593x_{B_1}y_{B_1} + 105.939x_{B_1} \\ -1.79781y_{B_1}^2 + 97.9667y_{B_1} - 1054.19 \end{pmatrix} & \begin{pmatrix} 4.89839x_{B_1}^2 + 3.59562x_{B_1}y_{B_1} - 99.4003x_{B_1} \\ +0.702454y_{B_1}^2 - 46.3680y_{B_1} + 3034.54 \end{pmatrix} & \begin{pmatrix} -134.539x_{B_1}^2 - 105.091x_{B_1}y_{B_1} + 4331.14x_{B_1} \\ -17.7675y_{B_1}^2 - 664.276y_{B_1} - 29219.3 \end{pmatrix} \\ \begin{pmatrix} -53.2624x_{B_1}^2 + 116.772x_{B_1}y_{B_1} - 2776.09x_{B_1} \\ +51.8288y_{B_1}^2 - 3684.6y_{B_1} + 53181.1 \end{pmatrix} & \begin{pmatrix} -134.539x_{B_1}^2 - 105.091x_{B_1}y_{B_1} + 4331.14x_{B_1} \\ -17.7675y_{B_1}^2 - 664.276y_{B_1} - 29219.3 \end{pmatrix} & \begin{pmatrix} 4236.81x_{B_1}^2 + 1731.64x_{B_1}y_{B_1} - 122679.x_{B_1} \\ +2288.69y_{B_1}^2 - 33612.8y_{B_1} + 921452. \end{pmatrix} \end{bmatrix} \tag{2.38}$$

Substituting Eq. (2.38) into (2.20) presents

$$\begin{aligned} |\mathbf{M}| = {} & 39.1785x_{B_1}^6 - 185.13x_{B_1}^5y_{B_1} + 4608.05x_{B_1}^5 + 433.967x_{B_1}^4y_{B_1}^2 \\ & - 13925.2x_{B_1}^4y_{B_1} - 320.867x_{B_1}^4 - 370.259x_{B_1}^3y_{B_1}^3 + 1955.7x_{B_1}^3y_{B_1}^2 \\ & + 902479.0x_{B_1}^3y_{B_1} - 1.94144e7x_{B_1}^3 + 750.399x_{B_1}^2y_{B_1}^4 - 83039.4x_{B_1}^2y_{B_1}^3 \\ & + 3460755.0x_{B_1}^2y_{B_1}^2 - 6.40205e7x_{B_1}^2y_{B_1} + 4.9395e8x_{B_1}^2 - 185.13x_{B_1}y_{B_1}^5 \\ & - 2652.35x_{B_1}y_{B_1}^4 + 2919555.0x_{B_1}y_{B_1}^3 - 1.74407e8x_{B_1}y_{B_1}^2 \\ & + 3.29746e9x_{B_1}y_{B_1} - 1.48565e10x_{B_1} + 355.61y_{B_1}^6 - 69114.3y_{B_1}^5 \\ & + 5022288.0y_{B_1}^4 - 1.76564e8y_{B_1}^3 + 3.57256e9y_{B_1}^2 - 5.12775e10y_{B_1} \\ & + 4.50738 \times 10^{11} \\ = {} & 0 \end{aligned} \tag{2.39}$$

**Table 2.3** Four different solutions

| $x_{B_1}$ (mm) | $y_{B_1}$ (mm) | $x_A$ (mm) | $y_A$ (mm) |
|---|---|---|---|
| −30.568446 | 20.960137 | 191.467702 | −64.838556 |
| 26.545161 | 32.179730 | −44.146967 | −17.273592 |
| −0.818763 | 64.804165 | −13.399540 | −60.318106 |
| 38.477471 | 3.216875 | −25.700624 | 9.351623 |

There are four different real solutions for the circle point $B_1$ $(x_{B_1}, y_{B_1})$ for equation (2.39) which are shown in Table 2.3. Substituting these solutions into Eq. (2.34) allows one to obtain the coordinates of the center point. For example, substituting the first solution of point $B_1$ into Eq. (2.34) presents

$$\begin{bmatrix} 1.028114 & 3.686569 \\ -6.356135 & -28.006752 \\ -11.429996 & -73.117632 \\ -8.202628 & -39.230632 \end{bmatrix} \begin{bmatrix} x_A \\ y_A \end{bmatrix} = - \begin{bmatrix} -42.181186 \\ 598.922840 \\ 2552.366548 \\ 973.119227 \end{bmatrix} \tag{2.40}$$

Solving Eq. (2.40) with equation (2.35) yields

$$\begin{cases} x_A = 191.467702 \\ y_A = -64.838556 \end{cases}$$

Similarly, the other center points corresponding to the circle points can be gained. The 4 different solutions for the specified 5-position problem are listed in Table 2.3.

According to Burmester theory [19], planar revolute-jointed four-bar linkage could exactly trace five arbitrary positions with the specified poses at most. But the method proposed in this chapter could also present the best approximate result when the 5-position design has no precise solution. The same is true for more position problems which will be discussed next.

### 2.3.3 *Design of Planar Four-Bar Linkages with More Approximate Positions*

Table 2.4 lists six positions for a planar four-bar linkage. The solutions for the circle point $B_1$ and its center point $A$ and the circle point $C$ and its center point $D$ can be obtained with the process discussed in Sect. 2.2.

Generally, there are only approximate solutions for circle points and center points when more than five positions for the link are specified. Fortunately, the 6-position design prescribed in Table 2.4 has accurate solutions which could be directly gained with Eq. (2.20) or (2.31) through (2.35). The solutions for the center points and circle points are listed in Table 2.5 which demonstrates the effectiveness of the

**Table 2.4** Six specified positions

| Positions | $x_P$-coordinate | $y_P$-coordinate | $x_Q$-coordinate | $y_Q$-coordinate |
|---|---|---|---|---|
| 1 | 1.046800 | 1.382700 | 2.577900 | 1.777300 |
| 2 | 1.127700 | 1.812200 | 2.708300 | 1.855700 |
| 3 | 0.684600 | 1.987400 | 2.264400 | 1.921500 |
| 4 | 0.119700 | 1.786300 | 1.700700 | 1.764300 |
| 5 | −0.256500 | 1.325300 | 1.315600 | 1.494400 |
| 6 | −0.316000 | 0.867500 | 1.189100 | 1.352100 |

**Table 2.5** Exact solutions for the circle point and centre point of six prescribed positions

| Circle point $B_1$ | Circle point $C_1$ | Center point $A$ | Center point $D$ |
|---|---|---|---|
| (0.749678, −0.000165) | (3.017368, 1.466613) | (−0.000396, −0.000156) | (2.700123, −0.001168) |

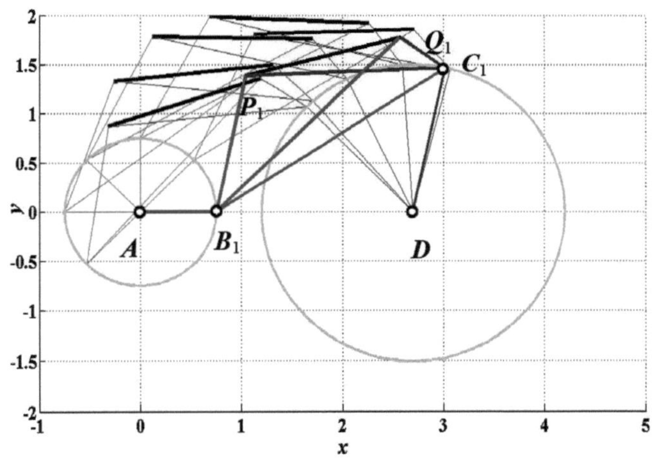


**Fig. 2.6** A mechanism exactly meets six different positions

method proposed in this chapter. The mechanism exactly satisfying the six positions prescribed in Table 2.4 is illustrated by Fig. 2.6. The quadrangle $B_1C_1Q_1P_1$ shown in Fig. 2.6 represents link $B_1C_1$.

There are only approximate solutions when these six positions are changed a little. Suppose the changed 6 positions are listed in Table 2.6 where only the second and fourth positions are altered near their original places.

The approximate circle point and center point corresponding to the positions in Table 2.6 are represented in Table 2.7, and the mechanism approximately satisfying the positions prescribed in Table 2.6 is shown in Fig. 2.7. The light bold lines represent the accurate positions specified by Table 2.6, and the dark bold lines are the real positions that the planar linkage could approximately trace.

**Table 2.6** Six arbitrary positions

| Positions | $x_P$-coordinate | $y_P$-coordinate | $x_Q$-coordinate | $y_Q$-coordinate |
|---|---|---|---|---|
| 1 | 1.046800 | 1.382700 | 2.577900 | 1.777300 |
| 2 | 1.227700 | 1.752200 | 2.808300 | 1.795700 |
| 3 | 0.684600 | 1.987400 | 2.264400 | 1.921500 |
| 4 | −0.080300 | 1.886300 | 1.500700 | 1.864300 |
| 5 | −0.256500 | 1.325300 | 1.315600 | 1.494400 |
| 6 | −0.316000 | 0.867500 | 1.189100 | 1.352100 |

**Table 2.7** Approximate solutions for the circle point and centre point of six arbitrary positions

| Circle point $B_1$ | Circle point $C_1$ | Center point $A$ | Center point $D$ |
|---|---|---|---|
| (0.819348, 0.055679) | (3.001470, 1.756439) | (0.064983, 0.073748) | (2.621502, 0.539307) |

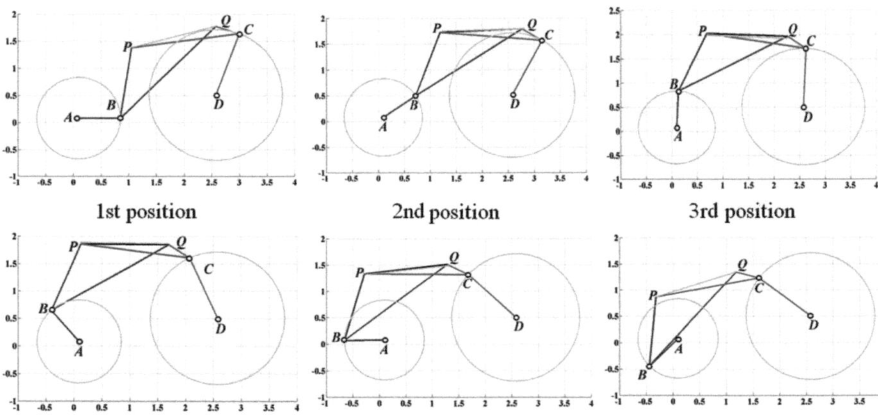


**Fig. 2.7** A mechanism approximately meets six different positions

When seven positions are specified, the solutions for the circle point $B_1$ and its center point $A$ and the circle point $C$ and its center point $D$ can be obtained with the similar process. Figure 2.8 illustrates the planar four-bar linkage that exactly satisfies the seven positions prescribed in Table 2.8. Figure 2.8 presents the case when the mechanism is in the second position for the specified seven positions. The solutions for the circle point and center point are listed in Table 2.9.

There are only approximate solutions when seven arbitrary positions are specified. Suppose the seven positions are listed in Table 2.10 where the second, fourth, and seventh positions are changed compared with those listed in Table 2.8.

The approximate circle point and center point corresponding to the positions in Table 2.10 are represented in Table 2.11, and the mechanism approximately satisfying the positions prescribed in Table 2.10 is shown in Fig. 2.9. The light bold lines represent the accurate positions specified by Table 2.10, and the dark bold lines are the real positions that the planar mechanism could approximately trace.

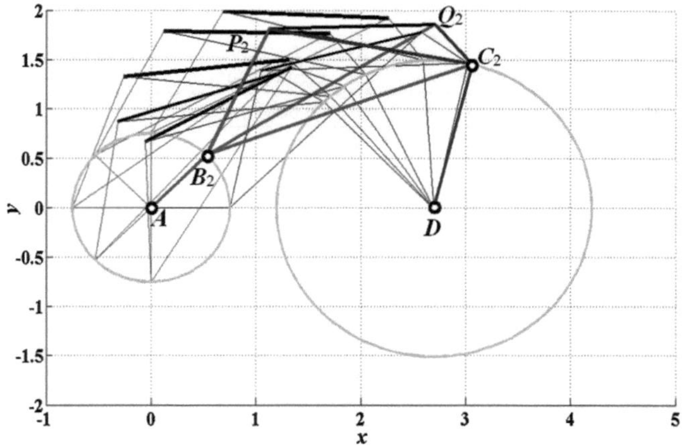


**Fig. 2.8** The mechanism in the second position

**Table 2.8** Seven specified positions

| Positions | $x_P$-coordinate | $y_P$-coordinate | $x_Q$-coordinate | $y_Q$-coordinate |
|---|---|---|---|---|
| 1 | 1.046800 | 1.382700 | 2.577900 | 1.777300 |
| 2 | 1.127700 | 1.812200 | 2.708300 | 1.855700 |
| 3 | 0.684600 | 1.987400 | 2.264400 | 1.921500 |
| 4 | 0.119700 | 1.786300 | 1.700700 | 1.764300 |
| 5 | −0.256500 | 1.325300 | 1.315600 | 1.494400 |
| 6 | −0.316000 | 0.867500 | 1.189100 | 1.352100 |
| 7 | −0.051700 | 0.663300 | 1.335700 | 1.421600 |

**Table 2.9** Precise solutions for the circle point and centre point of seven prescribed positions

| Circle point $B_1$ | Circle point $C_1$ | Center point $A$ | Center point $D$ |
|---|---|---|---|
| (0.749685, −0.000162) | (3.017259, 1.466166) | (−0.000386, −0.000153) | (2.6996, −0.004914) |

**Table 2.10** Seven arbitrary positions

| Positions | $x_P$-coordinate | $y_P$-coordinate | $x_Q$-coordinate | $y_Q$-coordinate |
|---|---|---|---|---|
| 1 | 1.046800 | 1.382700 | 2.577900 | 1.777300 |
| 2 | 1.227700 | 1.752200 | 2.808300 | 1.795700 |
| 3 | 0.684600 | 1.987400 | 2.264400 | 1.921500 |
| 4 | −0.080300 | 1.886300 | 1.500700 | 1.864300 |
| 5 | −0.256500 | 1.325300 | 1.315600 | 1.494400 |
| 6 | −0.316000 | 0.867500 | 1.189100 | 1.352100 |
| 7 | −0.001700 | 0.663300 | 1.385700 | 1.421600 |

**Table 2.11** Approximate solutions for the circle point and centre point of seven arbitrary positions

| Circle point $B_1$ | Circle point $C_1$ | Center point $A$ | Center point $D$ |
|---|---|---|---|
| (0.819348, 0.055679) | (3.001470, 1.756439) | (0.064983, 0.073748) | (2.621502, 0.539307) |

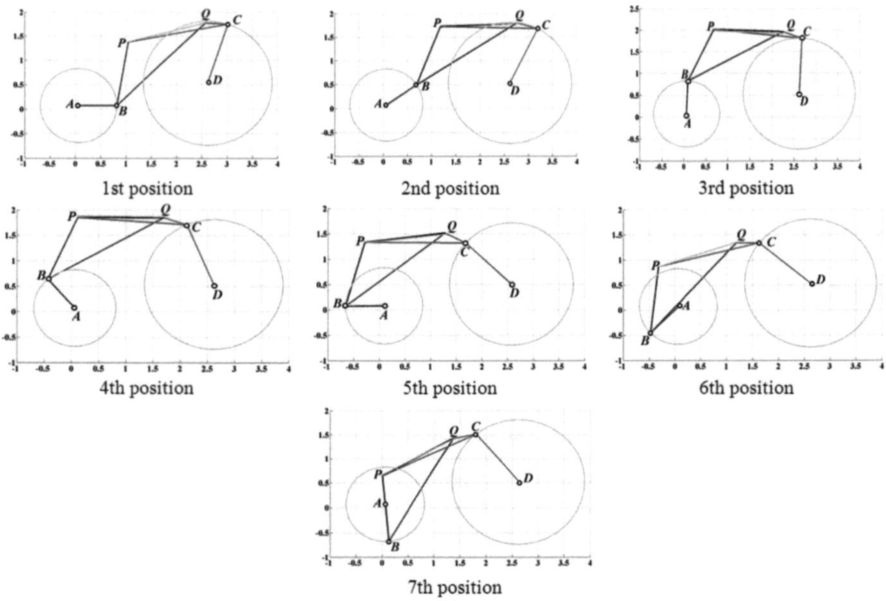


**Fig. 2.9** A planar four-bar linkage approximately meets seven different positions

**Table 2.12** Eight specified positions

| Positions | $x_P$-coordinate | $y_P$-coordinate | $x_Q$-coordinate | $y_Q$-coordinate |
|---|---|---|---|---|
| 1 | 1.046800 | 1.382700 | 2.577900 | 1.777300 |
| 2 | 1.127700 | 1.812200 | 2.708300 | 1.855700 |
| 3 | 0.684600 | 1.987400 | 2.264400 | 1.921500 |
| 4 | 0.119700 | 1.786300 | 1.700700 | 1.764300 |
| 5 | −0.256500 | 1.325300 | 1.315600 | 1.494400 |
| 6 | −0.316000 | 0.867500 | 1.189100 | 1.352100 |
| 7 | −0.051700 | 0.663300 | 1.335700 | 1.421600 |
| 8 | 0.469100 | 0.882600 | 1.851300 | 1.650300 |

When eight accurate positions are prescribed in Table 2.12, the solutions for the circle point $B_1$ and its center point $A$ and the circle point $C_1$ and its center point $D$ can be obtained with the same process.

The precise solutions for the specified eight positions shown in Table 2.12 can be immediately obtained with the process proposed in Sect. 2.2. The solutions are represented by Table 2.13.

**Table 2.13** Precise solutions for the circle point and centre point of eight prescribed positions

| Circle point $B_1$ | Circle point $C_1$ | Center point $A$ | Center point $D$ |
|---|---|---|---|
| (0.749755, −0.000061) | (3.017255, 1.466168) | (−0.000271, −0.000030) | (2.699542, −0.005128) |

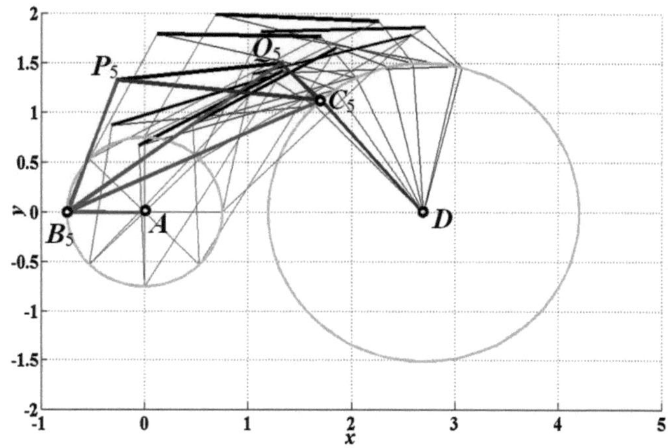


**Fig. 2.10** The mechanism in the fifth position

**Table 2.14** Eight arbitrary positions

| Positions | $x_P$-coordinate | $y_P$-coordinate | $x_Q$-coordinate | $y_Q$-coordinate |
|---|---|---|---|---|
| 1 | 1.046800 | 1.382700 | 2.577900 | 1.777300 |
| 2 | 1.227700 | 1.752200 | 2.808300 | 1.795700 |
| 3 | 0.684600 | 1.987400 | 2.264400 | 1.921500 |
| 4 | −0.080300 | 1.886300 | 1.500700 | 1.864300 |
| 5 | −0.256500 | 1.325300 | 1.315600 | 1.494400 |
| 6 | −0.316000 | 0.867500 | 1.189100 | 1.352100 |
| 7 | −0.001700 | 0.663300 | 1.385700 | 1.421600 |
| 8 | 0.489100 | 0.852600 | 2.808300 | 1.795700 |

Figure 2.10 shows the case when the mechanism moves to the fifth position for the specified 8-position problem.

There are also only approximate solutions when these eight arbitrary positions are specified. Table 2.14 shows the eight positions of a planar four-bar linkage where the second, fourth, seventh, and eighth positions are different from those listed in Table 2.12.

The approximate circle point and center point corresponding to the positions in Table 2.14 are represented in Table 2.15, and the mechanism approximately satisfying the positions prescribed in Table 2.14 is shown in Fig. 2.11. The light bold lines represent the accurate positions specified by Table 2.12, and the dark bold lines are the real positions that the planar mechanism could approximately trace.

**Table 2.15** Approximate solutions for the circle point and centre point of eight arbitrary positions

| Circle point $B_1$ | Circle point $C_1$ | Center point $A$ | Center point $D$ |
|---|---|---|---|
| (0.852813, 0.092069) | (3.032621, 1.789466) | (0.102542, 0.108504) | (2.683970, 0.458268) |

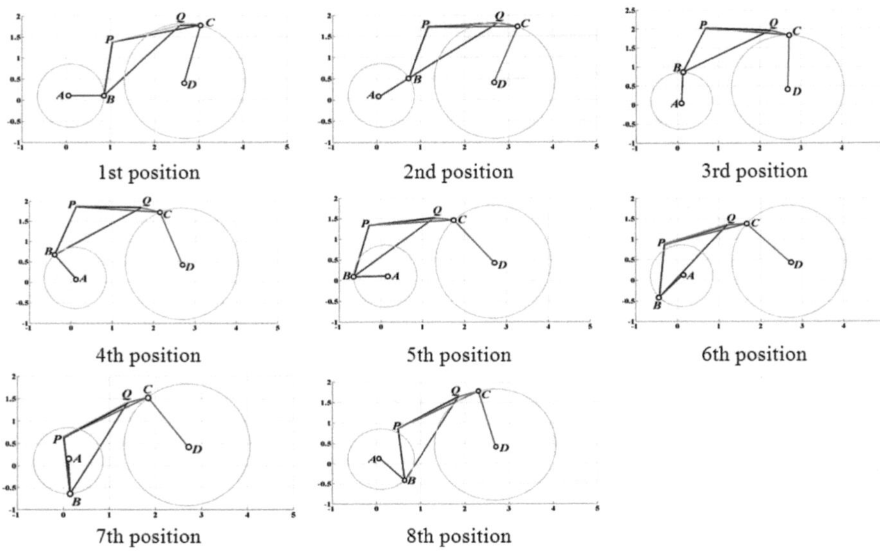


**Fig. 2.11** A mechanism approximately meets eight different positions

The method can also be directly used when more positions are required. The examples illustrate that the uniform equation is very convenient in engineering applications for the design of planar four-bar linkages with $n$ specified positions. It can search the accurate results when there are such solutions for the specified $n$ precise positions; otherwise, the best approximate solutions will be found. The merits of the method proposed in this chapter is that the engineers could directly obtain the planar four-bar linkages when the specified positions are known but need not intersect any two center/circle-point curves of four-position by combining the positions.

## 2.4 Conclusions

This chapter has proposed a uniform equation for the design of planar four-bar linkages with $n$ specified positions. The coordinates of circle point at the first position are used as the design variables to develop the distance constraint equations for the following successive positions through matrix transformation. Expanding the quadratic equations and canceling the quadratic items of the coordinates of the center point, a set of linear equations are obtained. The augmented coefficient matrix consisting the coordinates of circle point at the first position is then utilized to form a $3 \times 3$-matrix, $\mathbf{M}$. Solutions of the circle points could be represented as $|\mathbf{M}| = 0$ which provides a uniform expression for the design of planar four-bar linkages. The least-squares method is used to get the best approximate solution for the problem of more than four positions that might not have any precise solution.

## References


1. Hunt KH (1978) Kinematic geometry of mechanisms. Oxford University Press, New York
2. Suh CH, Radcliffe CW (1978) Kinematics and mechanisms design. Wiley, New York
3. Bottema O, Roth B (1979) Theoretical kinematics. North-Holland, New York
4. Angeles J (1982) Spatial kinematic chains: analysis, synthesis, and optimisation. Springer, Berlin
5. McCarthy JM, Soh GS (2011) Geometric design of linkages. Springer, New York
6. Shen Q, Al-Smadi YM, Martin PJ, Russell K, Sodhi RS (2009) An extension of mechanism design optimization for motion generation. Mech Mach Theory 44(9):1759–1767
7. Prentis JM (1991) The pole triangle, burmester theory and order and branching problems-I the order problem. Mech Mach Theory 26(1):19–30
8. Waldron KJ (1978) Location of burmester synthesis solutions with fully rotatable cranks. Mech Mach Theory 13(2):125–137
9. Waldron KJ, Strong RT (1978) Improved solutions of the branch and order problems of burmester linkage synthesis. Mech Mach Theory 13(2):199–207
10. Rao AC (1979) Synthesis of 4-bar function-generators using geometric programming. Mech Mach Theory 14(2):141–149
11. Ahmad A, Waldron KJ (1979) Synthesis of adjustable planar 4-bar mechanisms. Mech Mach Theory 14(6):405–411
12. Erdman AG (1981) Three and four precision point kinematic synthesis of planar linkages. Mech Mach Theory 16(3):227–245
13. Koetsier T (1989) The centenary of ludwig Burmester's "lehrbuch der kinematik". Mech Mach Theory 24(1):37–38
14. Nolle H (1974) Linkage coupler curve synthesis: a historical review – I. Developments up to 1875. Mech Mach Theory 9(2):147–168
15. Nolle H (1974) Linkage coupler curve synthesis: a historical review – II. Developments after 1875. Mech Mach Theory 9(3–4):325–348
16. Nolle H (1975) Linkage coupler curve synthesis: a historical review – III. Spatial synthesis and optimization. Mech Mach Theory 10(1):41–55
17. Chen Y, Fu J (1999) A computational approach for determining location of burmester solutions with fully rotatable cranks. Mech Mach Theory 34(4):549–558
18. Chen C, Angeles J (2008) A novel family of linkages for advanced motion synthesis. Mech Mach Theory 43(7):882–890
19. Burmester L (1888) Lehrbuch der kinematik. Arthur Felix Verlag, Leipzig
20. Davies TH, Baker JE, Thompson AGR (1979) A finite 3-dimensional atlas of 4-bar linkages. Mech Mach Theory 14(6):389–403
21. Chen J-S, Chiang CH (1992) Coordinations of coupler-line positions with input-link rotations by a planar four-bar linkage. Mech Mach Theory 27(5):555–561
22. Lin C-S, Erdman AG (1987) Dimensional synthesis of planar triads: motion generation with prescribed timing for six precision positions. Mech Mach Theory 22(5):411–419
23. Yang T, Han J, Yin L (2011) A unified synthesis method based on solution regions for four finitely separated and mixed "point-order" positions. Mech Mach Theory 46(11):1719–1731

24. Kim K-Y, Kwak B-M (1995) A modified selective precision synthesis technique for planar four-Bar rigid-body guidance. Proc Inst Mech Eng Part C J Mech Eng Sci 209(1):39–47
25. Han J (2004) Advanced mechanism design: analysis and synthesis. China Machine Press, Beijing, In Chinese
26. Han J, Qian W (2009) On the solution of region-based planar four-bar motion generation. Mech Mach Theory 44(2):457–465

# Chapter 3
# Design of Two Particular Positions for a Foldable Stair

This chapter proposes a foldable stair that is easily deployed for use and folded for storage. It consists of a number of identical deployable scissorlike elements which form the staircases when expanded. In addition to use, the folded stair can be used for hanging clothes and acting as a decoration. The mechanism of the stair is firstly synthesized in line with the common stairs between two horizontal levels. The actuating mechanism is then synthesized in accordance with two extreme positions of the stair, folded and unfolded. Because the stair can be folded after use, it is convenient in use and will witness a wide application both indoors and outdoors where there is no enough space for a fixed stair. In addition, this deployable stair is also particularly useful in evacuating people from a building when disasters such as earthquakes occur.

## 3.1 Introduction to the Foldable Structure

The first deployable structure was designed and constructed by Pinero in the 1960s [1, 2]. The need for mobile and reusable structures that are characterized by fast and easy erection procedures has existed for a long period [1]. From the viewpoint of theory of mechanism, the deployable structures were probed as early as the nineteenth century by Sarrus and Bricard and in the early twentieth century by Bennett et al. [3]. Deployable structures were widely surveyed and utilized in space structures with the increasing needs of complex space missions in 1970s [4]. Recently, substantial contributions to the general design of geometric and kinematic behaviors of deployable structures have been made by You and Pellegrino [5], Dai and Rees Jones [6], Gantes et al. [7], Chen and You [8], and Nagaraj et al. [9].

Deployable structures consist of units, which are composed of pairs of bars connected at joint that allows a compact and deployed configuration [10]. The configurations and even the mobility of these structures are often changing in the workspace [11]. These structures can be stored in a compact configuration and

J. Zhao et al., *Design of Special Planar Linkages*, Springer Tracts in Mechanical Engineering, DOI 10.1007/978-3-642-38448-6_3,

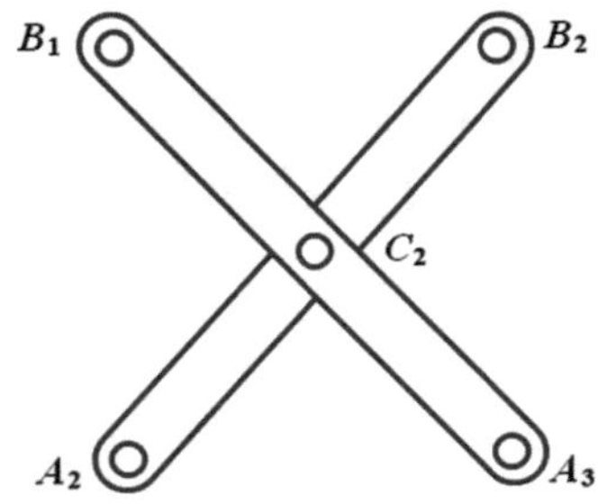


**Fig. 3.1** A scissorlike element

are designed to expand into stable structures that are capable of carrying loads after deployment [6, 12]. They are often used in the cases requiring minimal damage to structural components during repeated deployment/undeployment [13] in temporary construction industry and quick sheltering after natural disasters and in the aerospace industry [1]. The statics [14], kinematics [15, 16], and dynamics [17] of deployable structures are widely researched. Chen and You [8, 18] discussed the creations of 6R linkages by combining Goldberg 5R linkages [19]. Liu and Chen [20] proposed an assembly method to form a deployable structure using Myard linkage. The most outstanding characteristics of these deployable structures are that any two adjacent bars are all revolute jointed. This brings a great convenience to engineering applications. A fundamental design requirement of this type is that the structure is self-standing and stress-free when fully closed or fully deployed [7].

One of the most important foldable/deployable structures used in engineering is the scissorlike structure [7]. Typically, a scissorlike element (SLE) has a pair of equal length bars connected to each other at an intermediate point with a revolute joint. The joint allows the bars to rotate freely about an axis perpendicular to their common plane [12]. An SLE shown in Fig. 3.1, in mechanism theory, is one plane II-Assur group which will not induce any additional constraints to the mechanism [21]. And therefore the mobility of a mechanism to which it attaches will not change. When one or more such elements are added to the mechanism, the number of actuators needed to control the mechanism will not change [21] either.

Many applications of deployable structures can be found from a wide range of industry engineering [22, 23]. However, the foldable stairs are seldom found in civil engineering. It is well known that a staircase is a very important passage connecting two adjacent floors. Therefore, it should have enough stiffness and strength. Currently, the steel-reinforced concrete stairs are most popular in the buildings. Nevertheless, the foldable staircases are becoming feasible in engineering with the advent of high-stiffness, high-strength, and low-weight alloy materials. Consequently, this chapter proposes a foldable stair that can be folded after use for the sake of saving occupied space. The staircase adopts the SLE which has very good dexterity during folding and unfolding [2]. Therefore, this kind of foldable stair belongs among deployable structures. Next, the mobility of the foldable stair will be first investigated.

## 3.2 Mobility of the Foldable Stair

A planar four-bar linkage, $C_1A_2C_2B_1$, shown in Fig. 3.2, is an equilateral quadrangle mechanism. Therefore, when another SLE is added to connect this four-bar linkage, $C_1A_2C_2B_1$, through joints $B_2$ and $A_3$, individually, the mobility of the mechanism will not change [21].

When one or more identical SLEs shown in Fig. 3.1 are added in turn to connect the four-bar linkage in the mode discussed above, the mobility of the mechanism will not change either and the center points of all SLEs will always keep in the line $C_1C_2$ [24]. This is illustrated by the deployment process of a six-bar linkage shown in Fig. 3.3.

Suppose that the subtended angle between the radial line $B_1C_1$ and link $C_1A_2$ is denoted by $\alpha$ and the subtended angle from the radial line $B_1C_1$ to link $C_1C_2$ is denoted by $\beta$. Because link $A_2B_2$ is always parallel to the frame line $B_1C_1$, one

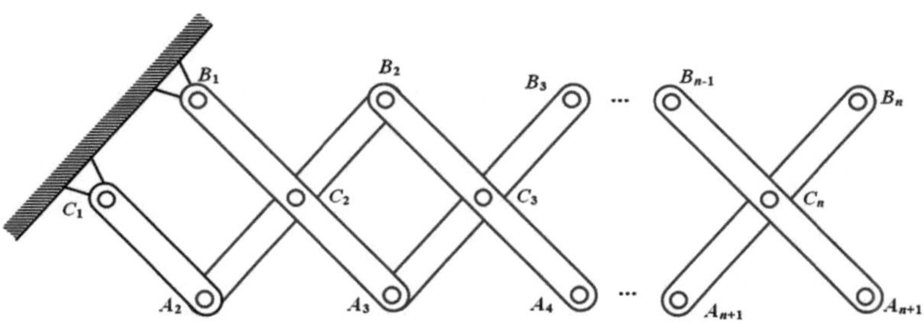


Fig. 3.2 A planar four-bar linkage

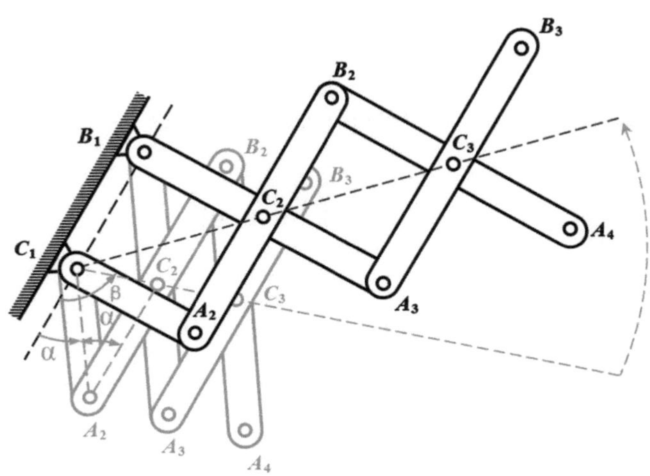


Fig. 3.3 Deployment geometry of a six-bar mechanism

immediately obtains that $\angle C_1A_2C_2 = \alpha$, and $\Delta C_1A_2C_2$ is an isosceles triangle, the two equal sides of which are $A_2C_1$ and $A_2C_2$. So,

$$\angle A_2C_1C_2 = \angle A_2C_2C_1 = \frac{1}{2}(\pi - \alpha) \tag{3.1}$$

and

$$\beta = \alpha + \angle A_2C_1C_2 = \frac{1}{2}(\pi + \alpha) \tag{3.2}$$

Links $B_1C_2$ and $B_1C_1$ will be perpendicular to each other when $\alpha = \frac{\pi}{2}$. Substituting $\alpha = \frac{\pi}{2}$ into Eq. (3.2) yields

$$\beta = \frac{3}{4}\pi \tag{3.3}$$

Figure 3.3 shows the process that the six-bar linkage unfolds to the case when any two adjacent links are perpendicular mutually. In this way, a foldable stair comes into being when a number of SLEs are added to the mechanism in series. As is discussed above, the number of actuators needed to control the folding and unfolding is still one. In order to completely and independently control the deployment of the stair, an RRR-Assur group, $A_1DA_2$, is particularly applied. Figure 3.4 shows a foldable stair with 5 staircases.

This foldable stair is a complicated parallel structure connected in series. The mobility can be analyzed with the methods proposed in [25]. In what follows, only one degree of freedom of the staircase will be checked in line with the theory of degree of freedom [26–28]:

$$\$_{A_1DA_2} = \begin{bmatrix} \$_{A_1} & \$_D & \$_{A_2} \end{bmatrix} \tag{3.4}$$

where $\$_{A_1}$, $\$_D$, and $\$_{A_2}$ denote the kinematic screws of joints $A_1$, $D$, and $A_2$, respectively, and

$$\$_{A_1} = \begin{pmatrix} 0 & 0 & 1 & 0 & 1 & 0 \end{pmatrix}^{\mathrm{T}},$$

$$\$_D = \begin{pmatrix} 0 & 0 & 1 & y_D & -x_D & 0 \end{pmatrix}^{\mathrm{T}},$$

$$\$_{A_2} = \begin{pmatrix} 0 & 0 & 1 & -l\sin\theta & l\cos\theta & 0 \end{pmatrix}^{\mathrm{T}},$$

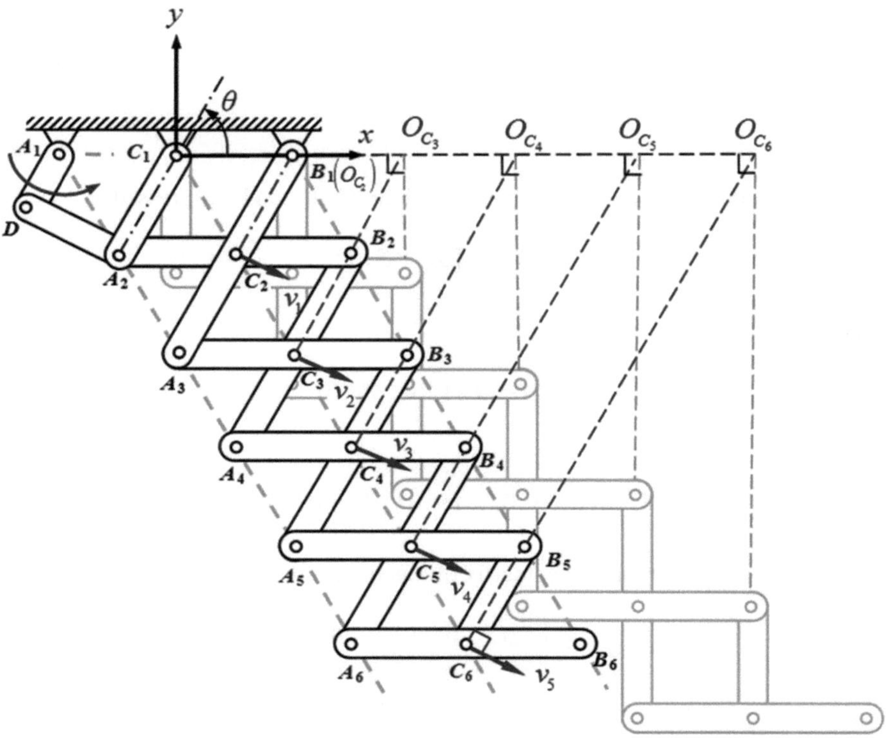


**Fig. 3.4** A foldable stair mechanism

where $l$ represents the length of $C_1A_2$ and the quadrilateral, $C_1A_2C_2B_1$, is a rhombus. Therefore, according to the screw theory, the terminal constraint screw matrix of kinematic chain, $A_1DA_2$, is immediately obtained by solving the reciprocal screw equation

$$\mathbf{\$}^{\mathrm{T}}\mathbf{\Delta}\mathbf{\$}^{\tau} = \mathbf{0} \tag{3.5}$$

where $\mathbf{\$}$ is a kinematic screw matrix, $\mathbf{\$}^{\tau}$ is the terminal constraint screw, and $\mathbf{\Delta} = \begin{bmatrix} \mathbf{0} & \mathbf{I} \\ \mathbf{I} & \mathbf{0} \end{bmatrix}$, in which $\mathbf{0}$ and $\mathbf{I}$ are $3 \times 3$-zero and unit matrices, respectively.

According to Eq. (3.5), one can obtain the terminal constraint screw matrix for kinematic chain $A_1DA_2$. This matrix is composed of a set of primary screws of the terminal constraints:

$$\$^{\tau}_{A_1DA_2} = \begin{bmatrix} 0 & 0 & 1 & 0 & 0 & 0 \\ 0 & 0 & 0 & 1 & 0 & 0 \\ 0 & 0 & 0 & 0 & 1 & 0 \end{bmatrix}^{\mathrm{T}} \tag{3.6}$$

Similarly, the screw matrix of kinematic chain $C_1A_2$ can be expressed as

$$\$_{C_1A_2} = [\$_{C_1} \quad \$_{A_2}] \tag{3.7}$$

where

$$\$_{C_1} = (0 \quad 0 \quad 1 \quad 0 \quad 0 \quad 0)^{\mathrm{T}},$$

$$\$_{A_2} = (0 \quad 0 \quad 1 \quad -l\sin\theta \quad l\cos\theta \quad 0)^{\mathrm{T}}.$$

The terminal constraint screw matrix of kinematic chain $C_1A_2$ is

$$\$^{\tau}_{C_1A_2} = \begin{bmatrix} 0 & 0 & 1 & 0 & 0 & 0 \\ 0 & 0 & 0 & 1 & 0 & 0 \\ 0 & 0 & 0 & 0 & 1 & 0 \\ \cot\theta & 1 & 0 & 0 & 0 & 0 \end{bmatrix}^{\mathrm{T}} \tag{3.8}$$

The screw matrix of kinematic chain $B_1C_2$ can be expressed as

$$\$_{B_1C_2} = [\$_{B_1} \quad \$_{C_2}] \tag{3.9}$$

where

$$\$_{B_1} = (0 \quad 0 \quad 1 \quad 0 \quad -l \quad 0)^{\mathrm{T}},$$

$$\$_{C_2} = (0 \quad 0 \quad 1 \quad -l\sin\theta \quad l\cos\theta - l \quad 0)^{\mathrm{T}}.$$

The terminal constraint screw matrix of kinematic chain $B_1C_2$ is

$$\$^{T}_{B_1C_2} = \begin{bmatrix} 0 & 0 & 1 & 0 & 0 & 0 \\ 0 & 0 & 0 & 1 & 0 & 0 \\ 0 & 0 & 0 & 0 & 1 & 0 \\ \cos\theta & \sin\theta & 0 & 0 & 0 & l\sin\theta \end{bmatrix}^{\mathrm{T}} \tag{3.10}$$

The constraint screw system exerted on $A_2C_2$ depends entirely on Eqs. (3.6), (3.8), and (3.10). Consequently, one can obtain the terminal constraint screw matrix for $A_2C_2$:

$$\$^{T}_{A_2C_2} = \begin{bmatrix} 0 & 0 & 1 & 0 & 0 & 0 \\ 0 & 0 & 0 & 1 & 0 & 0 \\ 0 & 0 & 0 & 0 & 1 & 0 \\ \cos\theta & \sin\theta & 0 & 0 & 0 & 0 \\ \cos\theta & \sin\theta & 0 & 0 & 0 & l\sin\theta \end{bmatrix}^{\mathrm{T}} \tag{3.11}$$

According to Eq. (3.5) again, one obtains the free motion screw of the first stair beam, $A_2C_2B_2$:

$$\$_{A_2C_2B_2} = \$_{A_2C_2} = \begin{pmatrix} 0 & 0 & 0 & \sin\theta & -\cos\theta & 0 \end{pmatrix}^{\mathrm{T}} \tag{3.12}$$

The second stair beam, $A_3C_3B_3$, suffers constraints from the kinematic chains $A_2C_2B_2-C_3$ and $B_1-A_3$. The kinematic screw matrix of kinematic chain $A_2C_2B_2-C_3$ can be expressed as

$$\$_{A_2C_2B_2-C_3} = \begin{bmatrix} \$_{A_2C_2B_2} & \$_{B_2} & \$_{C_3} \end{bmatrix} \tag{3.13}$$

where

$$\$_{A_2C_2B_2} = \begin{pmatrix} 0 & 0 & 0 & \sin\theta & -\cos\theta & 0 \end{pmatrix}^{\mathrm{T}},$$

$$\$_{B_2} = \begin{pmatrix} 0 & 0 & 1 & -l\sin\theta & l\cos\theta-2l & 0 \end{pmatrix}^{\mathrm{T}},$$

$$\$_{C_3} = \begin{pmatrix} 0 & 0 & 1 & -2l\sin\theta & 2l\cos\theta-2l & 0 \end{pmatrix}^{\mathrm{T}}.$$

The terminal constraint screw matrix of kinematic chain $A_2C_2B_2-C_3$ is

$$\mathbf{\$}^{\tau}_{A_2C_2B_2-C_3} = \begin{bmatrix} 0 & 0 & 1 & 0 & 0 & 0 \\ 0 & 0 & 0 & 1 & 0 & 0 \\ 0 & 0 & 0 & 0 & 1 & 0 \\ \cos\theta & \sin\theta & 0 & 0 & 0 & 2l\sin\theta \end{bmatrix}^{\mathrm{T}} \tag{3.14}$$

The screw matrix of kinematic chain $B_1 - A_3$ can be expressed as

$$\mathbf{\$}_{B_1-A_3} = \begin{bmatrix} \mathbf{\$}_{B_1} & \mathbf{\$}_{A_3} \end{bmatrix} \tag{3.15}$$

where

$$\mathbf{\$}_{B_1} = \begin{pmatrix} 0 & 0 & 1 & 0 & -l & 0 \end{pmatrix}^{\mathrm{T}},$$

$$\mathbf{\$}_{A_3} = \begin{pmatrix} 0 & 0 & 1 & -2l\sin\theta & 2l\cos\theta - l & 0 \end{pmatrix}^{\mathrm{T}}.$$

The terminal constraint screw matrix for the kinematic chain $B_1 - A_3$ is

$$\mathbf{\$}^{\tau}_{B_1-A_3} = \begin{bmatrix} 0 & 0 & 1 & 0 & 0 & 0 \\ 0 & 0 & 0 & 1 & 0 & 0 \\ 0 & 0 & 0 & 0 & 1 & 0 \\ \cos\theta & \sin\theta & 0 & 0 & 0 & l\sin\theta \end{bmatrix}^{\mathrm{T}} \tag{3.16}$$

The constraint screw system applied to $A_3C_3B_3$ depends completely on Eqs. (3.14) and (3.16). Consequently, one immediately obtains the terminal constraint screw matrix for $A_3C_3B_3$:

$$\mathbf{\$}^{\tau}_{A_3B_3C_3} = \begin{bmatrix} 0 & 0 & 1 & 0 & 0 & 0 \\ 0 & 0 & 0 & 1 & 0 & 0 \\ 0 & 0 & 0 & 0 & 1 & 0 \\ \cos\theta & \sin\theta & 0 & 0 & 0 & l\sin\theta \\ \cos\theta & \sin\theta & 0 & 0 & 0 & 2l\sin\theta \end{bmatrix}^{\mathrm{T}} \tag{3.17}$$

Therefore, the free motion screw of $A_3C_3B_3$ can be obtained in accordance to Eq. (3.15)

$$\mathbf{\$}_{A_3B_3C_3} = \begin{pmatrix} 0 & 0 & 0 & \sin\theta & -\cos\theta & 0 \end{pmatrix}^{\mathrm{T}} \tag{3.18}$$

Equation (3.18) indicates that the second stair beam, $A_3C_3B_3$, has the same motion with that of the first stair beam, $A_2C_2B_2$, represented by Eq. (3.12). With a similar process discussed above, one can find that the free motion of every stair beam can be expressed by Eq. (3.12) or (3.18). Therefore, each staircase of the structure has the same translational degree of freedom.

Next, one further analyzes the motions of the points on the vertical links. For example, one can investigate the motion of joint $C_3$. The joint $C_3$ suffers constraints from the kinematic chains $A_2C_2B_2 - B_2C_3$ and $C_2 - A_3 - C_3$ (or $B_1 - A_3 - C_3$ which can be similarly analyzed to be identical to $C_2 - A_3 - C_3$ in the following process). The screw matrix of kinematic chain $A_2C_2B_2 - B_2C_3$ can be expressed by

$$\$_{A_2C_2B_2-B_2C_3} = \begin{bmatrix} \$_{A_2C_2B_2} & \$_{B_2} & \$_{C_3} \end{bmatrix}$$

where

$$\$_{A_2C_2B_2} = \begin{pmatrix} 0 & 0 & 0 & \sin\theta & -\cos\theta & 0 \end{pmatrix}^{\mathrm{T}},$$

$$\$_{B_2} = \begin{pmatrix} 0 & 0 & 1 & -l\sin\theta & l\cos\theta - 2l & 0 \end{pmatrix}^{\mathrm{T}},$$

$$\$_{C_3} = \begin{pmatrix} 0 & 0 & 1 & -2l\sin\theta & 2l\cos\theta - 2l & 0 \end{pmatrix}^{\mathrm{T}}.$$

The terminal constraint screw matrix of kinematic chain $A_2C_2B_2 - B_2C_3$ is

$$\$^{r}_{A_2C_2B_2-B_2} = \begin{bmatrix} 0 & 0 & 1 & 0 & 0 & 0 \\ 0 & 0 & 0 & 1 & 0 & 0 \\ 0 & 0 & 0 & 0 & 1 & 0 \\ \cos\theta & \sin\theta & 0 & 0 & 0 & 2l\sin\theta \end{bmatrix}^{\mathrm{T}} \tag{3.19}$$

Similarly, the kinematic screw matrix of kinematic chain $C_2 - A_3 - C_3$ can be expressed as

$$\$_{C_2-A_3-C_3} = [\$_{A_2C_2B_2} \quad \$_{A_3} \quad \$_{C_3}] \tag{3.20}$$

where

$$\$_{A_3} = \begin{pmatrix} 0 & 0 & 1 & -2l\sin\theta & 2l\cos\theta - l & 0 \end{pmatrix}^{\mathrm{T}},$$

$$\$_{C_3} = \begin{pmatrix} 0 & 0 & 1 & -2l\sin\theta & 2l\cos\theta - 2l & 0 \end{pmatrix}^{\mathrm{T}}.$$

The terminal constraint screw matrix of kinematic chain $C_2 - A_3 - C_3$ is

$$\mathbf{\$}^{\tau}_{B_1-A_3-C_3} = \begin{bmatrix} 0 & 0 & 1 & 0 & 0 & 0 \\ 0 & 0 & 0 & 1 & 0 & 0 \\ 0 & 0 & 0 & 0 & 1 & 0 \end{bmatrix}^{\mathrm{T}} \tag{3.21}$$

The constraint screw system exerted on joint $C_3$ depends entirely on Eqs. (3.19) and (3.21). Consequently, one obtains the terminal constraint screw matrix for $C_3$:

$$\mathbf{\$}^{\tau}_{B_2C_3A_4} = \begin{bmatrix} 0 & 0 & 1 & 0 & 0 & 0 \\ 0 & 0 & 0 & 1 & 0 & 0 \\ 0 & 0 & 0 & 0 & 1 & 0 \\ \cos\theta & \sin\theta & 0 & 0 & 0 & 2l\sin\theta \end{bmatrix}^{\mathrm{T}} \tag{3.22}$$

According to Eq. (3.5) again, the screw matrix of $B_2C_3A_4$ can be obtained:

$$\mathbf{\$}_{B_2C_3A_4} = \begin{bmatrix} 0 & 0 & 1 & 0 & -2l & 0 \\ 0 & 0 & 0 & \sin\theta & -\cos\theta & 0 \end{bmatrix}^{\mathrm{T}} \tag{3.23}$$

The first screw is a rotation about the virtual fixed joint, $O_{C_3}$, shown in Fig. 3.4, when the angular coordinate is $\theta$, while the second screw is an instantaneous translation along the direction $\left(\sin\theta \;\; -\cos\theta \;\; 0\right)^{\mathrm{T}}$. Therefore, the two screws represented in Eq. (3.23) are not independent but restricted by the actuating parameter, $\theta$. This can be interpreted by the geometry shown in Fig. 3.4. The velocity vectors $v_i$ $(i = 1, 2, 3 \ldots,)$ shown in Fig. 3.4 denote the velocity directions of each vertical stair beam, and the motions can be considered as an instantaneous rotation around a number of fixed points, $O_{C_3}, O_{C_4}, O_{C_5}, \ldots,$ on the $x$-axis.

## 3.3 Synthesis of the Foldable Stair

In order to synthesize the foldable stair structure, two extreme positions should be provided. This, as a matter of fact, is the size synthesis of a planar four-bar mechanism shown in Fig. 3.5.

Suppose that the four-bar linkage, $A_1DA_2C_1$, is a crank-rocker mechanism and the length of $A_1C_1$ is identical to that of rocker $C_1A_2$. Because the crank-rocker mechanism acts as an actuator to the foldable stair, the primary parameters of $A_1DA_2C_1$ should satisfy the two extreme positions of the stair, namely, completely

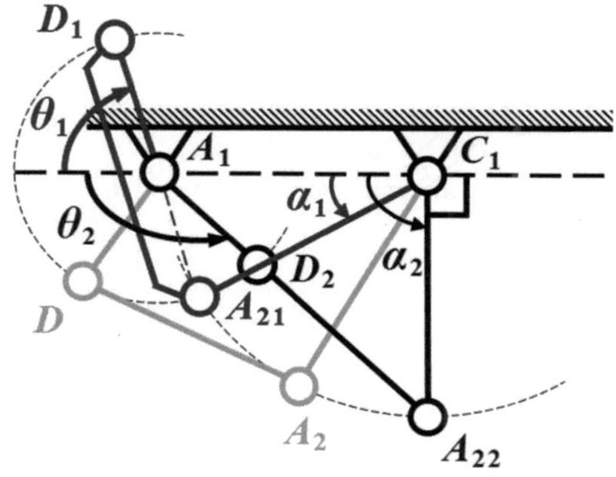


**Fig. 3.5** Synthesis of an actuating mechanism

folded and completely unfolded cases. For rocker $C_1A_2$, the two extreme positions are corresponding to the cases when the stair mechanism is wholly folded, $\alpha_1 = \frac{\pi}{6}$, and $\alpha_2 = \frac{\pi}{2}$ when wholly unfolded.

Suppose the length of link $A_1C_1$ is $l$. The following relationship holds:

$$l_{C_1A_1} = l_{C_1A_{21}} = l_{C_1A_{22}} = l \tag{3.24}$$

where $l_{C_1A_1}, l_{C_1A_{21}}$, and $l_{C_1A_{22}}$ represent the lengths of links $C_1A_1, C_1A_{21}$, $C_1A_{22}$, and $C_1A_{22}$, respectively.

In triangle $C_1A_1A_{21}$, the length of $A_1A_{21}$ is obtained in line with the law of cosine

$$l_{A_1A_{21}} = \sqrt{l^2 - 2l^2\cos\alpha_1 + l^2} = \sqrt{2-\sqrt{3}}l \tag{3.25}$$

where $l_{A_1A_{21}}$ denotes the length of $A_1A_{21}$.

Similarly, in the right triangle, $C_1A_1A_{22}$, the length of $A_1A_{22}$ is obtained in accordance to Pythagorean theorem

$$l_{A_1A_{22}} = \sqrt{l^2 + l^2} = \sqrt{2}l \tag{3.26}$$

where $l_{A_1A_{22}}$ denotes the length of $A_1A_{22}$.

If the two extreme positions are self-locked, then the crank $A_1D$ and the link $DA_2$ will locate in a line. As a result, one immediately obtains

$$\begin{cases} l_{A_2D} - l_{A_1D} = l_{A_1A_{21}} \\ l_{A_2D} + l_{A_1D} = l_{A_1A_{22}} \end{cases} \tag{3.27}$$

where $l_{A_2D}$ and $l_{A_1D}$ represent the lengths of $A_2D$ and $A_1D$, respectively. Substituting Eqs. (3.25) and (3.26) into (3.27) yields

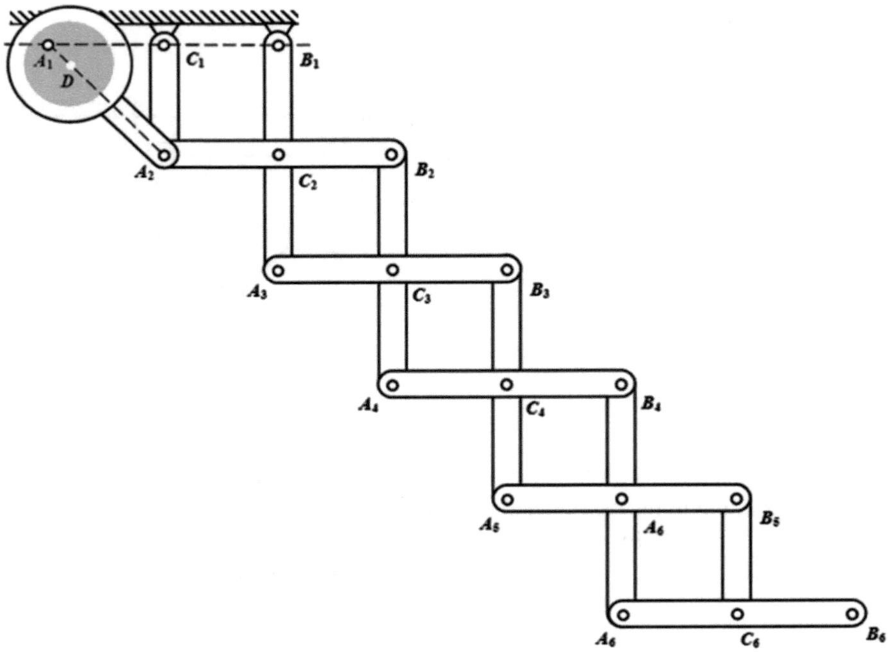


**Fig. 3.6** Unfolded stair mechanism

$$\begin{cases} l_{A_2D} - l_{A_1D} = \sqrt{2-\sqrt{3}}l \\ l_{A_2D} + l_{A_1D} = \sqrt{2}l \end{cases} \tag{3.28}$$

Solving Eq. (3.28), one obtains

$$\begin{cases} l_{A_1D} = \frac{1}{2}\left(\sqrt{2} - \sqrt{2-\sqrt{3}}\right)l \\ l_{A_2D} = \frac{1}{2}\left(\sqrt{2} + \sqrt{2-\sqrt{3}}\right)l \end{cases} \tag{3.29}$$

Considering the self-lock reliability and the strength of a short crank, one can use eccentric wheel instead of a short link. The deployed stair is shown in Fig. 3.6. The eccentric wheel jointed at point $A_1$ is the actuator. The distance between the geometry center of the eccentric wheel, $D$, and the pivoted point, $A_1$, is identical to the length of crank calculated in equation (3.29).

When the stair is folded, one or more clamp holders can be designed to hang the folded stair. Figure 3.7 shows the cases when the deployable stair is folded and clamped by two holders.

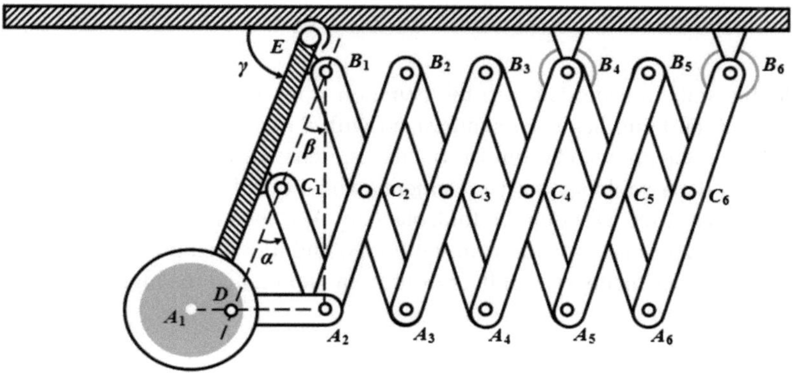


**Fig. 3.7** Folded stair

**Fig. 3.8** The folded and deployed stair model

Connecting joints $B_1$ and $A_2$, one can find that $B_1C_1A_2$ is an isosceles triangle, the two identical sides of which are $C_1B_1$ and $C_1A_2$, individually. One immediately finds out that

$$\beta = \frac{1}{2}\alpha = \frac{\pi}{12} \tag{3.30}$$

Generally, it is expected that line $A_2B_1$ is perpendicular to the ceiling plane. Therefore,

$$\gamma = \frac{\pi}{2} - \beta = \frac{5\pi}{12} \tag{3.31}$$

where $\gamma$ denotes the angular displacement of the support bard $ED$.

With the above parameters, a foldable stair model shown in Fig. 3.8 is built and experimented. The motion dexterity and flexibility are all very well within the desired space. The stair model demonstrates that the foldable stair is completely feasible in mechanism.

The foldable stair is very convenient to fold and unfold. It is available to substitute a concrete stair between two adjacent floors of a general building.

## 3.4 Conclusions

A foldable stair in accordance to the common stairs between two adjacent floors of a general building has been synthesized with a number of identical deployable scissorlike elements which could form the staircases when expanded. A number of identical deployable scissorlike elements which could form the staircases when expanded were used to set up the foldable stair. And then the actuating mechanism was synthesized in accordance with the folded and unfolded positions of the stair. Stair model experiments proved the dexterity in motion and the high strength of the deployed structure.

## References

1. Kaveht A, Davaranl A (1996) Analysis of pantograph foldable structures. Comput Struct 59(1):131–140
2. Escrig F, Valcarcel P (1996) Geometry of expandable space structures. Int J Space Struct 11(1):257–274
3. Chen Y, You Z, Tarnai T (2005) Threefold-symmetric bricard linkages for deployable structures. Int J Solids Struct 42(8):2287–2301
4. Chen W-J, Luo L-Y et al (2001) A study on space masts basedon octahedral truss family. Int J Space Struct 16(1):75–81
5. You Z, Pellegrino S (1997) Foldable bar structures. Int J Solids Struct 34(15):1825–1847
6. Dai JS, Jones JR (1999) Mobility in metamorphic mechanisms of foldable/erectable kinds. Trans ASME J Mech Des 121(3):375–382
7. Gantes C, Giakoumakis A, Vousvounis P (1997) Symbolic manipulation as a tool for design of deployable domes. Comput Struct 64(1–4):865–878
8. Chen Y, You Z (2009) Two-fold symmetrical 6R foldable frame and its bifurcations. Int J Solids Struct 46(25–26):4504–4514
9. Nagaraj BP, Pandiyan R, Ghosal A (2009) Kinematics of pantograph masts. Mech Mach Theory 44(4):822–834
10. Langbecker T (1999) Kinematic analysis of deployable scissor structures. Int J Space Struct 14(1):1–15
11. Wei G, Ding X, Dai JS (2010) Mobility and geometric analysis of the hoberman switch-pitch ball and its variant. Trans ASME J Mech Robot 2:031010
12. Nagaraj BP, Pandiyan R, Ghosal A (2010) A constraint jacobian based approach for static analysis of pantograph masts. Comput Struct 88(1–2):95–104
13. Hanaor A, Levy R (2001) Evaluation of deployable structures for space enclosures. Int J Space Struct 16(4):211–229
14. Mirats Tur JM, Juan SH (2008) Tensegrity frameworks: static analysis review. Mech Mach Theory 43(7):859–881
15. Mirats Tur JM, Juan SH (2009) Tensegrity frameworks: dynamic analysis review and open problems. Mech Mach Theory 44(1):1–18
16. Gan WW, Pellegrino S (2006) Numerical approach to the kinematic analysis of deployable structures forming a closed loop. Proc Inst Mech Eng Part C J Mech Eng Sci 220(7):1045–1056
17. Xu LJ, Tian GY, Duan Y, Yang SX (2001) Inverse kinematic analysis for triple-octahedron variable-geometry truss manipulators. Proc Inst Mech Eng Part C J Mech Eng Sci 215(2):247–251

18. Chen Y, You Z (2007) Spatial 6R linkages based on the combination of two goldberg 5R linkages. Mech Mach Theory 42(11):1484–1489
19. Wohlhart K (1991) Merging two general goldberg 5R linkages to obtain a new 6R space mechanism. Mech Mach Theory 26(7):659–668
20. Liu SY, Chen Y (2009) Myard linkage and its mobile assemblies. Mech Mach Theory 44(10):1950–1963
21. Shigley JE, Uicher JJ (1980) Theory of machines and mechanisms. McGraw-Hill Companies, Inc., New York
22. Rosenfeld Y, Logcher RD (1988) New concepts for deployable collapsable structures. Int J Space Struct 3(1):20–32
23. Escrig F, Valcarcel JP, Sanchez J (1996) Deployable cover on a swimming pool in seville. J Int Assoc Shell Spat Struct 37(1):39–70
24. Zhao J-S, Wang J-Y, Chu F, Feng Z-J, Dai JS (2011) Structure synthesis and statics analysis of a foldable stair. Mech Mach Theory 46(7):998–1015
25. Dai JS, Li D, Zhang QX, Jin GG (2004) Mobility analysis of a complex structured ball based on mechanism decomposition and equivalent screw system analysis. Mech Mach Theory 39(4):445–458
26. Zhao J-S, Zhou K, Feng Z-J (2004) A theory of degrees of freedom for mechanisms. Mech Mach Theory 39(6):621–643
27. Zhao J-S, Feng Z-J, Wang L-P, Dong J-X (2006) The free mobility of a parallel manipulator. Robotica 24(5):635–641
28. Zhao J-S, Feng Z-J, Dong J-X (2006) Computation of the configuration degree of freedom of a spatial parallel mechanism by using reciprocal screw theory. Mech Mach Theory 41(12):1486–1504

# Chapter 4
# Kinematic Design of Geared Five-Bar Linkage

This chapter starts from the discussion of the number of points that a planar four-bar linkage can precisely trace and then proposes a noncircular gear coupled steering mechanism which can strictly follow the Ackermann steering rule compared with the existing four-bar steering linkages.

## 4.1 Introduction to Vehicle Steering Mechanisms

The generally used steering mechanisms for four-wheel vehicles are four-bar linkages [1, 2] which are often called Ackermann-type steering mechanisms [1–6]. The input motion from the driver to the steering wheel is transmitted via a steering box and the steering control linkage to one of the steering knuckles and then transmitted to the other one through the Ackermann steering linkage [6]. The main kinematic requirement of the steering linkage of a vehicle is to provide the steerable wheels a correlated pivot such that their axes intersect at a point on the rear wheel axis [7]. The objective for the synthesis of a steering mechanism is to minimize the difference between the steering centers over the full range of steering angle inputs while fitting into a reasonable space [8]. To reach the target, several conflicting requirements should be simultaneously considered [5]. A possible formulation of the optimization-based synthesis problem is to search for the values of parameters [7].

Most trucks and off-highway vehicles have rigid steering axles equipped with Ackermann steering linkage [6]. So synthesis of function generators is a common mechanism design problem for steering systems of vehicles. The deviation between the desired and the real pivoting angles given to the wheels by the steering mechanism is called the steering error [7]. It requires finding the geometric parameters for which the input–output relationship of the mechanism best approximates a specified function [4, 9]. Error minimization studies in steering linkages have been attempted

J. Zhao et al., *Design of Special Planar Linkages*, Springer Tracts in Mechanical Engineering, DOI 10.1007/978-3-642-38448-6_4,

by many researchers [10]. The primary goals are ensuring minimum wheel-slip and symmetrical steering control for left and right turns, ensuring minimum cross-coupling between steering and axle oscillation, maintaining favorable pressure angles in the joints, and avoiding interference between the moving parts of the mechanism and between them and the body of the vehicle [5].

For the purpose of improving maneuverability at low speed and controllability and stability at high speed [11], a general method of optimal synthesis of steering mechanisms is developed by displacement matrix and constraints [3]. Design charts are presented which considerably facilitates the work of engineers in determining the optimum geometry of the central-lever steering mechanism of a vehicle [7]. Also, the minimum of the maximum steering errors of the steering linkage are considered as one of the optimization targets [8]. The sensitivity analysis is investigated with the design charts on which how parametric errors affect the performance of the steering mechanism [12]. A kinematic model of a rack-and-pinion-type steering linkage is developed, on the basis of which the synthesis of the mechanism is performed using the criteria of correct turning of the wheels and good transmissibility of the motion [13]. The practical cases of the rack-and-pinion and central-lever steering linkages are examined as planar mechanisms, and a maximum number of only three and four independent design parameters, respectively, are proven to exist [14]. A vehicle model is developed employing a detailed nonlinear kinematic representation of a MacPherson strut front independent automotive suspension with a rack-and-pinion steering system [15]. In the kinematic study of the McPherson-type steering suspension, the departure in the suspension geometry is expressed as a function of the travel of the strut and of the turn of the steering wheel [16].

In addition, differential mechanism coupled to a higher pair resolves the single input coming from the steering wheel into two outputs satisfying the condition of correct steering [17]. Structure classification, number, and dimensional synthesis of independent suspension mechanisms were investigated through the atlas of in-parallel suspension linkages [18]. Similarly the steering mechanism, which is often proposed as planar linkages [4, 10, 19], can be synthesized to satisfy the needs of number and size in engineering. However, the synthesis of a steering mechanism that meets the requirements of Ackermann turning geometry is far more complicated than the structure only. It is an exact trajectory tracing mechanism. Therefore, the Ackermann criteria should be put into the dimension synthesis of the steering mechanism. Noncircular gears make it possible to synthesize a steering mechanism that satisfies the Ackerman criteria [20, 21]. However, the new steering mechanism [20] and the eight-link mechanism [21] are much more complicated than the traditional four-bar steering systems.

This chapter starts from the discussion of the number of points that a planar four-bar linkage can precisely trace and then proposes a noncircular gear coupled steering mechanism which can strictly follow the Ackermann steering rule compared with its counterparts.

## 4.2 Restrictions of a Four-Bar Steering Linkage

The existing mechanical steering systems used in the automobiles all belong to four-bar linkages. To comprehensively understand the characteristics of the four-bar steering mechanism, the ideal turning criteria for a vehicle should be first discussed.

### 4.2.1 *Ackermann Criteria for Turning Wheels*

Figure 4.1 illustrates the ideal geometry conditions for an automobile to turn right when the four wheels only make pure rotations about an instant center $o$.

This geometry requirement shown in Fig. 4.1 can be expressed as

$$l\cot\theta_1 - l\cot\theta_2 = b \tag{4.1}$$

where $l$ denotes the wheelbase, $b$ represents the track of front wheels, and $\theta_1$ and $\theta_2$ indicate the due angles of the front left and front right wheels.

Rearranging Eq. (4.1) presents

$$\cot\theta_1 - \cot\theta_2 = \frac{b}{l} \tag{4.2}$$

Equation (4.2) is called the ideal Ackermann turning criteria [2, 7, 10]. Suppose that the turning angles shown in Fig. 4.1 are the upper limits when turning right. To keep the front wheels always satisfying this ideal constraint, the instant centers of the four wheels will shift from point $o$ to infinity.

However, if the turning angles do not satisfy the Ackermann turning criteria, the rotation centers of front left wheel and front right wheel will not coincide with each

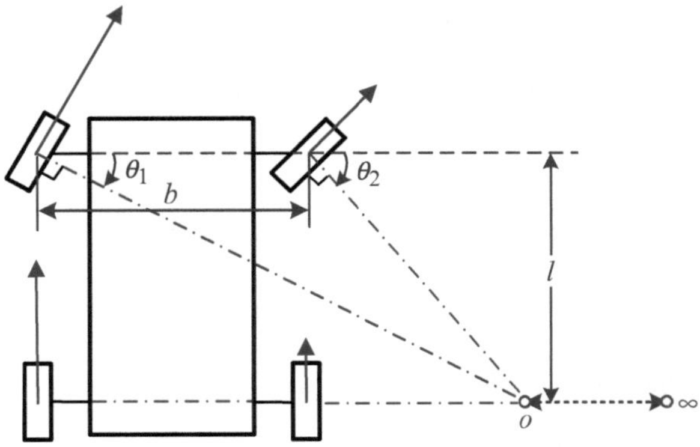


**Fig. 4.1** Ideal Ackermann turning geometry

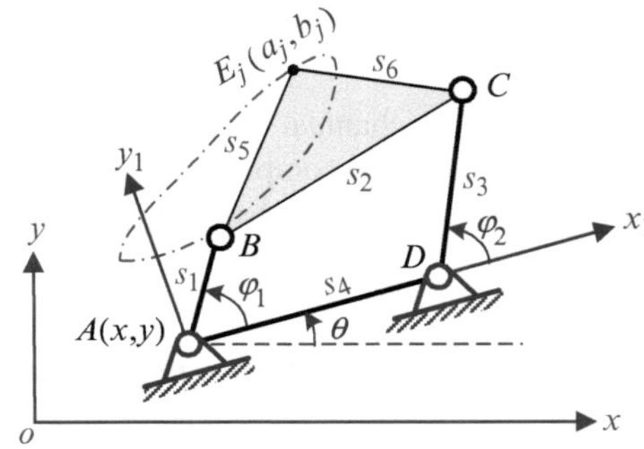


**Fig. 4.2** Trace of a four-bar linkage

other. In this instance, the wheels are not only rotating but also sliding. Hence, the steering error can significantly affect the tire wear, especially in high speed [22]. As a result, a steering mechanism meeting the requirements of Ackermann turning criteria is very meaningful in applications.

Can the steering four-bar linkage meet the needs of ideal Ackermann turning criteria? What follows will first discuss the maximal number of points that a general planar four-bar linkage can exactly trace.

### 4.2.2 *The Maximum Number of Points a Planar Four-Bar Linkage Could Trace*

Suppose that there is a planar four-bar mechanism shown in Fig. 4.2. Its end effector is triangle $BCE$.

In the planar four-bar mechanism system shown in Fig. 4.2, there are nine independent variables, namely, $x, y, \theta$, and $s_i$ $(i = 1, 2, \ldots, 6)$. Now, a pair of variables $\varphi_{1i}$ and $\varphi_{2i}$ are introduced for a specified point $E_j$ $(j = 1, 2, \ldots)$. In the coordinate frame $x_1Ay_1$, the coordinates of the four joints are

$$\boldsymbol{r}_{A_1} = \begin{bmatrix} 0 \\ 0 \end{bmatrix}, \quad \boldsymbol{r}_{B_1} = s_1 \begin{bmatrix} \cos\varphi_1 \\ \sin\varphi_1 \end{bmatrix}, \quad \boldsymbol{r}_{C_1} = \begin{bmatrix} s_4 + s_3\cos\varphi_2 \\ s_3\sin\varphi_2 \end{bmatrix}, \quad \boldsymbol{r}_{D_1} = \begin{bmatrix} s_4 \\ 0 \end{bmatrix} \tag{4.3}$$

The absolute coordinates of these four joints can be expressed as

$$\boldsymbol{r}_X = \boldsymbol{r}_A + \mathbf{R}\boldsymbol{r}_{X_1} \tag{4.4}$$

where $X = A, B, C, D$, and therefore, $\boldsymbol{r}_{X_1}$ represents the vectors in Eq. (4.3), $\boldsymbol{r}_A = \begin{bmatrix} x \\ y \end{bmatrix}$ and $\mathbf{R} = \begin{bmatrix} \cos\theta & -\sin\theta \\ \sin\theta & \cos\theta \end{bmatrix}$.

Substituting Eq. (4.3) in (4.4), we obtain

$$
\begin{aligned}
\boldsymbol{r}_A &= \begin{bmatrix} x \\ y \end{bmatrix}, \quad \boldsymbol{r}_B = \begin{bmatrix} x + s_1 \cos(\varphi_1 + \theta) \\ y + s_1 \sin(\varphi_1 + \theta) \end{bmatrix}, \\
\boldsymbol{r}_{C_1} &= \begin{bmatrix} x + s_4 \cos\theta + s_3 \cos(\varphi_2 + \theta) \\ y + s_4 \sin\theta + s_3 \sin(\varphi_2 + \theta) \end{bmatrix}, \quad \boldsymbol{r}_{D_1} = \begin{bmatrix} x + s_4 \cos\theta \\ s_4 \sin\theta \end{bmatrix}
\end{aligned}
\tag{4.5}
$$

For $E_j$ $(j = 1, 2, \ldots)$ are known points, we assume that

$$
\boldsymbol{r}_{E_j} = \begin{bmatrix} x_j \ y_j \end{bmatrix}^{\mathrm{T}} \tag{4.6}
$$

The following equations will always hold:

$$
\begin{cases}
\|\boldsymbol{r}_C - \boldsymbol{r}_B\| = s_2 \\
\|\boldsymbol{r}_{E_i} - \boldsymbol{r}_B\| = s_5 \\
\|\boldsymbol{r}_{E_i} - \boldsymbol{r}_C\| = s_6
\end{cases}
\tag{4.7}
$$

Therefore, one additional independent equation should be provided whenever a new point $E_i$ is introduced into the constraint system besides the pair of variables $\varphi_1$ and $\varphi_2$ correspondingly. As a result, the nine independent variables need and only need nine different points to be uniquely determined. So a planar four-bar linkage can only exactly trace nine points which are corresponding to five different position constraints discussed in Chap. 2. For a trapezoid steering four-bar linkage, the five constraints are as follows: (1) isosceles sides of the quadrangle linkage, (2) the status that the vehicle runs at a straight line is the second "position" constraint, and (3) three "points" on the rear axis which are corresponding to three different positions of the link. In this regard, the four-bar steering linkage can only precisely trace three specified positions when turning. Anyway, the trapezoid steering four-bar mechanism cannot always follow the Ackermann criteria from point $o$ to infinity. Therefore, the existing mechanical steering mechanisms could not always satisfy the Ackermann criteria when steering. Although the deformations resulting from the elastic links and rubber bushing can modify the Ackermann function, the method to synthesize such function-driven mechanism is urgently necessary in engineering applications and mechanism theory.

## 4.3 Ackermann-Type Steering Mechanism

This section focuses on the synthesis of Ackermann steering geometry and proposes a steering mechanism that can satisfy the ideal turning requirements.

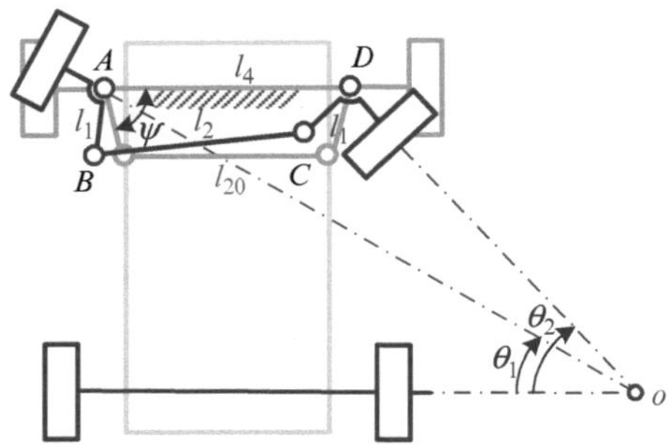


**Fig. 4.3** Ideal steering mechanism for Ackermann turning geometry

## *4.3.1 Requirements for an Ackermann-Type Steering Mechanism*

Suppose the bottom angle of a trapezoid linkage $ABCD$, shown in Fig. 4.3, is denoted by $\psi$, then when the turning angle equals zero, the initial length of link $BC$, $l_{20}$, can be expressed as

$$l_{20} = l_4 - 2l_1 \cos\psi \tag{4.8}$$

where $l_1$ denotes the length of link AB and $l_4$ denotes the length of link AD.

The steering linkage that meets the needs of Ackermann turning geometry is a particular path-tracing mechanism. During turning right, the left wheel turned $\theta_1$ and the right wheel turned $\theta_2$ correspondingly. To satisfy the Ackermann turning geometry, the length of the link should be

$$\begin{aligned} l_2 &= \sqrt{[l_4 - l_1\cos(\psi-\theta_2) - l_1\cos(\psi+\theta_1)]^2 + [l_1\sin(\psi-\theta_2) - l_1\sin(\psi+\theta_1)]^2} \\ &= \sqrt{2l_1^2 + l_4^2 + 2l_1^2\cos(2\psi+\theta_1-\theta_2) - 2l_1l_4[\cos(\psi-\theta_2)+\cos(\psi+\theta_1)]} \end{aligned} \tag{4.9}$$

where $l_2$ represents the length of link $BC$ that satisfies the Ackermann turning criterion.

From Eq. (4.2), there is

$$\theta_2 = \operatorname{arc\,cot}\left(\cot\theta_1 - \frac{b}{l}\right) \tag{4.10}$$

Substituting Eq. (4.10) into (4.9) yields

$$\begin{aligned} l_2 = l_4 &\sqrt{1 + 2\xi^2 + 2\xi^2\cos[2\psi + \theta_1 - \operatorname{arc\,cot}(\cot\theta_1 - \varsigma)]} \\ &\times \sqrt{-2\xi\{\cos(\psi+\theta_1) + \cos[\psi - \operatorname{arc\,cot}(\cot\theta_1 - \varsigma)]\}} \end{aligned} \tag{4.11}$$

where $\xi = \frac{l_1}{l_4}$, $\varsigma = \frac{b}{l}$.

Equation (4.11) indicates that the length of the link, $l_2$, should change with the change of the turning angle, $\theta_1$, if the Ackermann criteria were satisfied.

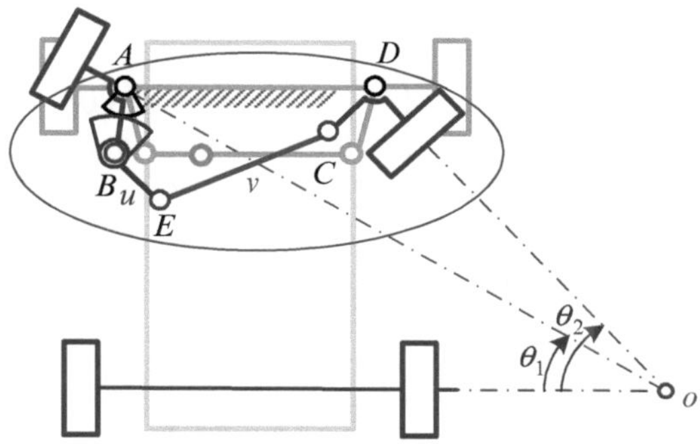


**Fig. 4.4** Geared steering mechanism to meet Ackermann turning geometry

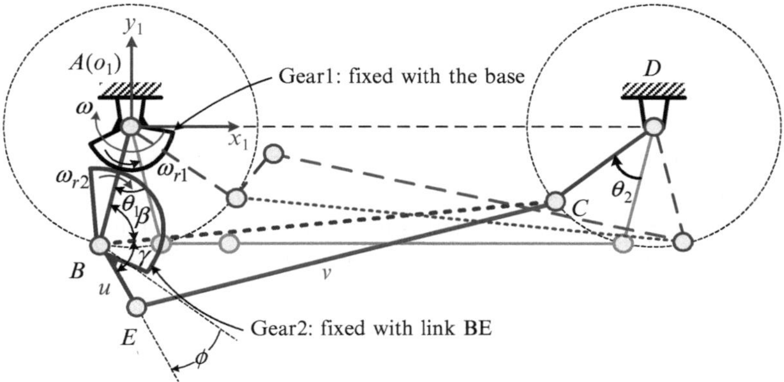


**Fig. 4.5** Geared five-bar steering mechanism

## 4.3.2 *Synthesis of the Ackermann-Type Steering Mechanism*

Suppose the original fixed link $BC$ are replaced by an Assur II-group linkage [23] $BEC$ shown in Fig. 4.4. It is an RRR-group linkage. Simultaneously, the link $BE$ is welded with a noncircular incomplete gear which meshes another noncircular incomplete gear fixed with the chassis. So, the synthesis of the steering mechanism is now transformed into the design of a pair of noncircular gears and an Assur II-group of revolute-jointed linkage $BEC$ shown in the circled section of Fig. 4.4 to satisfy Eq. (4.11). The link $BE$ and link $EC$ keep in one horizontal line at the position of zero turning which is shown by the light gray lines in Fig. 4.4.

Suppose that the length of link $BE$ is represented by $u$ and the length of link $EC$ is represented by $v$. The detailed structure of the geared five-bar steering mechanism is shown in Fig. 4.5 where $\omega$ is the angular velocity of link $AB$, $\omega_{r1}$ is the relative angular velocity of noncircular gear 1 with respect to link $AB$, $\omega_{r2}$ is the relative angular velocity of noncircular gear 2 with respect to link $AB$, $\beta$ is the angle between $AB$ and $BC$, and $\gamma$ is the angle between $BC$ and $BE$. Assume that the vehicle is turning from left to right; then the angular velocity directions of $\omega_{r1}$ and $\omega_{r2}$ are shown in Fig. 4.5.

The coordinates of joints, $A$, $B$, and $C$, in the coordinate system $x_1 o_1 y_1$ are

$$\begin{cases} A\left(0\ 0\right) \\ B\left(l_1 \cos\left(\psi + \theta_1\right)\ -l_1 \sin\left(\psi + \theta_1\right)\right) \\ C\left(l_4 - l_1 \cos\left(\psi - \theta_2\right)\ -l_1 \sin\left(\psi - \theta_2\right)\right) \end{cases} \tag{4.12}$$

So the angle $\beta$ is indicated by

$$\begin{aligned} \beta &= \arccos\left(\frac{\overrightarrow{BA} \cdot \overrightarrow{BC}}{\left\|\overrightarrow{BA}\right\| \left\|\overrightarrow{BC}\right\|}\right) \\ &= \arccos\left\{\frac{l_1 \left[1 + \cos\left(2\psi + \theta_1 - \theta_2\right)\right] - l_4 \cos\left(\psi + \theta_1\right)}{l_2}\right\} \end{aligned} \tag{4.13}$$

where $\overrightarrow{BA}$ stands for a vector from point $B$ to point $A$, "·" indicates the inner product of two vectors, and $\left\|\overrightarrow{BA}\right\|$ represents the Euclidean norm of vector $\overrightarrow{BA}$. In addition, the angle $\beta$ is always in the interval of $[0, \pi]$ considering the actual steering situations. Because $\arccos(x)$ is monotonic and real within the interval of $[0, \pi]$ in accordance with the definition of the inverse cosine, Eq. (4.13) has a unique solution.

Substituting Eq. (4.11) into (4.13) yields

$$\begin{aligned} \beta = \arccos &\left(\frac{\xi \left\{1 + \cos\left[2\psi + \theta_1 - \operatorname{arc\,cot}\left(\cot\theta_1 - \varsigma\right)\right]\right\}}{\sqrt{1 + 2\xi^2 + 2\xi^2 \cos\left[2\psi + \theta_1 - \operatorname{arc\,cot}\left(\cot\theta_1 - \varsigma\right)\right]}}\right) \\ &\times \frac{-\cos\left(\psi + \theta_1\right)}{\sqrt{-2\xi \left\{\cos\left(\psi + \theta_1\right) + \cos\left[\psi - \operatorname{arc\,cot}\left(\cot\theta_1 - \varsigma\right)\right]\right\}}} \end{aligned} \tag{4.14}$$

At the zero turning position, there is

$$v = l_{20} - u \tag{4.15}$$

So the angle $\gamma$ is represented by

$$\gamma = \arccos\left(\frac{u^2 + l_2^2 - v^2}{2ul_2}\right) = \arccos\left(\frac{l_2^2 - l_{20}^2 + 2ul_{20}}{2ul_2}\right) \tag{4.16}$$

Similarly, because the angle $\gamma$ is also defined within the interval of $[0, \pi]$, Eq. (4.16) has one solution.

Then the relative rotational angle $\phi$ between $\overrightarrow{BA}$ and $\overrightarrow{BE}$ is

$$\begin{cases} \phi = \beta + \gamma + \psi - \pi, & \theta_1 \geq 0 \\ \phi = \beta - \gamma + \psi - \pi, & \theta_1 < 0 \end{cases} \tag{4.17}$$

where $\theta_1 \geq 0$ indicates that the vehicle turns right; vice versa, $\theta_1 < 0$ indicates that the vehicle turns left.

The relative angular velocity of link $BE$ with respect to link $BA$ is indicated by

$$\omega_{r2} = \frac{\mathrm{d}\phi}{\mathrm{d}t} \tag{4.18}$$

The absolute angular velocity of link $BA$ is

$$\omega = \frac{\mathrm{d}\theta_1}{\mathrm{d}t} \tag{4.19}$$

Because the fact that gear 1 is fixed with the chassis, the relative angular velocity of noncircular gear 1 is

$$\omega_{r1} = -\omega \tag{4.20}$$

The relative transmission ratio of noncircular gears 1 and 2 with respect to link $AB$ is

$$i_{21}^{r} = \frac{\omega_{r2}}{\omega_{r1}} = \frac{\frac{\mathrm{d}\phi}{\mathrm{d}t}}{-\frac{\mathrm{d}\theta_1}{\mathrm{d}t}} = -\frac{\mathrm{d}\phi}{\mathrm{d}\theta_1} = \begin{cases} -\left(\frac{\mathrm{d}\beta}{\mathrm{d}\theta_1} + \frac{\mathrm{d}\gamma}{\mathrm{d}\theta_1}\right), & \theta_1 \geq 0 \\ -\left(\frac{\mathrm{d}\beta}{\mathrm{d}\theta_1} - \frac{\mathrm{d}\gamma}{\mathrm{d}\theta_1}\right), & \theta_1 < 0 \end{cases} \tag{4.21}$$

where "–" represents that the rotations of noncircular gear 1 and noncircular gear 2 are in opposite directions.

$\frac{\mathrm{d}\beta}{\mathrm{d}\theta_1}$ and $\frac{\mathrm{d}\gamma}{\mathrm{d}\theta_1}$ can be obtained from Eqs. (4.13) and (4.16), respectively:

$$\frac{\mathrm{d}\beta}{\mathrm{d}\theta_1} = \frac{l_1 l_2 \sin(1-\sigma_1)\left[2\psi + \theta_1 - \operatorname{arc\,cot}(\cot\theta_1 - \varsigma)\right] - l_2 l_4 \sin(\psi + \theta_1) + P\frac{\mathrm{d}l_2}{\mathrm{d}\theta_1}}{l_2\sqrt{l_2^2 - \sigma_2^2}} \tag{4.22}$$

where $\sigma_1 = \frac{\mathrm{d}\theta_2}{\mathrm{d}\theta_1} = \frac{1}{(\cos\theta_1 - \varsigma\sin\theta_1)^2 + \sin^2\theta_1}$, $\sigma_2 = l_1\{1 + \cos[2\psi + \theta_1 - \operatorname{arc\,cot}(\cot\theta_1 - \varsigma)]\} - l_4\cos(\psi + \theta_1)$,

$$\frac{\mathrm{d}l_2}{\mathrm{d}\theta_1} = \frac{l_4\xi\{\sin(\psi + \theta_1) - \sigma_1\sin[\psi - \operatorname{arc\,cot}(\cot\theta_1 - \varsigma)]}{\sqrt{1 + 2\xi^2 + 2\xi^2\cos[2\psi + \theta_1 - \operatorname{arc\,cot}(\cot\theta_1 - \varsigma)]}} \times \frac{-\xi(1-\sigma_1)\sin[2\psi + \theta_1 - \operatorname{arc\,cot}(\cot\theta_1 - \varsigma)]\}}{\sqrt{-2\xi[\cos(\psi + \theta_1) + \cos[\psi - \operatorname{arc\,cot}(\cot\theta_1 - \varsigma)]]}}$$

And

$$\frac{\mathrm{d}\gamma}{\mathrm{d}\theta_1} = -\frac{\left(l_2^2 + l_{20}^2 - 2ul_2\right)\frac{\mathrm{d}l_2}{\mathrm{d}\theta_1}}{l_2\sqrt{4u^2l_2^2 - \left(l_2^2 - l_{20}^2 + 2ul_{20}\right)^2}} \tag{4.23}$$

$i_{21}^r$ can also be expressed as

$$i_{21}^r = -\frac{r_1}{r_2} \tag{4.24}$$

where "–" represents that noncircular gear 1 rotates in an opposite direction as noncircular gear 2 does and

$$r_1 + r_2 = l_1 \tag{4.25}$$

From Eqs. (4.24) and (4.25), there are

$$\begin{cases} r_1 = -\dfrac{i_{21}^r l_1}{1 - i_{21}^r} \\ r_2 = \dfrac{l_1}{1 - i_{21}^r} \end{cases} \tag{4.26}$$

Equation (4.26) represents the pitch curves of the noncircular incomplete gears used to remedy the errors resulting from the fixed length of the link, denoted by $l_{20}$ in Fig. 4.3, which is used in the traditional trapezoid steering linkages. According to Eqs. (4.22) and (4.23), we obtain that $\frac{\mathrm{d}\beta}{\mathrm{d}\theta_1}$ and $\frac{\mathrm{d}\gamma}{\mathrm{d}\theta_1}$ are continuous about $\theta_1$, $\lim\limits_{\theta_1 \to 0} \frac{\mathrm{d}\gamma}{\mathrm{d}\theta_1} = 0$, and $\lim\limits_{\theta_1 \to 0} \frac{\mathrm{d}^2\gamma}{\mathrm{d}\theta_1^2} = 0$, so $i_{21}^r$ is continuous about $\theta_1$; then the functions of the pitch curves expressed by Eq. (4.26) are continuous about $\theta_1$.

### 4.3.3 Sensitivity Analysis

Without a doubt, the manufacturing errors and wears in joints, which are unavoidable in applications, could affect the accuracy of the mechanism. Hence, the sensitivity analysis is absolutely essential to the design.

Assume that the transmission ratio of the pair of noncircular gears is specified as $i_{21}^r$ which satisfies the Ackermann steering criterion. It is a continuous function about the steering angle, $\theta_1$. The relative rotational angle between $\overrightarrow{BA}$ and $\overrightarrow{BE}$, denoted by $\phi$, can be expressed as

$$\phi = \begin{cases} \int_0^{\theta_1} i_{21}^r \mathrm{d}\vartheta & \theta_1 \geq 0 \\ -\int_{\theta_1}^{0} i_{21}^r \mathrm{d}\vartheta & \theta_1 < 0 \end{cases}$$

where $\theta_1 \geq 0$ indicates that the vehicle turns right; vice versa, $\theta_1 < 0$ indicates that the vehicle turns left.

Accordingly, the subtended angle between $\overrightarrow{BA}$ and $\overrightarrow{BE}$ is

$$\varphi = \pi - \psi + \phi$$

The coordinates of joint $E$ in coordinate system $x_1 o_1 y_1$ shown in Fig. 4.5 are

$$\begin{cases} x_E = u\cos(\theta_1 + \phi) + l_1 \cos(\psi + \theta_1) \\ y_E = -u\sin(\theta_1 + \phi) - l_1 \sin(\psi + \theta_1) \end{cases} \tag{4.27}$$

Point $C$ is on the circle whose center is point $D$ and radius is $l_1$. Hence,

$$\begin{cases} x_C = l_4 - l_1 \cos(\psi - \theta_2) \\ y_C = -l_1 \sin(\psi - \theta_2) \end{cases} \tag{4.28}$$

Then the relationship between point $C$ and $E$ is

$$(x_E - x_C)^2 + (y_E - y_C)^2 = v^2 \tag{4.29}$$

Substituting Eqs. (4.27) and (4.28) into (4.29) presents the relationship between steering angle $\theta_2$ and the parameters $u$, $v$, $l_1$, $\phi$. The partial derivatives of steering angle related to different variable parameters interpret the sensitivity of links to the output steering angle.

$$\begin{cases} \dfrac{\partial\theta_2}{\partial u} = \dfrac{-\cos(\theta_1+\phi)(x_E - x_C) + \sin(\theta_1+\phi)(y_E - y_C)}{l_1\sin(\psi-\theta_2)(x_E - x_C) - l_1\cos(\psi-\theta_2)(y_E - y_C)} \\ \dfrac{\partial\theta_2}{\partial v} = \dfrac{v}{l_1\sin(\psi-\theta_2)(x_E - x_C) - l_1\cos(\psi-\theta_2)(y_E - y_C)} \\ \dfrac{\partial\theta_2}{\partial l_1} = -\dfrac{[\cos(\psi+\theta_1) + \cos(\psi-\theta_2)](x_E - x_C)}{l_1\sin(\psi-\theta_2)(x_E - x_C)} \\ \qquad \times \dfrac{+[\sin(\psi-\theta_2) - \sin(\psi+\theta_1)](y_E - y_C)}{-l_1\cos(\psi-\theta_2)(y_E - y_C)} \\ \dfrac{\partial\theta_2}{\partial \phi} = \dfrac{u\sin(\theta_1+\phi)(x_E - x_C) + u\cos(\theta_1+\phi)(y_E - y_C)}{l_1\sin(\psi-\theta_2)(x_E - x_C) - l_1\cos(\psi-\theta_2)(y_E - y_C)} \end{cases} \tag{4.30}$$

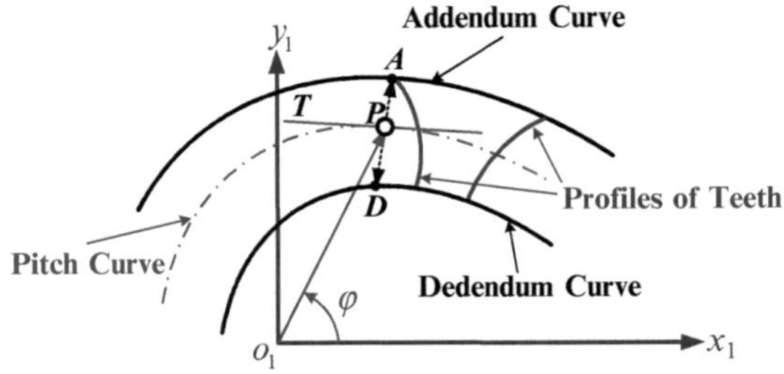


**Fig. 4.6** Addendum and dedendum curves of a noncircular incomplete gear

Therefore, combining with numerical simulation, the sensitivity values could be computed. Positive sensitivity values indicate that the output steering angle increases with the increase of the corresponding parameters, and vice versa.

### *4.3.4 Noncircular Profile Design*

To facilitate the analysis of pitch curves of the noncircular incomplete gears, a local coordinate frame, $x_1 o_1 y_1$ shown in Fig. 4.5, is established by letting the origin, $o_1$, be on the center of gear 1 which is also the fixed revolute joint of link $AB$. The $x_1$-axis is along the link $AD$. The $y_1$-axis is pointing up. The coordinates of a point on the pitch curve can be expressed as

$$\begin{cases} x_1 = r_1 \cos\varphi \\ y_1 = r_1 \sin\varphi \end{cases} \tag{4.31}$$

where $\varphi$ represents the polar angle of point $P$ shown in Fig. 4.6.

Specifying the parameters, $b$, $l$, $l_i (i = 1, 2, 4)$, $u$, and $v$, the pitch curve of the noncircular gear 1 can be obtained by associating Eqs. (4.26) and (4.31). After knowing the pitch curves of the two noncircular incomplete gears, the addendum and dedendum curves can be obtained in line with the equidistance line equation. In the coordinate frame, $x_1 o_1 y_1$, shown in Fig. 4.6, the addendum curve equation of the fixed noncircular incomplete gear can be represented by

$$\begin{cases} x_1^{\mathrm{a}} = x_1 + \dfrac{(\mathrm{d}y_1/\mathrm{d}\varphi)}{\sqrt{\left(\frac{\mathrm{d}x_1}{\mathrm{d}\varphi}\right)^2 + \left(\frac{\mathrm{d}y_1}{\mathrm{d}\varphi}\right)^2}} h_{\mathrm{a}}^* m \\ y_1^{\mathrm{a}} = y_1 - \dfrac{(\mathrm{d}x_1/\mathrm{d}\varphi)}{\sqrt{\left(\frac{\mathrm{d}x_1}{\mathrm{d}\varphi}\right)^2 + \left(\frac{\mathrm{d}y_1}{\mathrm{d}\varphi}\right)^2}} h_{\mathrm{a}}^* m \end{cases} \tag{4.32a}$$

where $x_1$ and $y_1$ are prescribed by Eq. (4.31), $h_{\mathrm{a}}^*$ is the addendum coefficient, $m$ is the gear module, $\frac{\mathrm{d}x_1}{\mathrm{d}\varphi}$ is the derivative of $x_1$ with respect to $\varphi$, and $\frac{\mathrm{d}y_1}{\mathrm{d}\varphi}$ is the derivative of $y_1$ with respect to $\varphi$.

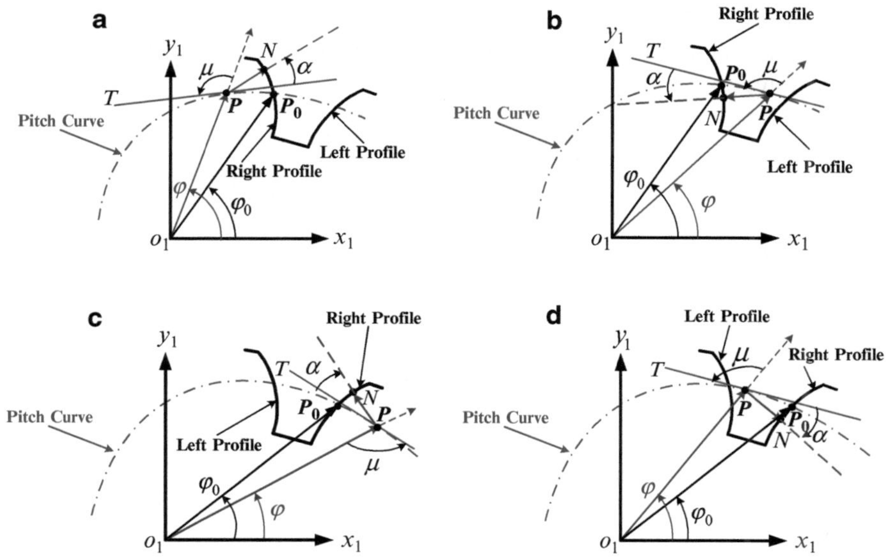


**Fig. 4.7** Tooth profiles of a noncircular incomplete gear

Similarly, the dedendum curve of the bottom noncircular gear can also be obtained as

$$\begin{cases} x_1^{\mathrm{d}} = x_1 - \dfrac{(\mathrm{d}y_1/\mathrm{d}\varphi)}{\sqrt{\left(\frac{\mathrm{d}x_1}{\mathrm{d}\varphi}\right)^2 + \left(\frac{\mathrm{d}y_1}{\mathrm{d}\varphi}\right)^2}} h_{\mathrm{f}}^* m \\ y_1^{\mathrm{d}} = y_1 + \dfrac{(\mathrm{d}x_1/\mathrm{d}\varphi)}{\sqrt{\left(\frac{\mathrm{d}x_1}{\mathrm{d}\varphi}\right)^2 + \left(\frac{\mathrm{d}y_1}{\mathrm{d}\varphi}\right)^2}} h_{\mathrm{f}}^* m \end{cases} \tag{4.32b}$$

where $h_{\mathrm{f}}^*$ is the dedendum coefficient.

The functions of the addendum and dedendum curves are continuous because the pitch curves are continuous as mentioned above.

For any point, $N$, on the gear profile shown in Fig. 4.7, it either locates outside the pitch curve or inside (including on) it. Assume that the inward or outward normal of the profile on $N$ intersects the pitch curve at a point, $P$. The directional vectors, $\boldsymbol{r}_1$, and its tangent line, $\boldsymbol{r}'_1$, will be

$$\boldsymbol{r}_1 = [r_1 \cos\varphi \ r_1 \sin\varphi]^{\mathrm{T}}, \quad \boldsymbol{r}'_1 = \left[r_1' \cos\varphi - r_1 \sin\varphi \ r_1' \sin\varphi + r_1 \cos\varphi\right]^{\mathrm{T}} \tag{4.33}$$

The tangent of the pitch curve at point $P$ is indicated by $PT$ in Fig. 4.7, and the angle turned from $OP$ to $PT$ anticlockwise about point $P$ is denoted by $\mu \in (0, \pi)$.

Therefore,

$$\cos \mu = \frac{\boldsymbol{r}_1 \cdot \boldsymbol{r'}_1}{\|\boldsymbol{r}_1\| \, \|\boldsymbol{r}'_1\|} = \frac{r'_1(\varphi)}{\sqrt{r_1^2(\varphi) + \left[r'_1(\varphi)\right]^2}} \tag{4.34a}$$

where $\mu \in (0, \pi)$, $r_1(\varphi)$ is represented by Eq. (4.26), and $r'_1(\varphi)$ is the derivative of $r_1(\varphi)$ with respect to $\varphi$, the polar angle of radial line $OP$. Alternatively, $\mu$ can also be expressed as

$$\tan \mu = \frac{r_1(\varphi)}{r'_1(\varphi)} \tag{4.34b}$$

The polar angle of vector $\overline{PN}$ shown in Fig. 4.7a is

$$\psi_{\mathrm{Ro}} = \varphi + \mu + \alpha - \pi \tag{4.35a}$$

where $\psi_{\mathrm{Ro}}$ represents the polar angle of the vector $\overline{PN}$ when point $N$ locates on the right profile of the tooth and is outside of the pitch curve.

When point $N$ locates on the right profile of the tooth but is inside of the pitch curve shown in Fig. 4.7b, there is

$$\psi_{\mathrm{Ri}} = \varphi + \mu + \alpha \tag{4.35b}$$

where $\psi_{\mathrm{Ri}}$ represents the polar angle of the vector $\overline{PN}$ when point $N$ locates on the right profile of the tooth and is outside of the pitch curve.

Similarly, we can find that the polar angles of the vector $\overline{PN}$ when point $N$ locates on the left profile of the tooth, shown in Fig. 4.7(c, d), are

$$\begin{cases} \psi_{\mathrm{Lo}} = \varphi + \mu - \alpha \\ \psi_{\mathrm{Li}} = \varphi + \mu - \alpha - \pi \end{cases} \tag{4.35c}$$

where $\psi_{\mathrm{Lo}}$ represents the polar angle of the vector $\overline{PN}$ when point $N$ locates on the left profile of the tooth and is outside of the pitch curve, and $\psi_{\mathrm{Li}}$ represents the polar angle of the vector $\overline{PN}$ when point $N$ locates on the left profile of the tooth and is outside of the pitch curve.

The coordinates of the profile can be expressed segment by segment as follows:

$$\begin{cases} x_{1l} = r_1 \cos\varphi \pm \displaystyle\int_{\varphi_0}^{\varphi} r_1(\varphi) \cos\alpha \cos(\varphi + \mu - \alpha)\, \mathrm{d}\varphi \\ y_{1l} = r_1 \sin\varphi \pm \displaystyle\int_{\varphi_0}^{\varphi} r_1(\varphi) \cos\alpha \sin(\varphi + \mu - \alpha)\, \mathrm{d}\varphi \\ x_{1r} = r_1 \cos\varphi \mp \displaystyle\int_{\varphi_0}^{\varphi} r_1(\varphi) \cos\alpha \cos(\varphi + \mu + \alpha)\, \mathrm{d}\varphi \\ y_{1r} = r_1 \sin\varphi \mp \displaystyle\int_{\varphi_0}^{\varphi} r_1(\varphi) \cos\alpha \sin(\varphi + \mu + \alpha)\, \mathrm{d}\varphi \end{cases} \tag{4.36}$$

where the subscript $1l$ denotes the left profile of gear 1, subscript $1r$ denotes the right profile of gear 1, $r_1$ is the polar radius of point $P$ on gear 1, $\varphi_0$ is the polar angle of vector $\overrightarrow{o_1 P_0}$ shown in Fig. 4.7, the upper sign in $\pm$ or $\mp$ represents the profiles outside the pitch curve, while the lower sign represents the profiles inside the pitch curve, $P_0$ is the intersection points of the profiles and the pitch curve, $\mu \in (0, \pi)$ which can be worked out using Eq. (4.24), and $\alpha$ is the specified pressure angle.

Theoretically, the profiles are defined precisely in accordance to Eq. (4.36); however, it is impossible to calculate the whole accurate points on the profiles. Therefore, the calculation precision of the profiles depends on the interval of polar angle, $\varphi$; the smaller the interval is, the higher the precision will be.

For a standard spur gear, the pitch $p$ and the tooth thickness $s$ are both $\pi m/2$ on the pitch curve, where $m$ is the gear module. Therefore, the arc length determined by the two adjacent crossover points of the profiles and the pitch curve is $\pi m/2$. Then the polar angles of other intersection points, $\varphi_0$, can be determined by solving the integral equation

$$\int_0^{\varphi_0} r_{1L}(\varphi)\mathrm{d}\varphi = \frac{n\pi m}{2} \tag{4.37}$$

For a rack cutter, the transitional curves are shaped by the envelope of the tip fillet of rack teeth. As shown in Fig. 4.8, the rack cutter has tip fillets of identical radii of $\rho$. The points $S$ and $E$ represent the start point and end point of the fillet curve, and $O_\mathrm{f}$ is the center of the fillet. Point $A$ is any point on the tip fillet. Line $AO_\mathrm{f}$ and the pitch line intersect at point $C$. Since $O_\mathrm{f}$ is the center of the fillet, line $AO_\mathrm{f}$, in fact, is the inward normal of the tip fillet. $\alpha_A$ is the acute angle subtended by $AC$ and the pitch line, so $\alpha < \alpha_A < \frac{\pi}{2}$ where $\alpha$ is the standard pressure angle. Consequently, there is

$$\left\| \overrightarrow{CO_\mathrm{f}} \right\| = \frac{h_{\mathrm{a0}}^{*} m - \rho \sin\alpha}{\sin\alpha_A} \tag{4.38}$$

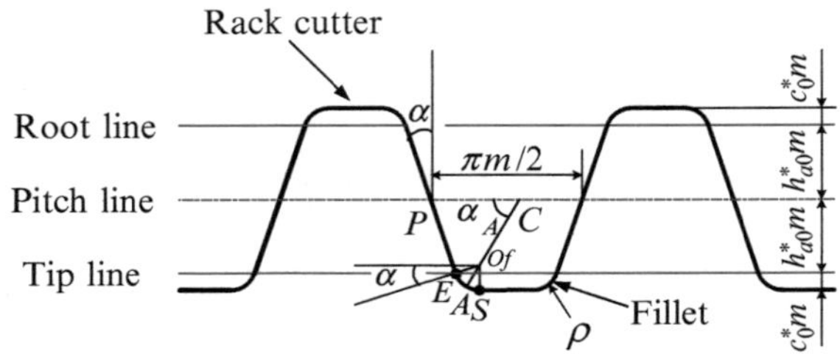


**Fig. 4.8** Rack cutter and tip fillet

where $\left\|\overrightarrow{C O_{\mathrm{f}}}\right\|$ is the length of the line segment $C O_{\mathrm{f}}$, $h_{\mathrm{a0}}^*$ is the addendum coefficient of the rack and generally equals 1, and $m$ is the module. Therefore,

$$\left\|\overrightarrow{AC}\right\| = \left\|\overrightarrow{C O_{\mathrm{f}}}\right\| + \rho \tag{4.39}$$

and

$$\left\|\overrightarrow{CP}\right\| = h_{\mathrm{a0}}^* m \tan\alpha + \rho\cos\alpha + \left\|\overrightarrow{C O_{\mathrm{f}}}\right\| \cos\alpha_A \tag{4.40}$$

where $\left\|\overrightarrow{AC}\right\|$ is the length of the line segment $AC$, and $\rho$ is the radius of the fillets shown in Fig. 4.8.

Since line $AC$ is the inward normal of the tip fillet, the rack cutter tip fillet and the transition curve of the noncircular gear should be tangent at point $A$ if the pitch line of the rack cutter and the pitch curve of the noncircular gear are tangent at point $C$. Therefore, the corresponding point on the transition curve can be determined when the position of point $A$ is specified.

Assume that the polar angle of $\overrightarrow{OC}$ is denoted by $\varphi_C$ and $\mu_C \in (0, \pi)$ is the subtended angle between the point vector and its tangent line at point $C$. Hence, $\varphi_C$ can be worked out using

$$\int_{\varphi_0}^{\varphi_C} r_{1L}(\varphi)\mathrm{d}\varphi = \left\|\overrightarrow{CP}\right\| \tag{4.41}$$

where $\varphi_0$ is calculated by Eq. (4.37), and $\left\|\overrightarrow{CP}\right\|$ is represented with Eq. (4.40). Then $\mu_C$ can be worked out using Eq. (4.34a) or Eq. (4.34b) where $\varphi_{1L} = \varphi_C$.

Thus, after specifying the angle $\alpha_A$ shown in Fig. 4.9, the polar angle of $\overrightarrow{CA}$ is

$$\varphi_{CA} = \varphi_C + \mu_C + \alpha_A - 2\pi \tag{4.42}$$

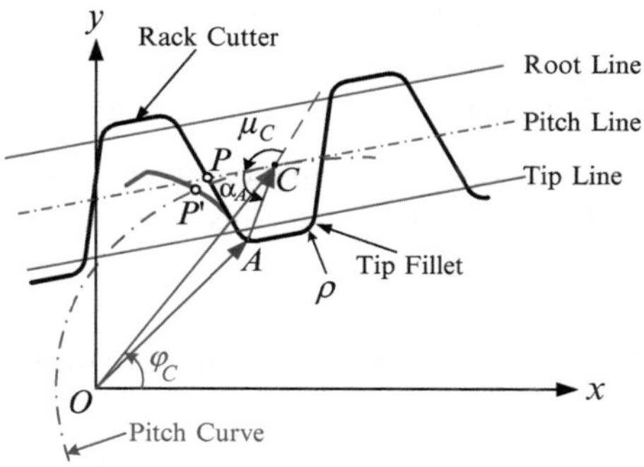


**Fig. 4.9** Coordinates of the transition curve

Therefore, the coordinates of point $A$ are now represented by

$$\begin{cases} x_A = r_{1L}(\varphi_C)\cos\varphi_C + \left\|\overrightarrow{AC}\right\| \cos\varphi_{CA} \\ y_A = r_{1L}(\varphi_C)\sin\varphi_C + \left\|\overrightarrow{AC}\right\| \sin\varphi_{CA} \end{cases} \tag{4.43}$$

where $\varphi_C$ can be worked out using Eq. (4.41), and $\varphi_{CA}$ is represented by Eq. (4.42).

When $\alpha_A$ changes from $\alpha$ to $\frac{\pi}{2}$, the position of point $A$ will correspondingly change on the tip fillet. To avoid undercutting by rack cutters, there should be

$$m_{\max} \le \frac{\rho_{\min}\sin^2\alpha}{h^*_{a0}} \tag{4.44}$$

where $m_{\max}$ is the maximum gear module allowed, $\rho_{\min}$ is the minimal radius of curvature of the pitch curve, $h^*_{a0}$ is the addendum coefficient of the cutter and is generally equal to 1.

With theoretical analysis we can get the numerical method for gear profile as follows:

1. Substitute the structure parameters into Eq. (4.26) and get the pitch curves.
2. Calculate the addendum and dedendum curves in accordance with Eqs. (4.32a) and (4.32b).
3. Calculate the maximum gear module allowed in accordance to Eq. (4.44), and choose an appropriate module $m$.
4. Assume that the polar angle of the starting point $P_0$ on the pitch curve, $\varphi_0$, is specified and the corresponding profile is a right(left) profile.
5. Prescribe the interval of the polar angle $\varphi$, $\Delta\varphi$, such as $\Delta\varphi = 0.01$ rad, then the polar angle of the $i$th ($i = 1, 2, 3, \ldots$) point on the profile is $\varphi = \varphi_0 \pm i\Delta\varphi$.
6. Calculate the $\mu$ in line with Eq. (4.34b), then get the coordinates of the $i$th ($i = 1, 2, 3, \ldots$) point on the profile according to Eq. (4.36).

7. Calculate the polar angle of the $i$th ($i = 1, 2, 3, \ldots$) point $P_n$ on the pitch curve, $\varphi_n$, by solving the integral equation $\int_{\varphi_0}^{\varphi_n} r_1(\varphi)\mathrm{d}\varphi = n\pi m/2$.
8. Repeat steps 5∽7 to complete the whole gear profile.

Moreover, the calculation precision of the pitch curves and the profiles respectively depend on the intervals of $\theta_1$ and $\varphi$.

## 4.4 Parametric Optimization and Numerical Simulation

In general, the maximum turning angle allowed by front wheels is a certain value between $\frac{7\pi}{36} \sim \frac{\pi}{4}$ or $35° \sim 45°$ in degrees. For the steering mechanism shown in Fig. 4.5, only the parameters $l_1$, $u$, and $\psi$ could be optimized because the structure parameters $b$, $l$, and $l_4$ are specified by the vehicle. To reach this target, the following process can be followed.

The steering mechanism should satisfy the request of the maximum turning angle of the front wheels which is no more than $\frac{\pi}{4}$. Consequently the maximum turning angle interval of the steering mechanism is $\theta_1 \in \left[\mathrm{arc\,cot}\left[\cot\left(-\frac{\pi}{4}\right) + \frac{b}{l}\right], \frac{\pi}{4}\right]$ in line with Eq. (4.10). Analyzing the profile calculation method, we can find that the maximum polar angle $\theta_{1\max}$ should meet that $\theta_{1\max} > \frac{\pi}{4}$; otherwise it is impossible to get the last segment of the left profile, as Fig. 4.10a shows. Similarly, the minimum polar angle $\theta_{1\min}$ should be smaller than the minimum turning angle which is represented in Fig. 4.10b. As a result, there should be $\theta_1 \in \left[\mathrm{arc\,cot}\left[\cot\left(-\frac{\pi}{4}\right) + \frac{b}{l}\right] - \Delta\theta_2, \frac{\pi}{4} + \Delta\theta_1\right]$ where $\Delta\theta_1$ and $\Delta\theta_2$ are determined by the pitch curves of the two gears.

Obviously, the coordinates of point $P$ satisfy both Eqs (4.32b) and (4.36). That is to say, we can obtain $\Delta\theta_1$ and $\Delta\theta_2$ by associating Eqs. (4.32b) and (4.36). Generally, it is safe to assume $\Delta\theta_1 = \Delta\theta_2 = 10°$. For instance, assuming that $b = 1{,}600$ mm, $l = 2{,}800$ mm, then $\theta_1 \in [-76.8°, 55°]$.

What follows will investigate the optimization of the structure parameters of the steering mechanism.

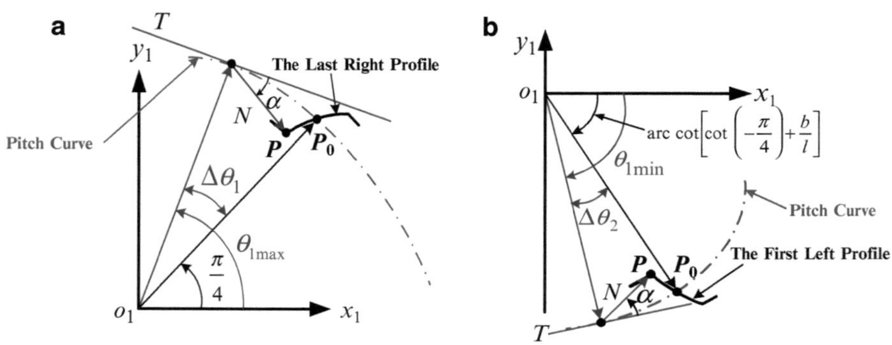


**Fig. 4.10** The extreme values of polar angle $\theta_1$

### 4.4.1 The Optimization Variables and Constraint Conditions

As mentioned above, the optimization variables, in vector form, can be expressed as

$$\boldsymbol{x} = [l_1 \; u \; \psi]^{\mathrm{T}} \tag{4.45}$$

Firstly, the parameters should be neither too big nor too small so long as they meet the space arrangement of chassis. The upper and lower limits of the variable vectors can be expressed with

$$\begin{cases} \boldsymbol{x}_U = [l_{1\max} \; u_{\max} \; \psi_{\max}]^{\mathrm{T}} \\ \boldsymbol{x}_L = [l_{1\min} \; u_{\min} \; \psi_{\min}]^{\mathrm{T}} \end{cases} \tag{4.46}$$

The minimum radius of pitch curves should not be too small, otherwise structure interference might occur. Assume that the minimum radius of pitch curve of gear 1 is $\rho_1$ and the minimum radius of pitch curve of gear 2 is $\rho_2$. Associating Eq. (4.26) presents

$$\begin{cases} r_1 = -\dfrac{i_{21}^r(\boldsymbol{x})l_1}{1 - i_{21}^r(\boldsymbol{x})} \geq \rho_1 \\ r_2 = \dfrac{l_1}{1 - i_{21}^r(\boldsymbol{x})} \geq \rho_2 \end{cases} \tag{4.47}$$

As a result, the transmission ratio satisfies that

$$\frac{\rho_1}{l_1 - \rho_1} \leq -i_{21}^r(\boldsymbol{x}) \leq \frac{l_1 - \rho_2}{\rho_2} \tag{4.48}$$

Inequality (4.48) always holds for $\theta_1 \in [\theta_{1\min}, \theta_{1\max}]$.

### 4.4.2 Object Function and the Optimization Analysis

The particular characteristic of noncircular gear is that the transmission ratio is variable within a certain interval. To keep a balanced design for meshing a noncircular pair, the width of the interval should be as narrow as possible. Suppose

$$E(\boldsymbol{x}) = \frac{\int_{\theta_{1\min}}^{\theta_{1\max}} i_{21}^r(\theta_1, \boldsymbol{x})\,\mathrm{d}\theta_1}{\theta_{1\max} - \theta_{1\min}} = \frac{\phi(\theta_{1\max}, \boldsymbol{x}) - \phi(\theta_{1\min}, \boldsymbol{x})}{\theta_{1\max} - \theta_{1\min}} \tag{4.49}$$

where $E(\boldsymbol{x})$ denotes the mean value of $i_{21}^r(\theta_1, \boldsymbol{x})$ within the interval $[\theta_{1\min}, \theta_{1\max}]$.

**Table 4.1** Primary parameters of a vehicle

| Sequence | $l_4$ (mm) | $l$ (mm) | $b$ (mm) | $m$ (mm) | $\alpha$ (°) | $h_a^*$ | $h_f^*$ |
|---|---|---|---|---|---|---|---|
| 1 | 1,200 | 2,800 | 1,600 | 2 | 20 | 1 | 1.25 |
| 2 | 1,350 | 2,800 | 1,600 | 2 | 20 | 1 | 1.25 |
| 3 | 1,350 | 2,950 | 1,600 | 2 | 20 | 1 | 1.25 |
| 4 | 1,350 | 2,950 | 1,750 | 2 | 20 | 1 | 1.25 |

The variance of transmission ratio can be subsequently defined as

$$f(\boldsymbol{x}) = \frac{\int_{\theta_{1\min}}^{\theta_{1\max}} \left[i_{21}^{r}(\theta_1, \boldsymbol{x}) - E(\boldsymbol{x})\right]^2 \mathrm{d}\theta_1}{\theta_{1\max} - \theta_{1\min}} \tag{4.50}$$

Substituting the Eq. (4.49) to (4.50) and simplifying it presents

$$f(\boldsymbol{x}) = \frac{\int_{\theta_{1\min}}^{\theta_{1\max}} \left[i_{21}^{r}(\theta_1, \boldsymbol{x})\right]^2 \mathrm{d}\theta_1}{\theta_{1\max} - \theta_{1\min}} - E(\boldsymbol{x})^2 \tag{4.51}$$

where $f(\boldsymbol{x})$ measures the variable amplitude of the transmission ratio. Thus, the object function, $F(\boldsymbol{x})$, can be defined as:

$$F(\boldsymbol{x}) = \min\left[f(\boldsymbol{x})\right] \tag{4.52}$$

where $\boldsymbol{x}$ is subjected to inequality (4.48) while $\boldsymbol{x}_L \le \boldsymbol{x} \le \boldsymbol{x}_U$.

Because $i_{21}^{r}(\boldsymbol{x})$ and $\phi(\boldsymbol{x})$ are continuous functions about $\boldsymbol{x}$, so $f(x)$ is also a continuous function. As a result, the minimum value of $f(\boldsymbol{x})$ must occur at one of the three points, the upper boundary of the $\boldsymbol{x}$-variable interval, the lower boundary of the $\boldsymbol{x}$-variable interval and the point where the partial derivative of $f(\boldsymbol{x})$ with respect to $\boldsymbol{x}$ equals zero. The partial derivative of $f(\boldsymbol{x})$ can be expressed as

$$\frac{\partial f(\boldsymbol{x})}{\partial \boldsymbol{x}} = \frac{2\left\{\int_{\theta_{1\min}}^{\theta_{1\max}} i_{21}^{r}(\theta_1, \boldsymbol{x})\frac{\partial i_{21}^{r}(\theta_1, x)}{\partial x}\mathrm{d}\theta_1 - \left[\phi(\theta_{1\max}, \boldsymbol{x}) - \phi(\theta_{1\min}, \boldsymbol{x})\right]\frac{\partial E(x)}{\partial x}\right\}}{\theta_{1\max} - \theta_{1\min}} \tag{4.53}$$

The numerical algorithm such as trapezoid integration equation can be used to obtain an enough precise solution of the above problem by programming at MATLAB.

Table 4.1 lists the primary parameters of a vehicle. With these parameters, numerical simulations based on the optimization method proposed above are executed. Suppose that

$$\begin{cases} \boldsymbol{x}_U = [200\ 80\ 75^\circ]^{\mathrm{T}} \\ \boldsymbol{x}_L = [100\ 20\ 105^\circ]^{\mathrm{T}'} \end{cases} \quad \rho_1 = \rho_2 = 20\ \mathrm{mm}$$

**Table 4.2** Optimization results

| Sequence | $l_1$ (mm) | $u$ (mm) | $\psi$ (rad)mm |
|---|---|---|---|
| 1 | 100.00 | 26.24 | $\frac{7\pi}{12}$ |
| 2 | 100.00 | 26.24 | $\frac{7\pi}{12}$ |
| 3 | 100.00 | 25.96 | $\frac{7\pi}{12}$ |
| 4 | 100.00 | 26.42 | $\frac{7\pi}{12}$ |

Four sets of vehicle parameters shown in Table 4.1 are used to demonstrate the effectiveness of the method. The optimization results are listed in Table 4.2.

The pitch curves of noncircular gears corresponding to the parameters listed in Tables 4.1 and 4.2 in polar coordinate are presented in Fig. 4.11. And the gear profile can be obtained by the aforementioned numerical method.

Accordingly, the sensitivity values can be computed, as Fig. 4.12 shows, by substituting the parameters in sequence 1 into Eq. (4.30). From Fig. 4.12, we can find that the variation range of the sensitivity value of link $BE$ is the largest, but the absolute sensitivity value of link $CE$ is the largest. Therefore, the wear resistance ability of joints connected to link $BE$ and $CE$ should be improved in applications, or the joints with preload function can be used in this system to reduce the influence of wear in joints. What's more, the steering system design should allow the vehicle to tend to understeer once the abrasion occurs.

With any set of parameters of Tables 4.1 and 4.2, the first set of parameters, for example, an Ackermann-type steering mechanism can be designed and assembled by software Pro/Engineer 4.0 as Fig. 4.12 shows.

Figure 4.13 indicates that this kind of steering mechanism is similar in structural compactness to the trapezoid four-bar mechanism but can precisely satisfy the Ackermann turning requirements. Figure 4.14 verifies the steering motion of the proposed mechanism. In applications, the pitch curve might be modified when taking into account the tires and other elastic elements. However, the method proposed in this chapter can be used to design an ideal Ackermann-type steering mechanism.

## 4.5 Conclusions

This chapter has proposed a steering mechanism that precisely satisfies the needs of Ackermann turning geometry. After pointing out that a planar four-bar linkage can exactly trace 9 points at most, it put the Ackermann criteria into the size synthesis for steering mechanism by using the incomplete noncircular gears. Therefore, the synthesis of the steering mechanism becomes the design of the pitch curves, addendum curves, dedendum curves, tooth profiles, and transition curves of a pair of noncircular gears. Kinematic simulations show that the target

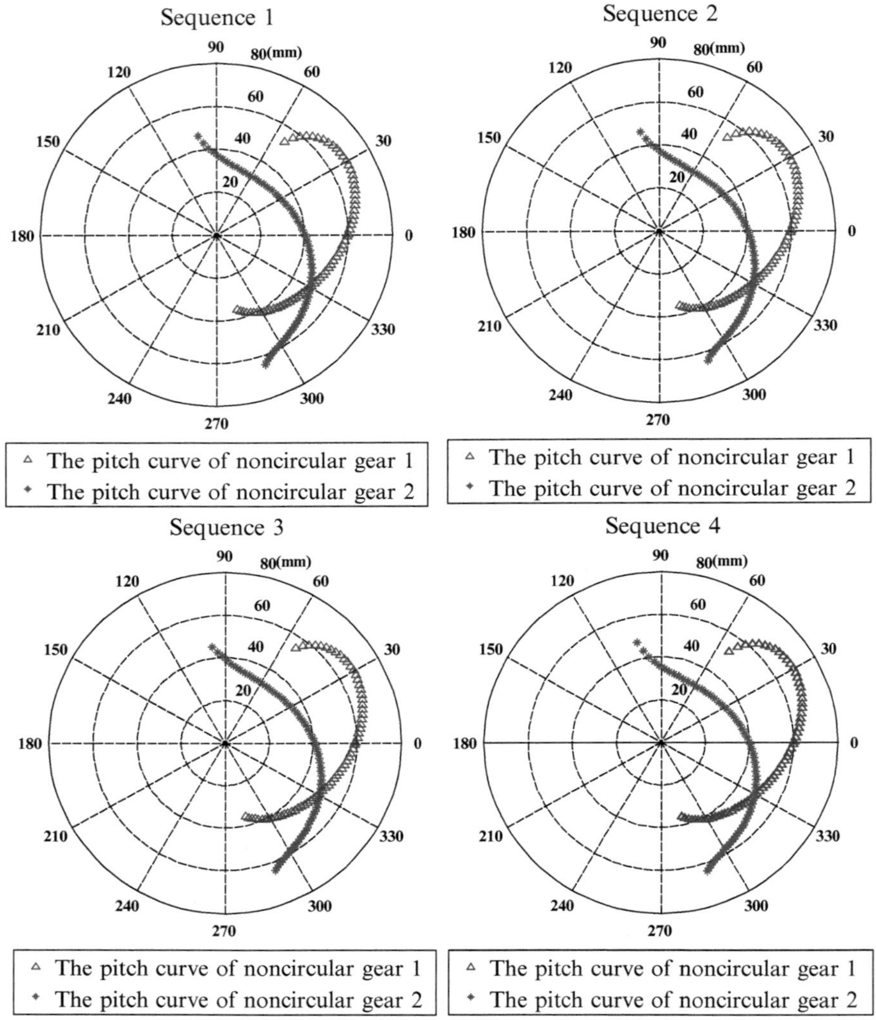


**Fig. 4.11** The pitch curves of noncircular gears in polar coordinate

of design can be completely reached by the incomplete noncircular gear coupled five-link mechanism. The steering mechanism should improve the vehicle steering performance but has similar structure to the existing steering four-bar linkages; as a result, it has a very wide application prospect, especially in the light carriages.

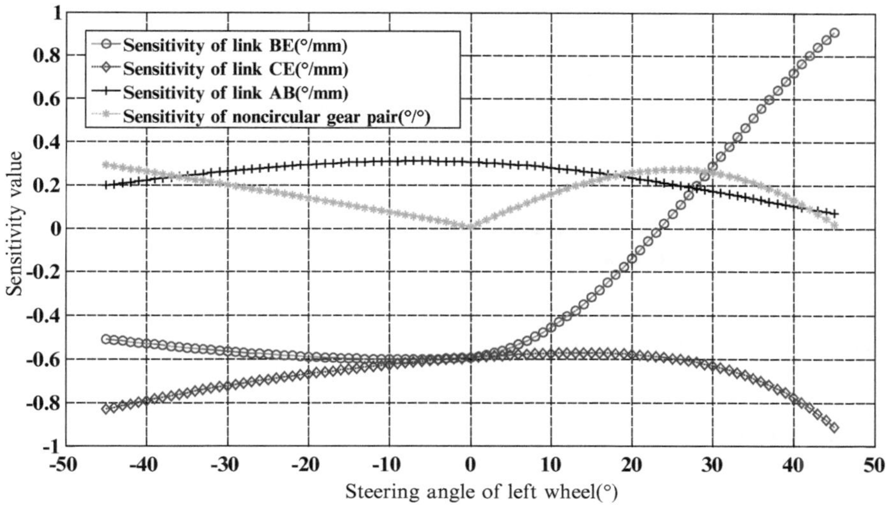


**Fig. 4.12** Sensitivity values with parameters in sequence 1

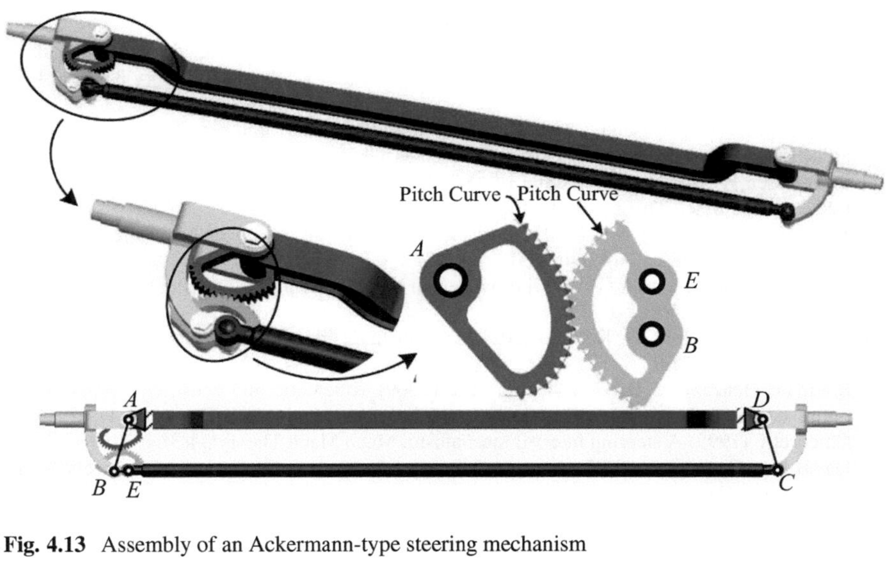


**Fig. 4.13** Assembly of an Ackermann-type steering mechanism

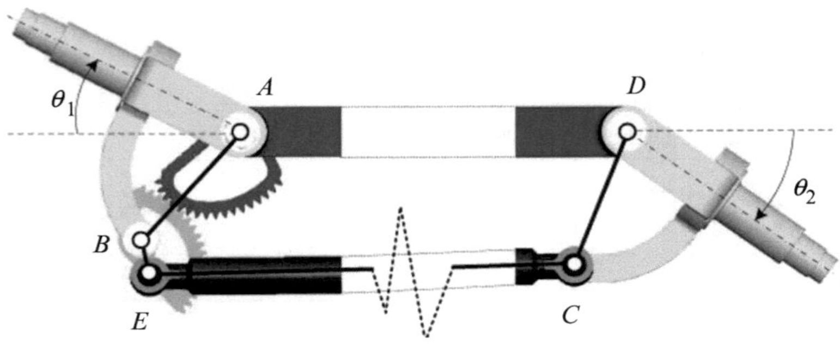


**Fig. 4.14** Steering motion of the mechanism

# References


1. Pramanik S (2002) Kinematic synthesis of a six-member mechanism for automotive steering. J Mech Des 124(4):642–645
2. Reimpell H, Stoll H, Betzler JW (2002) The automotive chassis: engineering principles, 2nd edn. Butterworth Heinemann, Oxford
3. Suh CH, Mecklenburg AW (1973) Optimal design of mechanisms with the use of matrices and least squares. Mech Mach Theory 8:479–495
4. Simionescu PA, Beale D (2002) Optimum synthesis of the four-bar function generator in its symmetric embodiment: the Ackermann steering linkage. Mech Mach Theory 37(12):1487–1504
5. Simionescu PA, Talpasanu I (2007) Synthesis and analysis of the steering system of an adjustable tread-width four-wheel tractor. Mech Mach Theory 42(5):526–540
6. Simionescu PA, Beale D, Talpasanu I (2007) Dynamic effect of the bump steer in a wheeled tractor. Mech Mach Theory 42(10):1352–1361
7. Simionescu PA, Smith MR (2002) Initial estimates in the design of central-lever steering linkages. J Mech Des 124(4):646–651
8. Zarak CE, Townsend MA (1983) Optimal design of rack-and-pinion steering linkages. J Mech Trans Auto Des 105(2):220–226
9. Erdman AG, Sandor GN, Kota S (2001) Mechanism design: analysis and synthesis. Prentice Hall, Englewood Cliffs
10. Rahmani Hanzaki A, Rao PVM, Saha SK (2009) Kinematic and sensitivity analysis and optimization of planar rack-and-pinion steering linkages. Mech Mach Theory 44(1):42–56
11. Chicurel E (1999) A steering interval mechanism. Mech Mach Theory 34(3):421–436
12. David DA, Danwen Q (1987) Analytical design of seven joint spatial steering mechanisms. Mech Mach Theory 22(4):315–319
13. Simionescu PA, Smith MR, Tempea I (2000) Synthesis and analysis of the two loop translational input steering mechanism. Mech Mach Theory 35(7):927–944
14. Simionescu PA, Smith MR (2000) Applications of watt II function generator cognates. Mech Mach Theory 35(11):1535–1549
15. Felzien ML, Cronin DL (1985) Steering error optimization of the Macpherson strut automotive front suspension. Mech Mach Theory 20(1):17–26
16. Mántaras DA, Pablo L, Carlos V (2004) Development and validation of a three-dimensional kinematic model for the McPherson steering and suspension mechanisms. Mech Mach Theory 39(6):603–619

17. Carcaterra A, D'Ambrogio W (1998) A function generating differential mechanism for an exact solution of the steering problem. Mech Mach Theory 33(5):535–549
18. Raghavan M (1996) Number and dimensional synthesis of independent suspension mechanisms. Mech Mach Theory 31(8):999–1195
19. Huang XZ, Zhang YM (2010) Reliability sensitivity analysis for rack-and-pinion steering linkages. J Mech Des 132(7):071012
20. Emura T, Arakawa A (1992) A new steering mechanism using noncircular gears. JSME Int J 35(4):604–610
21. Donner DB (2001) Function generation utilizing an eight-link mechanism and optimized non-circular gear elements with application to automotive steering. Proc Inst Mech Eng Part C J Mech Eng Sci 215(7):847–857
22. Miller G, Reed R, Wheeler F (1991) Optimum Ackerman for improved steering axle tire wear on trucks, vol 912694, SAE technical paper. SAE, Warrendale
23. Hunt KH (1978) Kinematic geometry of mechanisms. Oxford University Press, Oxford

# Chapter 5
# Workspace and Singularity of Planar Parallel Linkage

The layout of this chapter is that the theoretical foundation for the workspace of planar parallel mechanism under rotational actuations is firstly investigated, and then the algorithms to search the reachable and the dexterous workspaces of such mechanisms are discussed with application examples.

## 5.1 Workspace of a Mechanism

The workspace of a mechanism is the reachable domain of its end effector and is bounded in the three-dimensional space [1], which is therefore called reachable workspace. Great interest has been addressed to workspace evaluation for both analysis and synthesis purposes in the past decades. Kumar and Waldron [2] discussed the workspace of mechanical mechanisms. Gosselin and Jean [3] and Merlet et al. [4] investigated the determination of the workspace of planar parallel manipulators with joint limits. Gosselin [5] and Merlet [6] addressed the determination and comparison of the workspace of 6-DoF parallel manipulators based on geometry properties of the workspace.

In the concept design phase, it is always necessary to determine the boundaries of the workspace of a new mechanism [7]. Snyman et al. [8] proposed an approach to compute the boundaries of the workspaces of planar manipulators. This numerical method focuses on finding a suitable radiating point in the output coordinate space and then determining the points of intersection of a representative pencil of rays with the boundary of the accessible set. Yang et al. [9] analyzed the workspace boundary of serial manipulators with non-unilateral constraints. Karim Abdel-Malek et al. [10, 11] discussed the boundary determination of manipulator workspaces and the interior and exterior boundaries of the workspace of mechanical manipulators. Bonev and Gosselin [12] analyzed the determination problems of the workspace of spherical symmetrical parallel mechanisms. However, the determination of the reachable workspace of the parallel mechanism whose actuations are rotational

J. Zhao et al., *Design of Special Planar Linkages*, Springer Tracts in Mechanical Engineering, DOI 10.1007/978-3-642-38448-6_5,

inputs is seldom systematically investigated. The constraints for the reachable workspace of spatial parallel mechanisms that resulted from the rotational inputs are much more complicated than those of translational actuations; this chapter therefore will first investigate the transformations of the constraint equations with additional restrictions from the angular inputs by applying trigonometric identities and then discuss the searching algorithms for some representative parallel mechanisms.

Many scholars have proposed numerous approaches to search the boundary curves or surfaces. Davidson and Chaney [13] presented an algebraic method and the design procedure for planar RPR robot. Merlet [14] investigated the designing problems of a parallel manipulator with a specific workspace. Cervantes-Sánchez and Rendón-Sánchez [15] addressed a simplified approach for obtaining the workspace of a class of 2-dof planar parallel mechanisms. Ceccarelli [16] presented a formulation for the workspace boundary of general $n$-revolute open chain mechanisms. However, the searching task for the boundary curves or boundary surfaces of the reachable workspace of a parallel mechanism is much more difficult than that of an open chain mechanism. Consequently, this chapter focuses on the general method of obtaining the reachable workspace of a parallel mechanism whose actuations are restricted to rotations.

## 5.2 Theoretical Foundation for the Workspace of Parallel Mechanism Under Rotational Actuations

The reachable workspace of a mechanism has also been defined in literature as the totality of positions that a particular identified point of the end effector of the mechanism can reach [4, 6, 13], which will be adopted in this chapter. Merlet [4] defined the reachable workspace of a mechanism as the region which can be reached by a reference point $C$ located on the end effector, and defined the dexterous workspace as the region that can be reached by the reference point with any orientation in a prescribed range. For a parallel mechanism, the reachable workspace and the dexterous workspace are often described by a set of rigid-body constraint equations with different variable ranges. Generally, the reachable workspace of a specified point $C$ may be expressed as

$$W_{\mathrm{R}} = \left\{(x, y, z) \,\middle|\, F(x, y, z, \theta_x, \theta_y, \theta_z) = 0\right\} \tag{5.1}$$

where $W_{\mathrm{R}}$ represents the reachable workspace and $F(\cdot)$ denotes a set of rigid-body constraint equations.

If the dexterous workspace of a mechanism is defined as the domain where the reference point $C$ of the end effector can reach with any orientation in a prescribed pose range, the dexterous workspace of the point can be expressed as

$$\begin{aligned} W_{\mathrm{D}} &= \left\{(x, y, z) \,\middle|\, F(x, y, z, \theta_x, \theta_y, \theta_z)\right. \\ &\left.= 0, \theta_x \in [\theta_{x_1}, \theta_{x_2}], \theta_y \in \left[\theta_{y_1}, \theta_{y_2}\right], \theta_z \in [\theta_{z_1}, \theta_{z_2}]\right\} \end{aligned} \tag{5.2}$$

where $W_D$ represents the desired workspace, $\theta_{x_1}$ denotes the lower limit of $\theta_x$, and $\theta_{x_2}$ denotes the upper limit of $\theta_x$ and the meanings of other symbols can be similarly deduced.

However, for the family of spatial parallel mechanisms under rotational actuations, the constraint equations of the reference point of the end effector often have the following general form:

$$F(\theta_1, \theta_2, \ldots, \theta_m) = \begin{bmatrix} [u_{11}(x) + a_{11}\cos\theta_1 + b_{11}\sin\theta_1]^2 + [v_{11}(y) + c_{11}\cos\theta_1 + d_{11}\sin\theta_1]^2 + [w_{11}(z) + e_{11}\cos\theta_1 + f_{11}\sin\theta_1]^2 - l_1^2 \\ [u_{21}(x) + a_{21}\cos\theta_2 + b_{21}\sin\theta_2]^2 + [v_{21}(y) + c_{21}\cos\theta_2 + d_{21}\sin\theta_2]^2 + [w_{21}(z) + e_{21}\cos\theta_2 + f_{21}\sin\theta_2]^2 - l_2^2 \\ \vdots \\ [u_{m1}(x) + a_{m1}\cos\theta_m + b_{m1}\sin\theta_m]^2 + [v_{m1}(y) + c_{m1}\cos\theta_m + d_{m1}\sin\theta_m]^2 + [w_{m1}(z) + e_{m1}\cos\theta_m + f_{m1}\sin\theta_m]^2 - l_m^2 \end{bmatrix} \tag{5.3}$$

where $u_{i1}(\cdot)$, $v_{i1}(\cdot)$, and $w_{i1}(\cdot)$ represent the functions of $x$, $y$, and $z$, respectively; $i = 1, 2, \ldots, m$, $m$ denotes the number of constraint equations; $a_{i1}, b_{i1}, \ldots, f_{i1}$ represent the constant coefficients; and $\theta_i$ represents the $i$th independent angular input.

Equation (5.3) can be equivalently transformed into a function of $m$ angular inputs:

$$F(\theta_1, \theta_2, \ldots, \theta_m) = \begin{bmatrix} p_{11} - p_{12}\cos\theta_1 - p_{13}\sin\theta_1 \\ p_{21} - p_{22}\cos\theta_2 - p_{23}\sin\theta_2 \\ \vdots \\ p_{m1} - p_{m2}\cos\theta_3 - p_{m3}\sin\theta_m \end{bmatrix} \tag{5.4}$$

where $p_{ij}(i = 1, 2, \ldots, m; j = 1, 2, 3)$ are the functions of position and pose parameters, $x$, $y$, $z$, $\theta_x$, $\theta_y$, $\theta z$, the independent variables.

For a set of angular inputs $(\theta_1\ \theta_2 \cdots \theta_m)$, there are at least one set of $x, y, z, \theta_x, \theta_y, \theta_z$ such that $F(\theta_1, \theta_2, \ldots, \theta_3) = 0$ holds; vice versa, for any point within the reachable workspace of the end effector, there must be a set of real parameters $(\theta_1\ \theta_2 \cdots \theta_m)$ such that the position and pose parameters $x, y, z, \theta_x, \theta_y, \theta_z$ of the point satisfy the equation $F(\theta_1, \theta_2, \ldots, \theta_3) = 0$. However, the additional constraints of $\cos\theta_i \in [-1, 1]$ and $\sin\theta_i \in [-1, 1]$ are very difficult to directly utilize to search the reachable workspace. The fact that $\tan\alpha \left(-\frac{\pi}{2} < \alpha < \frac{\pi}{2}\right)$ belongs to $(-\infty, \infty)$ might shed a light on the cracking of this difficult problem. So, we can try to obtain the criteria for the boundary surfaces of the reachable and dexterous workspace through the trigonometric transformations, such as $\cos\theta_i = \frac{1-\tan^2(\theta_i/2)}{1+\tan^2(\theta_i/2)}$ and $\sin\theta_i = \frac{2\tan(\theta_i/2)}{1+\tan^2(\theta_i/2)}$.

Suppose $t_i = \tan\frac{\theta_i}{2}(i = 1, 2, \ldots, m)$, we can immediately transform Eq. (5.4) into the following form:

$$F(\theta_1, \theta_2, \ldots, \theta_m) = \begin{bmatrix} (p_{11} + p_{12})t_1^2 - 2p_{13}t_1 + p_{11} - p_{12} \\ (p_{21} + p_{22})t_2^2 - 2p_{23}t_2 + p_{21} - p_{22} \\ \vdots \\ (p_{m1} + p_{m2})t_m^2 - 2p_{m3}t_3 + p_{m1} - p_{m2} \end{bmatrix} \quad (5.5)$$

The criteria that equation $F(\theta_1, \theta_2, \ldots, \theta_m) = 0$ must have real solutions are

$$\begin{cases} \Delta_1 = 4\left(p_{12}^2 + p_{13}^2 - p_{11}^2\right) \geq 0 \\ \Delta_2 = 4\left(p_{22}^2 + p_{23}^2 - p_{21}^2\right) \geq 0 \\ \vdots \\ \Delta_m = 4\left(p_{m2}^2 + p_{m3}^2 - p_{m1}^2\right) \geq 0 \end{cases} \quad (5.6)$$

where $p_{i1}, p_{i2}, p_{i3}$ are functions of the mere position and pose parameters $x, y, z, \theta_x, \theta_y, \theta_z$.

With the criteria of (5.6), we can obtain the reachable and dexterous workspaces of any parallel mechanisms whose actuations are rotations. In the following sections, we will illustrate the validity and efficiency of the above algorithm for the mechanisms under rotational actuations.

## 5.3 Workspace of a Planar 3-RRR Parallel Mechanism

A planar parallel mechanism shown in Fig. 5.1 is characterized with that the end effector $C_1C_2C_3$ is an equilateral triangle whose three vertexes are connected with the vertexes of the fixed equilateral triangle $A_1A_2A_3$ through three identical RRR

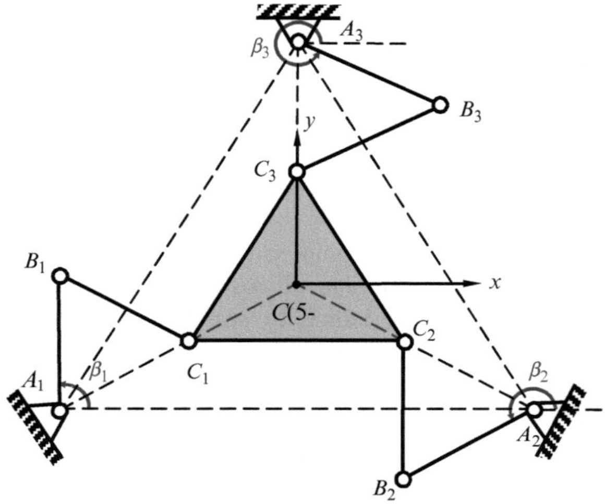


**Fig. 5.1** A planar 3-RRR parallel mechanism

kinematic chains, $A_iB_iC_i$, $i = 1, 2, 3$. Suppose that the reference point of the end effector $C_1C_2C_3$ is represented by its geometry center, $C$. We now can investigate the reachable and dexterous workspace of the mechanism.

Suppose the coordinates of the reference point of the end effector, denoted by $C$, in the fixed coordinate system $oxyz$ are $\boldsymbol{r} = (x\ y\ 0)^{\mathrm{T}}$, the angular displacement about the $z$-axis is $\theta$; then, the absolute coordinates of the three vertexes of the triangle end effector $C_1C_2C_3$ can be obtained, individually:

$$\boldsymbol{r}_{C_i} = \boldsymbol{r} + \mathbf{R}\boldsymbol{r}^L_{CC_i}, i = 1, 2, 3 \tag{5.7}$$

where

$$\mathbf{R} = \begin{bmatrix} \cos\theta & -\sin\theta & 0 \\ \sin\theta & \cos\theta & 0 \\ 0 & 0 & 1 \end{bmatrix}, \quad \boldsymbol{r}^L_{CC_1} = \begin{bmatrix} -\frac{\sqrt{3}}{2}r \\ -\frac{1}{2}r \\ 0 \end{bmatrix},$$

$$\boldsymbol{r}^L_{CC_1} = \begin{bmatrix} -\frac{\sqrt{3}}{2}r \\ -\frac{1}{2}r \\ 0 \end{bmatrix}, \quad \text{and} \quad \boldsymbol{r}^L_{CC_3} = \begin{bmatrix} 0 \\ r \\ 0 \end{bmatrix}$$

$r$ denotes the circumcircle radius of the triangle $C_1C_2C_3$, and $\boldsymbol{r}^L_{CC_i} (i = 1, 2, 3)$ denotes the vector coordinates of $\overrightarrow{CC}_i$.

The coordinates of points $B_1$, $B_2$, and $B_3$ can be expressed with the three angular displacements, $\beta_1$, $\beta_2$, and $\beta_3$:

$$\boldsymbol{r}_{B_i} = \boldsymbol{r}_{A_i} + l_1 \begin{bmatrix} \cos\beta_i \\ \sin\beta_i \\ 0 \end{bmatrix}, \quad i = 1, 2, 3 \tag{5.8}$$

So, the constraint equations are

$$F(x, y, \theta) = \begin{bmatrix} \left(x - \frac{\sqrt{3}}{2}r\cos\theta + \frac{1}{2}r\sin\theta - x_{A_1} - l_1\cos\theta_1\right)^2 + \left(y - \frac{\sqrt{3}}{2}r\sin\theta - \frac{1}{2}r\cos\theta - y_{A_1} - l_1\sin\theta_1\right)^2 - l_2^2 \\ \left(x + \frac{\sqrt{3}}{2}r\cos\theta + \frac{1}{2}r\sin\theta - x_{A_2} - l_1\cos\theta_2\right)^2 + \left(y + \frac{\sqrt{3}}{2}r\sin\theta - \frac{1}{2}r\cos\theta - y_{A_2} - l_1\sin\theta_2\right)^2 - l_2^2 \\ (x - r\sin\theta - x_{A_3} - l_1\cos\theta_3)^2 + (y + r\cos\theta - y_{A_3} - l_1\sin\theta_3)^2 - l_2^2 \end{bmatrix} \tag{5.9}$$

Therefore, from Eq. (5.1), the reachable workspace of the specified point $C$ is expressed as

$$W_{\mathrm{R}} = \{(x, y)\,|\,F(x, y, \theta) = 0\} \tag{5.10}$$

According to Eq. (5.2), the dexterous workspace of point $C$ can be represented by

$$W_D = \{(x, y) \,|\, F(x, y, \theta) = 0, \quad \theta \in [\theta_{z_1}, \theta_{z_2}]\} \tag{5.11}$$

Following, we will investigate the searching problems of the reachable workspace and the dexterous workspace. Firstly, the reachable workspace should be obtained. The primary steps to fulfill such a task might be the following:

**Step 1**: Establish parametric equations of polar coordinates of a point within the workspace:

$$\begin{cases} x = \rho \cos(i\Delta\gamma) \\ y = \rho \sin(i\Delta\gamma) \\ \theta = 0 \end{cases} \tag{5.12}$$

where $\rho$ is the searching polar radius with a polar angle $i\Delta\gamma$ from $x$-axis and $0 \le i \le \frac{2\pi}{\Delta\gamma}$ and $\Delta\gamma$ is the angular interval and $i$ is a natural number.

**Step 2**: Suppose the workspace is divided evenly by $k-1$ radial lines, where $k$ is the largest natural number that is less than or equal to $\frac{2\pi}{\Delta\gamma}$. And therefore, $0 \le i \le k$.

**Step 3**: Search the boundary point in the $i$th radial line. Let $\rho = \rho(i\Delta\gamma)$ increase from zero with a proper interval $\Delta\rho$, namely, $\rho(i\Delta\gamma) = j\Delta\rho,\ j = 0, 1, 2, \ldots$. Merlet [6, 14], Bonev and Ryu [17], and Dash et al. [18] proposed such a numerical algorithm for generating the reachable workspace of parallel manipulators. However, the stop criterion for such an increase is not obvious for the mechanism with rotational actuations. Therefore, we will discuss this problem in the following section.

The reachable workspace of the mechanism is represented with Eq. (5.10). Supposing the position and pose parameters $(x, y, \theta)$ of the end effector to be parametric coefficients, we can immediately transform the relationship constraint equation (5.9) into a function of $\theta_1$, $\theta_2$, and $\theta_3$ in accordance with Eq. (5.4):

$$F(\theta_1, \theta_2, \theta_3) = \begin{bmatrix} p_{11} - p_{12}\cos\theta_1 - p_{13}\sin\theta_1 \\ p_{21} - p_{22}\cos\theta_2 - p_{23}\sin\theta_2 \\ p_{31} - p_{32}\cos\theta_3 - p_{33}\sin\theta_3 \end{bmatrix} \tag{5.13}$$

where

$$p_{11} = \left(x - \frac{\sqrt{3}}{2}r\cos\theta + \frac{1}{2}r\sin\theta - x_{A_1}\right)^2 + \left(y - \frac{\sqrt{3}}{2}r\sin\theta - \frac{1}{2}r\cos\theta - y_{A_1}\right)^2 + l_1^2 - l_2^2,$$

$$p_{12} = 2l_1 \left( x - \frac{\sqrt{3}}{2} r \cos\theta + \frac{1}{2} r \sin\theta - x_{A_1} \right),$$

$$p_{13} = 2l_1 \left( y - \frac{\sqrt{3}}{2} r \sin\theta - \frac{1}{2} r \cos\theta - y_{A_1} \right);$$

$$p_{21} = \left( x + \frac{\sqrt{3}}{2} r \cos\theta + \frac{1}{2} r \sin\theta - x_{A_2} \right)^2 + \left( y + \frac{\sqrt{3}}{2} r \sin\theta - \frac{1}{2} r \cos\theta - y_{A_2} \right)^2 + l_1^2 - l_2^2,$$

$$p_{22} = 2l_1 \left( x + \frac{\sqrt{3}}{2} r \cos\theta + \frac{1}{2} r \sin\theta - x_{A_2} \right),$$

$$p_{23} = 2l_1 \left( y + \frac{\sqrt{3}}{2} r \sin\theta - \frac{1}{2} r \cos\theta - y_{A_2} \right);$$

$$p_{31} = (x - r\sin\theta - x_{A_3})^2 + (y + r\cos\theta - y_{A_3})^2 + l_1^2 - l_2^2,$$

$$p_{32} = 2l_1 (x - r\sin\theta - x_{A_3}),$$

$$p_{33} = 2l_1 (y + r\cos\theta - y_{A_3}).$$

Obviously, within the reachable workspace, the three angular inputs $\theta_1$, $\theta_2$, and $\theta_3$ should be real values. Equivalently, Eq. (5.13) should have real solutions for $\theta_1$, $\theta_2$, and $\theta_3$ if we let $F(\theta_1, \theta_2, \theta_3) = 0$. Assuming $t_1 = \tan\frac{\theta_1}{2}$, $t_2 = \tan\frac{\theta_2}{2}$, and $t_3 = \tan\frac{\theta_3}{2}$, we can immediately have

$$\begin{cases} (p_{11} + p_{12})t_1^2 - 2p_{13}t_1 + p_{11} - p_{12} = 0 \\ (p_{21} + p_{22})t_2^2 - 2p_{23}t_2 + p_{21} - p_{22} = 0 \\ (p_{31} + p_{32})t_3^2 - 2p_{33}t_3 + p_{31} - p_{32} = 0 \end{cases} \quad (5.14)$$

Consequently, the boundaries of the reachable workspace are now transformed into the criteria for real solutions of the three quadric equations of (5.14). With Eq. (5.6), the criteria for the boundaries of the reachable workspace of the mechanism are now expressed as

$$\begin{cases} \Delta_1 = 4\left(p_{12}^2 + p_{13}^2 - p_{11}^2\right) \geq 0 \\ \Delta_2 = 4\left(p_{22}^2 + p_{23}^2 - p_{21}^2\right) \geq 0 \\ \Delta_3 = 4\left(p_{32}^2 + p_{33}^2 - p_{31}^2\right) \geq 0 \end{cases} \quad (5.15)$$

Therefore, the stop criterion for the polar radius's increase in the radial direction can be guaranteed with inequality function set (5.15) while the free variables are

**Table 5.1** Structure parameters for the planar 3-RRR parallel mechanism shown in Fig. 5.1

| Item / Sequence | $r$ (5-mm) | $R$ (5-mm) | $l_1$ (5-mm) | $l_2$ (5-mm) | Range of pose angles | | |
|---|---|---|---|---|---|---|---|
| 1 | 100 | 800 | 300 | 600 | $\left[-\frac{\pi}{6},\frac{\pi}{6}\right]$ | $\left[-\frac{\pi}{3},\frac{\pi}{3}\right]$ | $\left[-\frac{2\pi}{3},\frac{2\pi}{3}\right]$ |
| 2 | 100 | 800 | 500 | 500 | $\left[-\frac{\pi}{6},\frac{\pi}{6}\right]$ | $\left[-\frac{\pi}{3},\frac{\pi}{3}\right]$ | $\left[-\frac{2\pi}{3},\frac{2\pi}{3}\right]$ |
| 3 | 100 | 800 | 800 | 600 | $\left[-\frac{\pi}{6},\frac{\pi}{6}\right]$ | $\left[-\frac{\pi}{3},\frac{\pi}{3}\right]$ | $\left[-\frac{2\pi}{3},\frac{2\pi}{3}\right]$ |

merely restricted in $(x, y, \theta)$, the three positions and pose parameters. Select the largest value of $\rho(i\Delta\gamma)$ as the radius of boundary point in the $i$th radial direction under the constraint of inequality function set (5.15). Consequently, the closed boundary curve of the reachable workspace will be obtained as $i$ increases from 0 to $k$.

Based on the above process, we can also get the following steps to obtain the dexterous workspace of the reference point of the end effector.

**Step 4**: Within step 3, let the orientation angle $\theta$ increase evenly from the lower limit to the upper limit at point $(\rho, i\Delta\gamma)$. The dexterous workspace boundary point in the radial direction occurs when at least one inequality equation set of (5.15) does not hold for a certain orientation within $[\theta_1, \theta_2]$ as $\rho(i\Delta\gamma)$ increases in the radial direction. Similarly, the closed boundary curve of the dexterous workspace is obtained as $i$ increases from 0 to $k$.

We here made some simulations on the mechanism shown in Fig. 5.1 with different parameters shown in Table 5.1. The workspaces and the corresponding dexterous workspaces defined by the prescribed range of pose angles are shown in Fig. 5.2. Simulations indicate that the algorithm proposed in this chapter is valid and effective for searching the reachable and dexterous workspaces.

## 5.4 Distribution of Singularities of a Planar 3-RRR Parallel Mechanism Within Its Reachable Workspace

A planar parallel mechanism shown in Fig. 5.1 is characterized with that its end effector, $C_1C_2C_3$, is an equilateral triangle whose three vertexes are connected with the vertexes of the fixed equilateral triangle, $A_1A_2A_3$, through three identical RRR

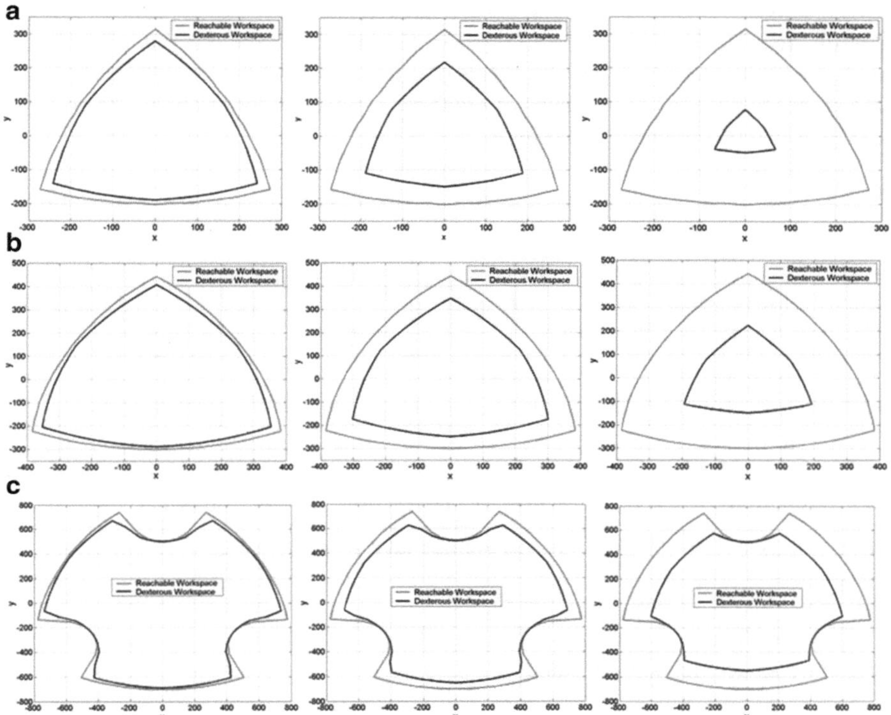


**Fig. 5.2** Different workspaces of the planar 3-RRR parallel mechanism. The reachable and dexterous workspaces of the mechanism with (**a**) the 1st set of parameters, (**b**) the 2nd set of parameters, and (**c**) the 3rd set of parameters

kinematic chains, $A_i B_i C_i$, $i = 1, 2, 3$, individually. Suppose that the end effector, $C_1C_2C_3$, is represented by its geometry center, $C$. The singularity distribution within its workspace will be investigated in what follows.

Obviously, the symmetries of the planar mechanism shown in Fig. 5.1 can be interpreted by a symmetrical group $S_3$, the elements of which are the permutations of a set $V = \{A_1, A_2, A_3\}$. Hence, $S_3 = \{\varphi_1, \varphi_2, \varphi_3, \varphi_4, \varphi_5, \varphi_6\}$, where $\varphi_1 = (1)$, $\varphi_2 = (123)$, $\varphi_3 = (132)$, $\varphi_4 = (23)$, $\varphi_5 = (13)$, and $\varphi_6 = (12)$. $S_3$ describes all symmetries of the mechanism structure shown in Fig. 5.1. From the viewpoint of geometry symmetry, the first three elements of $S_3$ represent the three rotations 0, $\frac{2\pi}{3}$, and $\frac{4\pi}{3}$ about $z$-axis which is perpendicular to the $A_1A_2A_3$-plane; and the last three denote three symmetries about $CA_1$, $CA_2$, and $CA_3$, respectively. Therefore, with the theorem proposed in this chapter, we can reduce the searching range of the polar angle to $\frac{1}{6}$ of that of the initial. However, the whole region is particularly searched to verify the validation of the theorem.

Suppose the coordinates of the reference point of the end effector, $C$, in the fixed coordinate system are $\boldsymbol{r} = (x\ y\ 0)^{\mathrm{T}}$, the rotational angle about $z$-axis with respect to $x$-axis is denoted by $\theta$, and the radius of circumcircle of the triangle, $C_1C_2C_3$, is $R_1$; then, the absolute coordinates of the three vertexes of the triangle end effector, $C_1C_2C_3$, can be obtained, individually:

$$\boldsymbol{r}_{C_i} = \boldsymbol{r} + \mathbf{R}\boldsymbol{r}_{CC_i}, \quad i = 1, 2, 3 \tag{5.16}$$

where $\mathbf{R} = \begin{bmatrix} \cos\theta & -\sin\theta & 0 \\ \sin\theta & \cos\theta & 0 \\ 0 & 0 & 1 \end{bmatrix}$, $\boldsymbol{r}_{CC_1} = \begin{bmatrix} -\frac{\sqrt{3}}{2}R_1 \\ -\frac{1}{2}R_1 \\ 0 \end{bmatrix}$, $\boldsymbol{r}_{CC_2} = \begin{bmatrix} \frac{\sqrt{3}}{2}R_1 \\ -\frac{1}{2}R_1 \\ 0 \end{bmatrix}$, and $\boldsymbol{r}_{CC_3} = \begin{bmatrix} 0 \\ R_1 \\ 0 \end{bmatrix}$ and $R_1$ denotes the circumcircle radius of the triangle $C_1C_2C_3$.

Suppose that the lengths of $A_iB_i$ and $B_iC_i$ are $l_1$ and $l_2$, respectively; the radius of circumcircle of the triangle $A_1A_2A_3$ is $R_2$; and the free rotational angle range about $z$-axis is $\theta_z = [\theta_L, \theta_U]$. Also presume that $\boldsymbol{r}_{A_1} = \left[-\frac{\sqrt{3}}{2}R_2\ -\frac{1}{2}R_2\ 0\right]^{\mathrm{T}}$, $\boldsymbol{r}_{A_2} = \left[\frac{\sqrt{3}}{2}R_2\ -\frac{1}{2}R_2\ 0\right]^{\mathrm{T}}$, and $\boldsymbol{r}_{A_3} = [0\ R_2\ 0]^{\mathrm{T}}$.

The coordinates of points $B_1$, $B_2$, and $B_3$ can be expressed with the three input angular displacements, $\theta_1$, $\theta_2$, and $\theta_3$:

$$\boldsymbol{r}_{B_i} = \boldsymbol{r}_{A_i} + l_1 \begin{bmatrix} \cos\theta_i \\ \sin\theta_i \\ 0 \end{bmatrix}, \quad i = 1, 2, 3 \tag{5.17}$$

The constraint equations for link $B_iC_i (i = 1, 2, 3)$ can be expressed by

$$\|r_{C_i} - r_{B_i}\|^2 - l_2^2 = 0 \tag{5.18}$$

Substituting Eqs. (5.16) and (5.17) into (5.18), we obtain that

$$\boldsymbol{F}(x, y, \theta) = \begin{bmatrix} \left(x - \frac{\sqrt{3}}{2}R_1\cos\theta + \frac{1}{2}R_1\sin\theta + \frac{\sqrt{3}}{2}R_2 - l_1\cos\theta_1\right)^2 + \left(y - \frac{\sqrt{3}}{2}R_1\sin\theta - \frac{1}{2}R_1\cos\theta + \frac{1}{2}R_2 - l_1\sin\theta_1\right)^2 - l_2^2 \\ \left(x + \frac{\sqrt{3}}{2}R_1\cos\theta + \frac{1}{2}R_1\sin\theta - \frac{\sqrt{3}}{2}R_2 - l_1\cos\theta_2\right)^2 + \left(y + \frac{\sqrt{3}}{2}R_1\sin\theta - \frac{1}{2}R_1\cos\theta + \frac{1}{2}R_2 - l_1\sin\theta_2\right)^2 - l_2^2 \\ (x - R_1\sin\theta - l_1\cos\theta_3)^2 + (y + R_1\cos\theta - R_2 - l_1\sin\theta_3)^2 - l_2^2 \end{bmatrix} \tag{5.19}$$

According to Eq. (5.6), the reachable workspace of the specified point $C$ is

$$\Omega = \{(x, y)\,|\,F(x, y, \theta) = 0\}$$

The Jacobian matrix of the constraint equation (5.19) can be expressed as

$$\frac{\partial \boldsymbol{F}(x,y,\theta)}{\partial(x,y,\theta)} = 2\begin{bmatrix} x - \frac{\sqrt{3}}{2}R_1\cos\theta + \frac{1}{2}R_1\sin\theta + \frac{\sqrt{3}}{2}R_2 - l_1\cos\theta_1 & y - \frac{\sqrt{3}}{2}R_1\sin\theta - \frac{1}{2}R_1\cos\theta + \frac{1}{2}R_2 - l_1\sin\theta_1 & t_1 \\ x + \frac{\sqrt{3}}{2}R_1\cos\theta + \frac{1}{2}R_1\sin\theta - \frac{\sqrt{3}}{2}R_2 - l_1\cos\theta_2 & y + \frac{\sqrt{3}}{2}R_1\sin\theta - \frac{1}{2}R_1\cos\theta + \frac{1}{2}R_2 - l_1\sin\theta_2 & t_2 \\ x - R_1\sin\theta - l_1\cos\theta_3 & y + R_1\cos\theta - R_2 - l_1\sin\theta_3 & t_3 \end{bmatrix} \tag{5.20}$$

where $t_1 = R_1x\sin\left(\theta + \frac{\pi}{6}\right) + R_1y\sin\left(\theta - \frac{\pi}{3}\right) + l_1R_1\sin\left(\theta_1 - \theta - \frac{\pi}{6}\right) + R_1R_2\sin\theta$,

$$t_2 = -R_1x\sin\left(\theta - \frac{\pi}{6}\right) + R_1y\sin\left(\theta + \frac{\pi}{3}\right) + l_1R_1\sin\left(\theta - \theta_2 - \frac{\pi}{6}\right) + R_1R_2\sin\theta,$$

$$t_3 = -R_1x\cos\theta - R_1y\sin\theta + l_1R_1\cos(\theta - \theta_3) + R_1R_2\sin\theta.$$

According to Eq. (5.7), the singularities within the workspace can be denoted with

$$S = \left\{(x,y)\left|\begin{array}{l} \boldsymbol{F}(x,y,\theta) = 0 \\ \left|\dfrac{\partial \boldsymbol{F}(x,y,\theta)}{\partial(x,y,\theta)}\right| = 0 \end{array}\right.\right\} \tag{5.21}$$

Next, the searching problem of singularities within the reachable workspace is investigated. Firstly, the reachable workspace should be obtained. The basic steps to fulfill such a task might be found in [17, 18], etc.

**Step 1**: Suppose that [·] represents the largest natural number that is less than or equal to "·". Establish parametric equations of polar coordinates for a point within the workspace

$$\begin{cases} x = \rho\cos(i\,\Delta\alpha) \\ y = \rho\sin(i\,\Delta\alpha) \\ \theta = j\Delta\theta \end{cases} \tag{5.22}$$

where $\rho$ is the searching polar radius with a polar angle $i\,\Delta\alpha$ from $x$-axis and $0 \le i \le \left[\frac{2\pi}{\Delta\alpha}\right]$ and $\Delta\alpha$ is the angular interval for the polar angle; $i$ is a natural number;

**Table 5.2** Structure parameters for another planar 3-RRR parallel mechanism shown in Fig. 5.1

| Item / Sequence | $R_1$ (mm) | $R_2$ (mm) | $l_1$ (mm) | $l_2$ (mm) |
|---|---|---|---|---|
| 1 | 100 | 700 | 200 | 600 |
| 2 | 100 | 700 | 200 | 800 |
| 3 | 100 | 800 | 300 | 600 |

$0 \leq j \leq \left[\frac{\theta_{\max}-\theta_{\min}}{\Delta\theta}\right]$ and $\Delta\theta$ is the angular interval from the minimum pose angle, $\theta_{\min}$, to the maximum pose angle, $\theta_{\max}$; and $j$ is also a natural number.

**Step 2**: Suppose that at the pose angle of $\theta = \theta_{\min} + j\Delta\theta$, the $xoy$-plane is sliced evenly with $k$ radial lines from the origin of the coordinate system, where $k = \left[\frac{2\pi}{\Delta\alpha}\right]$. And therefore, $0 \leq i \leq k$.

**Step 3**: To search the boundary points for the desired workspace and the reachable workspace in the $i$th radial line, we can let $\rho = \rho(i\Delta\alpha)$ increase from zero with a proper interval $\Delta\rho$, namely, $\rho(i\Delta\alpha) = m\Delta\rho,\ m = 0, 1, 2, \ldots$.

**Step 4**: For each point $(\rho(i\Delta\alpha), \Delta\alpha)$ in the radial line, let $j$ increase evenly from zero to $\left[\frac{\theta_{\max}-\theta_{\min}}{\Delta\theta}\right]$, and correspondingly, the pose angle, $\theta$, will also increase evenly from $\theta_{\min}$ to $\theta_{\max}$. If there is at least one $\theta_j$ such that when Eq. (5.22) is substituted into (5.19), there is at least one $\theta_i (i = 1, 2, 3)$ that has no real solution; the increase in the $i$th radial line should be terminated. Then, select the largest parameter $\rho(i\Delta\alpha)$ as the boundary point of the desired workspace in the $i$th radial line. It is not difficult to find that the searched point in the $i$th line will be the boundary point of the reachable workspace when we let $\theta$ change from $-\pi$ to $\pi$. Of course, we can also first solve the maximum and the minimum values of the pose angle $\theta$ within the range $[-\pi, \pi]$ at the geometry center of the mechanism structure, and then let $\theta$ change from the minimum limit to the maximum limit. It is not difficult to prove that the absolute values of the maximum and minimum pose angle $\theta$ must be identical for a mechanism with symmetrical structure.

**Step 5**: When $i$ increases from 0 to $k$, the polar angle can be correspondingly increased from 0 to $2\pi$, and the boundary curves of the desired workspace and the reachable workspace of the end effector are gained, respectively.

According to the theorem, the searching region of the polar angle can be reduced to $\frac{1}{6}$ of that of the initial one. In order to verify the validation of the theorem proposed in this chapter, the whole workspace is searched by computer. Some numerical simulations on the mechanism shown in Fig. 5.1 with different parameters shown in Table 5.1 are executed.

For the first set of parameters shown in Table 5.2, the reachable workspace of the center $C$ of the end effector, $C_1C_2C_3$, is the workspace when the pose angle range is set to any real value, which is shown by subfigure (a) in Fig. 5.3. The different desired workspaces with respect to the pose angle $\theta$ are shown by the interior blank region of subfigure (b) and the interior blank region of subfigure (c)

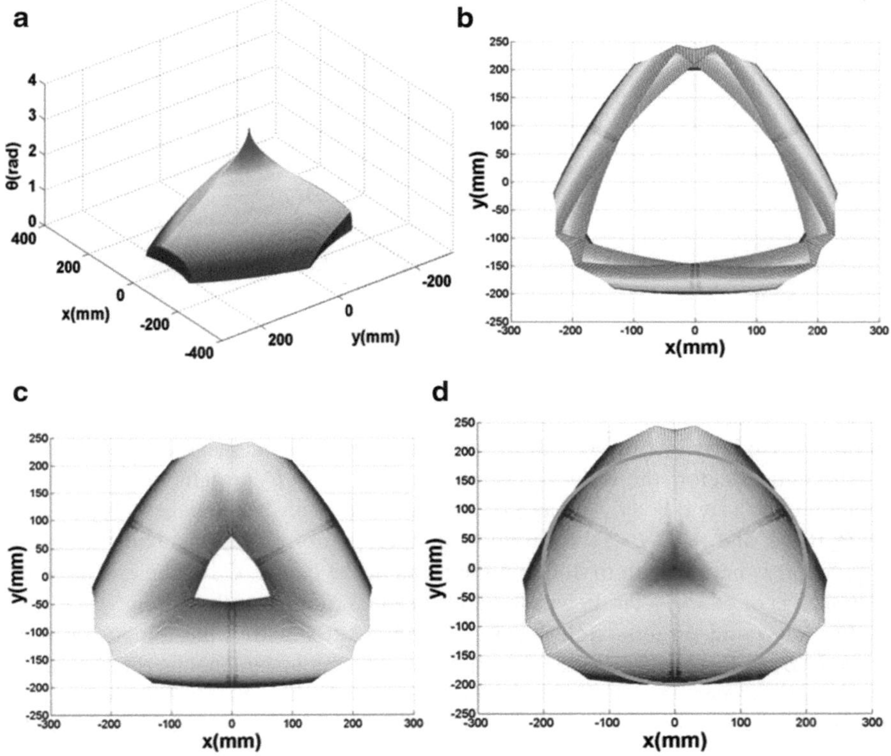


**Fig. 5.3** Desired and reachable workspaces of the mechanism with the 1st set of parameters (**a**) Reachable workspace to the pose angle $\theta$, (**b**) Desired workspace within $\left[-\frac{\pi}{3}, \frac{\pi}{3}\right]$, (**c**) Desired workspace within $\left[-\frac{2\pi}{3}, \frac{2\pi}{3}\right]$, and (**d**) Continuous singular curve within the workspace

in Fig. 5.3, respectively. Computer simulation indicates that the singularities within its reachable workspace only occur at $\theta = 0$ and the singularities form a continuous curve, which is shown by subfigure (d) in Fig. 5.3.

Next, we will investigate the physical meanings of the singular curve which is denoted by the circle in subfigure (d) of Fig. 5.3, and the radius is just the length of link $A_i B_i$.

Firstly, the singularities occur when the pose angle is zero. Next, we can study the geometry characteristics of the mechanism with the first set of parameters in Table 5.1 (refer to Table 5.1, sequence 1). The following relationship always holds:

$$R_2 - R_1 = l_2 \tag{5.23}$$

Under the condition that the pose angle $\theta = 0$, equation (5.23) indicates that the points $B_1$, $B_2$, and $B_3$ can form an equilateral triangle which is shown by the

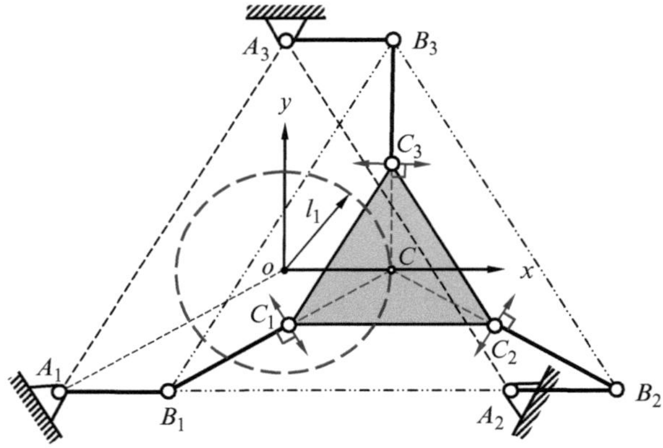


**Fig. 5.4** A singular configuration of the mechanism

double dashed triangle in Fig. 5.4. If we do not consider the instantaneous rotation of the end effector, the double dashed triangle, $B_1B_2B_3$, might be temporarily thought of as a rigid body. And as a result, the virtual four-bar linkages, $A_1B_1B_2A_2$, $A_2B_2B_3A_3$, and $A_1B_1B_3A_3$, form three parallelogram four-bar mechanisms. From Fig. 5.4, it is not difficult to find that the locus of the center, $C$, of the end effector, $C_1C_2C_3$, is of course a circle whose radius is equal to the length of $l_1$.

For any point on this circle locus, the end effector within the temporary rigid triangle $B_1B_2B_3$ (double dashed triangle) can make an instantaneous rotation about the geometry center point, $C$. The instantaneous velocities of the three points, $C_1$, $C_2$, and $C_3$ with respect to the instantaneous velocity center, $C$, are shown by the three double arrows in Fig. 5.4. Therefore, all points on the dashed circle in Fig. 5.4 are singularities, which demonstrate the correctness of the computer simulation. Consequently, the relationship (5.16) holds for the mechanism shown in Fig. 5.4.

For the second set of parameters shown in Table 5.1, the boundaries of the reachable workspace of the center, $C$, of the end effector, $C_1C_2C_3$, with respect to the pose angle, $\theta$, is shown by subfigure (a) in Fig. 5.5. The reachable workspace, as a matter of fact, can be obtained by letting the pose angle $\theta$ be any real value between $-\pi$ and $\pi$, which is shown by the colored area of subfigure (b) in Fig. 5.5. Computer simulation indicates that singularity of the mechanism is just the origin of the $xoy$-system, which is on the geometry center of the desired and reachable workspaces.

Next, the physical meanings of this singular point will be also investigated. It is not difficult to find that the following relationship always holds for the second set of parameters (refer to Table 5.1, sequence 2).

$$R_2 - R_1 = l_2 - l_1 \tag{5.24}$$

Equation (5.24) indicates that the points $A_i, B_i, C_i\, (i = 1, 2, 3)$ can keep in one line, which is coincident with a midline of the equilateral triangle, $C_1C_2C_3$, that

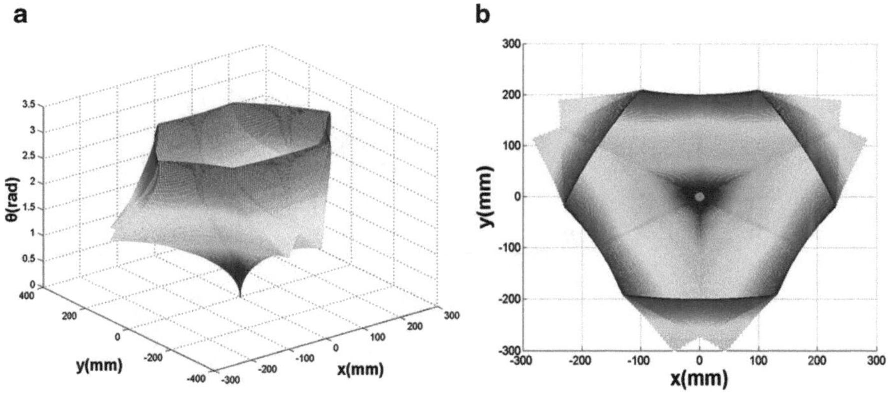


**Fig. 5.5** Reachable workspace and the singularity of the mechanism with the 2nd set of parameters (**a**) Boundaries of the reachable workspace and (**b**) Singularity within the workspace

**Fig. 5.6** A singular configuration of the mechanism within the reachable workspace

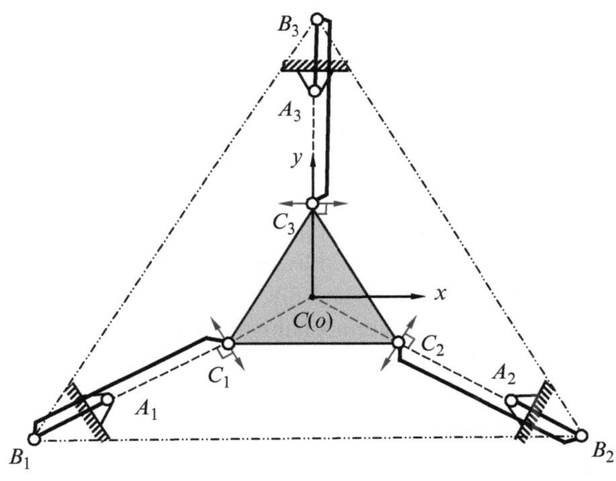


passes through the vertex, $C_i$. As is shown in Fig. 5.6, at the configuration where $x = y = \theta = 0$, the end effector can make an instantaneous rotation about its center point, $C$. The instantaneous velocities of the three points, $C_1$, $C_2$, and $C_3$ with respect to $C$ are shown by the three double arrows in Fig. 5.6. As a matter of fact, such a configuration is a singularity of the mechanism, which is illustrated by subfigure (b) in Fig. 5.5.

Consequently, the relationship (5.16) also holds for the mechanism shown in Fig. 5.6. For the third set of parameters (refer to Table 5.1, sequence 3), we can also make a numerical simulation. The increment in the radial direction is $\Delta\rho = 1$, angular increment for the polar angle is $\Delta\alpha = \frac{\pi}{144}$, and the angular increment for the pose angle is $\Delta\theta = \frac{\pi}{144}$. Computer simulation indicates that there is no singularity

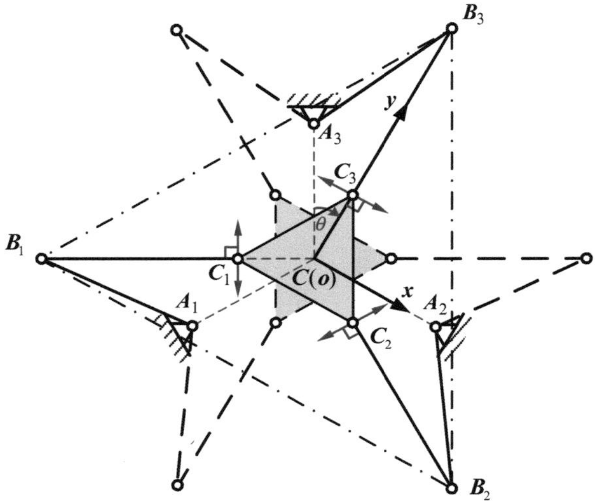


**Fig. 5.7** Two singular configurations of the mechanism when $\rho = 0$

within the whole reachable workspace except when $\rho = 0,\ \theta = \pm \arccos \frac{13}{14}$. These two configurations are shown in Fig. 5.7.

The reachable workspace in $oxy\theta$-space and in $xoy$-plane are shown in subfigures (a) and (b) in Fig. 5.8. The desired workspaces with respect to the pose angle, $\theta$, are shown in subfigures (c) and (d) in Fig. 5.8. The interior blank regions in subfigures (c) and (d) represent the desired workspaces with pose angle ranges $\left[-\frac{\pi}{3}, \frac{\pi}{3}\right]$ and $\left[-\frac{2\pi}{3}, \frac{2\pi}{3}\right]$, individually.

From Figs. 5.3, 5.5, and 5.8, we can draw a conclusion that the relationship of (5.16) holds for all the mechanisms described by the parameters in Table 5.1. And this verifies the theorem.

## 5.5 Conclusions

This chapter addresses the determinations of the reachable and dexterous workspaces of planar parallel linkages under rotational actuations. The relationship constraint equations established from geometry and mobility characteristics of a mechanism are transformed into the constraints for the boundary curves or boundary surfaces of the reachable and dexterous workspaces of a reference point of the end effector. To realize the algorithm, discretization methodology for searching the boundaries of the reachable and dexterous workspaces is investigated. This algorithm can be utilized to obtain the workspaces of any planar parallel mechanisms with rotational actuations. Theoretical investigation and the mechanism illustrations demonstrate the validity and efficiency of the algorithm.

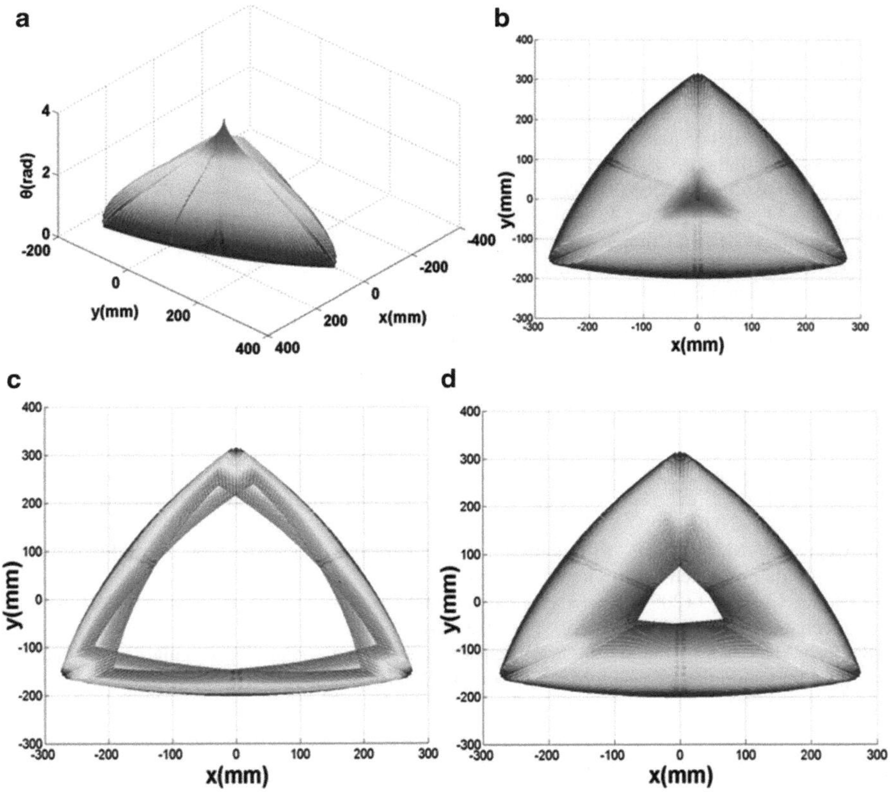


**Fig. 5.8** No singularity within the reachable and desired workspaces (**a**) Reachable workspace, (**b**) overview of the reachable workspace, (**c**) $\left[-\frac{\pi}{3}, \frac{\pi}{3}\right]$-desired workspace, and (**d**) $\left[-\frac{2\pi}{3}, \frac{2\pi}{3}\right]$-desired workspace

# References


1. Sen D, Mruthyunjaya TS (1998) A centro-based characterization of singularities in the workspace of planar closed-loop mechanisms. Mech Mach Theory 33(8):1091–1105
2. Kumar A, Waldron KJ (1981) The workspaces of a mechanical mechanism. J Mech Des 103:665–672
3. Gosselin CM, Jean M (1996) Determination of the workspace of planar parallel manipulators with joint limits. Robot Auton Syst 17(3):129–138
4. Merlet J-P, Gosselin CM, Mouly N (1998) Workspaces of planar parallel manipulators. Mech Mach Theory 33(1):7–20
5. Gosselin C (1990) Determination of the workspace of 6-DoF parallel manipulators. J Mech Des 112(3):331–336
6. Merlet J-P (1999) Determination of 6D workspaces of Gough-type parallel manipulator and comparison between different geometries. Int J Robot Res 18(9):902–916

7. Dibakar S, Mruthyunjaya TS (1999) A computational geometry approach for determination of boundary of workspaces of planar mechanisms with arbitrary topology. Mech Mach Theory 34(1):149–169
8. Snyman JA, du Plessis LJ, Duffy J (2000) An optimization approach to the determination of the boundaries of manipulator workspaces. J Mech Des 122(4):447–456
9. Yang J, Abdel-Malek K, Zhang Y (2008) On the workspace boundary determination of serial manipulators with non-unilateral constraints. Robot Comput Integr Manuf 24:60–76. doi:10.1016/j.rcim.2006.06.005
10. Abdel-Malek K, Adkins F et al (1997) On the determination of boundaries to manipulator workspaces. Robot Comput Integr Manuf 13(1):63–72
11. Abdel-Malek K, Yeh H-J, Othman S (2000) Interior and exterior boundaries to the workspace of mechanical manipulators. Robot Comput Integr Manuf 16(5):365–376
12. Bonev IA, Gosselin CM (2006) Analytical determination of the workspace of symmetrical spherical parallel mechanisms. IEEE Trans Robot 22(5):1011–1017
13. Davidson JK, Chaney KD (1999) A design procedure for RPR planar robotic workcells: an algebraic approach. Mech Mach Theory 34(2):193–203
14. Merlet J-P (1997) Designing a parallel manipulator for a specific workspace. Int J Robot Res 16(4):545–556
15. Cervantes-Sánchez JJ, Rendón-Sánchez JG (1999) A simplified approach for obtaining the workspace of a class of 2-dof planar parallel mechanisms. Mech Mach Theory 34(7):1057–1073
16. Ceccarelli M (1996) A formulation for the workspace boundary of general N-revolute manipulators. Mech Mach Theory 31(5):637–646
17. Bonev IA, Ryu J (2001) A new approach to orientation workspace analysis of 6-DOF parallel mechanisms. Mech Mach Theory 36(1):15–28
18. Dash AK, Chen I-M et al (2005) Workspace generation and planning singularity-free path for parallel manipulators. Mech Mach Theory 40(7):776–805

# Chapter 6
# Statics Analysis of Special Planar Linkages

After analyzing the geometry of the foldable stair mechanism in Chap. 3, the statics should be discussed. Statics analysis is one of the most important tasks in synthesizing a foldable structure. The statics [1, 2] and kinematics [3, 4] of the deployable structures are widely investigated. Nagaraj et al. [5] investigated the kinematics and statics of a kind of pantograph mast. They addressed the constraints of the joints by Lagrange multiplier method. In this chapter, the statics of rigid-body system of the stair mechanism is first developed, and then the inner forces of every link are analyzed.

## 6.1 Rigid Statics of the Foldable Stair

First, forces exerted on each footplate of the stair should be investigated. It might be supposed that the weight of a person is applied to a point on the footplate which is $a$ away from the right pivot bar shown in Fig. 6.1. For the distance between every two adjacent horizontal pivot bars is $l$, the forces exerted on the right and left pivot bars should satisfy the following moment equations:

$$\begin{cases} \sum M_{Bm}(F) = 0 \quad Ga - F_{Cm}l = 0 \\ \sum M_{Cm}(F) = 0 \quad -G(l-a) + F_{Bm}l = 0 \end{cases} \tag{6.1}$$

where $M_{Bm}(F)$ indicates that the moments in the equation are all about point $B_m$ and $M_{C_m}(F)$ indicates that the moments in the equation are all about point $C_m$.

Solving equation set (6.1) yields

$$\begin{cases} F_{Bm} = \left(1 - \dfrac{a}{l}\right) G \\ F_{Cm} = \dfrac{a}{l} G \end{cases} \tag{6.2}$$

J. Zhao et al., *Design of Special Planar Linkages*, Springer Tracts in Mechanical Engineering, DOI 10.1007/978-3-642-38448-6_6,

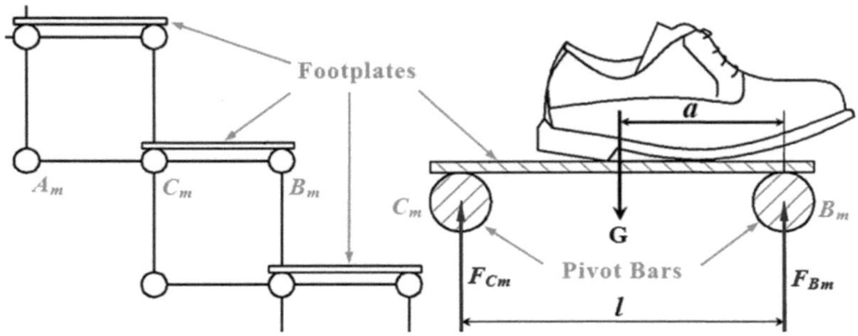


**Fig. 6.1** Forces exerted on the footplate of stairs

To simplify the analysis, we may as well suppose that $a = \frac{1}{2}l$. As a result, the following relationship from Eq. (6.2) can be immediately obtained:

$$F_{Bm} = F_{Cm} = \frac{1}{2}G \tag{6.3}$$

Equation (6.3) indicates that the weight of a person is equally supported by two joints, $Cm$ and $Bm$, of link $A_m C_m B_m$. Generally, we can suppose that there are three forces, $p_{A_n}$, $p_{C_n}$, and $p_{B_n}$, which exert on joints $A_n$, $C_n$, and $B_n$, individually. Without loss of generality, a horizontal disturbing force, $F$, is particularly assumed to act on link $A_n B_n$. Now the joint forces of the stair under the external forces shown in Fig. 6.2 can be investigated.

Because the stair mechanism works under the conditions that the vertical and horizontal links are perpendicular or near perpendicular, the following analysis adopts the orthogonal assumptions for every pair of adjacent links, namely, $\psi = \frac{\pi}{2}$. Now, the statics model of the links of a stair shown in Fig. 6.3 can be built.

For the vertical link in a stair shown in the left of Fig. 6.3, we immediately obtain the following equations:

$$\begin{cases} M(C_k) = 0 \\ M(B_{k-1}) = 0 \\ \sum F_y = 0 \end{cases} \tag{6.4}$$

Equation set (6.4) can be expanded to

$$\begin{cases} S_{B(k-1)} = S_{A(k+1)} \\ S_{Ck} = -2S_{A(k+1)} \\ T_{Ck} + T_{A(k+1)} - T_{Ck} - T_{A(k+1)} = 0 \end{cases} \tag{6.5}$$

where $S_{Ck}$ represents the horizontal force exerted to joint $C$ of the $k$th vertical link and $T_{Ck}$ represents the vertical force exerted to joint $C$ of the $k$th vertical link; the rest of the symbols can be similarly deduced.

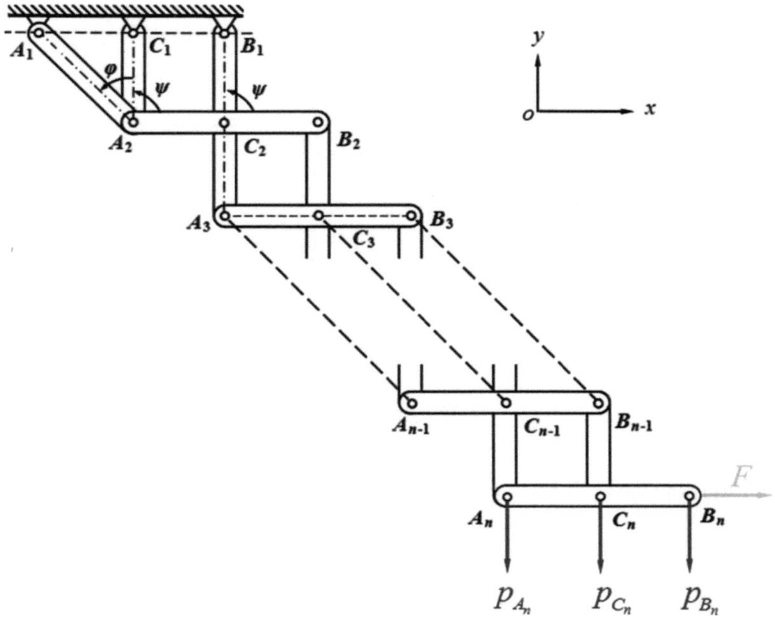


Fig. 6.2 Extreme conditions of the foldable stair mechanism

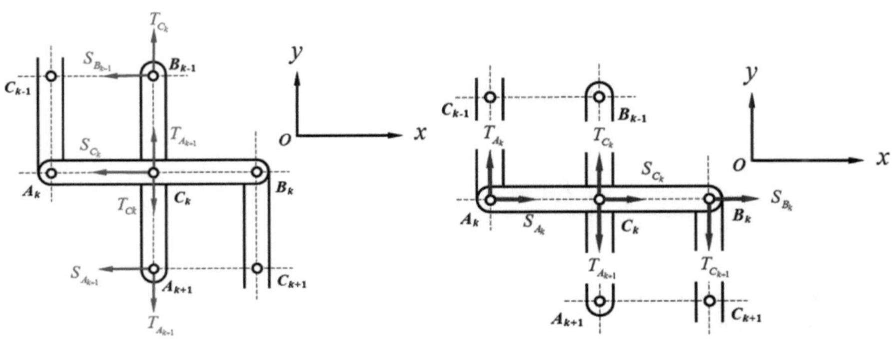


Fig. 6.3 Statics analysis of the links of a stair

For the horizontal link in a stair shown in the right of Fig. 6.3, we also have the following equations:

$$\begin{cases} M(C_k) = 0 \\ M(A_k) = 0 \\ \sum F_x = 0 \end{cases} \tag{6.6}$$

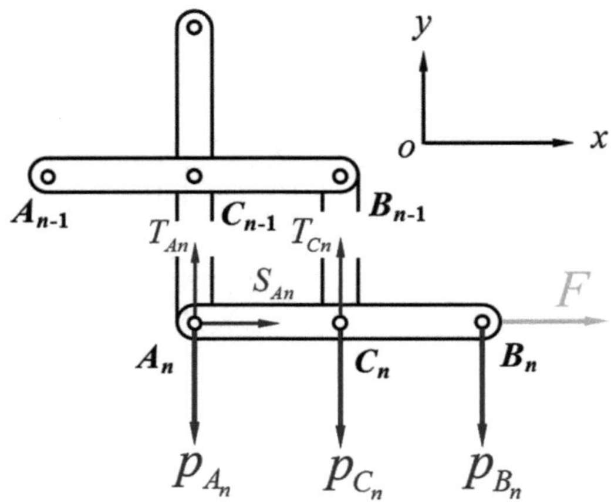


**Fig. 6.4** Statics analysis of the links of the last stair

Equation set (6.6) can be expanded to

$$\begin{cases} T_{Ak} = -T_{C(k+1)} \\ T_{Ck} = 2T_{C(k+1)} + T_{A(k+1)} \\ S_{Ak} + S_{Ck} + S_{Bk} = 0 \end{cases} \tag{6.7}$$

Combining Eqs. (6.5) and (6.7) yields

$$\begin{cases} S_{Ak} = -S_{Ck} - S_{Bk} \\ S_{B(k-1)} = S_{A(k+1)} \\ S_{Ck} = -2S_{A(k+1)} \\ T_{Ak} = -T_{C(k+1)} \\ T_{Ck} = 2T_{C(k+1)} + T_{A(k+1)} \end{cases} \tag{6.8}$$

Now we can analyze the inner forces of the links of the last stair shown in Fig. 6.4. The link $C_n B_{n-1}$ is a two-force component and

$$\begin{cases} S_{Cn} = 0 \\ S_{B(n-1)} = 0 \end{cases} \tag{6.9}$$

$$\begin{cases} M(C_n) = 0 \\ M(A_n) = 0 \\ \sum F_x = 0 \\ T_{Bn} = 0 \end{cases} \tag{6.10}$$

Equation set (6.10) can be expanded to

$$\begin{cases} p_{An}l - p_{Bn}l - T_{An}l = 0 \\ T_{Cn}l - p_{Cn}l - 2p_{Bn}l = 0 \\ S_{An} + F = 0 \\ T_{Bn} = 0 \end{cases} \tag{6.11}$$

Solving equation set (6.11) and combining equation set (6.9), we gain

$$\begin{cases} S_{An} = -F \\ S_{Cn} = 0 \\ S_{B(n-1)} = 0 \\ T_{Bn} = 0 \\ T_{An} = p_{An} - p_{Bn} \\ T_{Cn} = 2p_{Bn} + p_{Cn} \end{cases} \tag{6.12}$$

When $k = n - 1$, equation set (6.8) can be transformed into

$$\begin{cases} S_{A(n-1)} = -S_{C(n-1)} - S_{B(n-1)} \\ S_{B(n-2)} = S_{An} \\ S_{C(n-1)} = -2S_{An} \\ T_{A(n-1)} = -T_{Cn} \\ T_{C(n-1)} = 2T_{Cn} + T_{An} \end{cases} \tag{6.13}$$

Associating Eqs. (6.8), (6.12), and (6.13) yields

$$\begin{cases} S_{Ak} = -(n-k+1)F \\ S_{Bk} = -(n-k-1)F \\ S_{Ck} = 2(n-k)F \\ T_{Ak} = -(n-k-1)p_{An} - (n-k+1)p_{Bn} - (n-k)p_{Cn} \\ T_{Ck} = (n-k)p_{An} + (n-k+2)p_{Bn} + (n-k+1)p_{Cn} \end{cases} \tag{6.14}$$

From equation set (6.14), we obtain that

$$\begin{cases} T_{Ak} = -T_{C(k+1)} \\ S_{A(k+1)} = S_{B(k-1)} \end{cases} \tag{6.15}$$

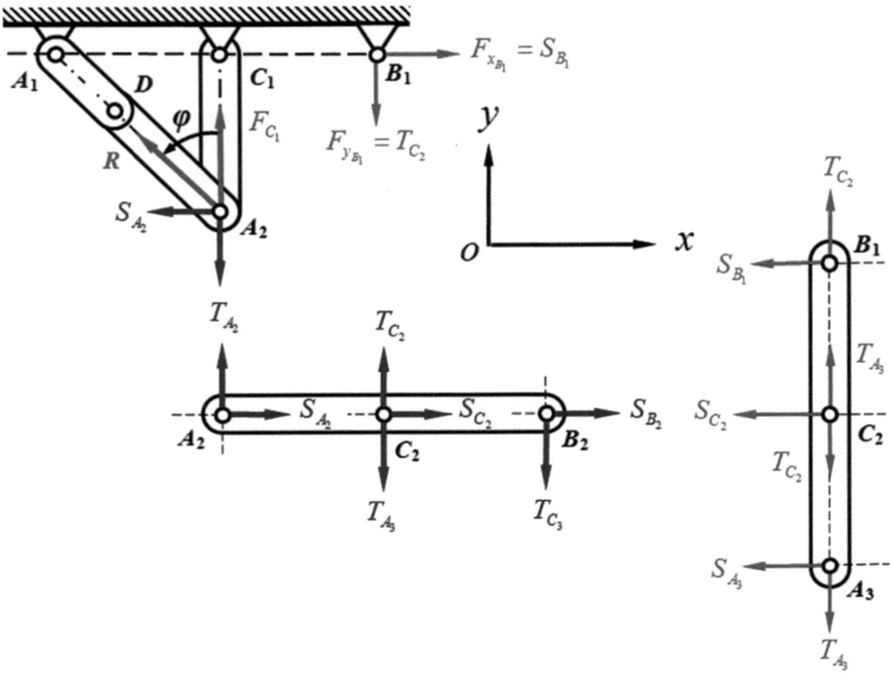


**Fig. 6.5** Statics analysis of the first stair unit

The minus indicates that force $T_{Ak}$ is opposite to the right of Fig. 6.3 in direction but identical in magnitude.

When $k = 2$, Eq. (6.14) can be expressed as

$$\begin{cases} S_{A_2} = -(n-1)F \\ S_{B_2} = -(n-3)F \\ S_{C_2} = 2(n-2)F \\ T_{A_2} = -(n-3)p_{An} - (n-1)p_{Bn} - (n-2)p_{Cn} \\ T_{C_2} = (n-2)p_{An} + np_{Bn} + (n-1)p_{Cn} \end{cases} \tag{6.16}$$

Under the working conditions, links $A_1D$ and $DA_2$ keep in a line. Therefore, $A_1DA_2$ can be thought of as a two-force member, and the force exerted to joint $A_2$ is denoted by $R$ which is shown in Fig. 6.5. Similarly, link $C_1A_2$ is also a two-force member, and the force exerted to joint $A_2$ is represented by $F_{C_1}$.

For joint $A_2$, we have the following equilibrium conditions:

$$\begin{cases} \sum F_x = 0, \quad -R\sin\varphi - S_{A_2} = 0 \\ \sum F_y = 0, \quad R\cos\varphi + F_{C_1} - T_{A_2} = 0 \end{cases} \tag{6.17}$$

With Eq. (6.16), we obtain

$$\begin{cases} R = (n-1)F\csc\varphi \\ F_{C_1} = -(n-3)p_{An} - (n-1)p_{Bn} - (n-2)p_{Cn} - (n-1)F\cot\varphi \end{cases} \tag{6.18}$$

From Eq. (6.14), we can also find that

$$S_{A_3} = -(n-2)F \tag{6.19}$$

For the vertical link, $B_1C_2A_3$, shown in Fig. 6.5, there is

$$\sum F_x = 0, \quad S_{B_1} + S_{C_2} + S_{A_3} = 0 \tag{6.20}$$

Associating Eqs. (6.16), (6.19), and (6.20) presents

$$S_{B_1} = -S_{C_2} - S_{A_3} = -2(n-2)F + (n-2)F = -(n-2)F \tag{6.21}$$

Therefore, the forces of the fixed joint $B_1$ can be obtained from Eqs. (6.16) and (6.21)

$$\begin{cases} F_{xB_1} = S_{B_1} = -(n-2)F \\ F_{yB_1} = T_{C_2} = (n-2)p_{An} + np_{Bn} + (n-1)p_{Cn} \end{cases} \tag{6.22}$$

where the minus indicates that the force $F_{xB_1}$ is opposite to that of Fig. 6.5 in direction but identical in magnitude.

The real working condition is that the stair mechanism is fully deployed when each pair of adjacent links are orthogonally connected. When $\varphi = \frac{\pi}{4}$, Eqs. (6.20) and (6.22) can be jointly simplified as

$$\begin{cases} F_{xB_1} = -(n-2)F \\ F_{yB_1} = (n-2)p_{An} + n\,p_{Bn} + (n-1)p_{Cn} \\ F_{C_1} = -(n-3)p_{An} - (n-1)p_{Bn} - (n-2)p_{Cn} - (n-1)F \\ R = \sqrt{2}(n-1)F \end{cases} \tag{6.23}$$

When $p_{A_n} = p_{C_n} = 0$, $p_{B_n} = G$ where $G$ represents the weight of a person, Eq. (6.23) is further simplified to

$$\begin{cases} F_{xB_1} = -(n-2)F \\ F_{yB_1} = nG \\ F_{C_1} = -(n-1)G - (n-1)F \\ R = \sqrt{2}(n-1)F \end{cases} \tag{6.24}$$

Equation (6.24) indicates that the inner forces of link $A_1A_2$ and the horizontal component of joint $B_1$ only result from the horizontal force, $F$, in real working. As a matter of fact, $F$ is just a disturbing component of a person's self-weight. So it is not large in value. As a result, Eq. (6.24) can be loosely enlarged to

$$\begin{cases} F_{xB_1} = -nF \\ F_{yB_1} = nG \\ F_{C_1} = -(n-1)G - (n-1)F \\ R = \sqrt{2}(n-1)F \end{cases} \tag{6.25}$$

Equation (6.25) indicates that the largest tensions of links $B_1C_2$ and $C_1A_2$ are proportional to the number of stairs. Considering the general height of one floor is about 3 m and the height of one stair is 0.2 m, the number of stairs needed is 15 and the weight of one person is not more than 100 kg, and the disturbing force $F = 0.2\,G$, so the inner forces of links $B_1C_2$ and $C_1A_2$ are $15 \times 100 \times 9.8 \times \sqrt{1+0.2^2} = 1,4991$ N and $-14 \times 100 \times 9.8 \times (1+0.2) = -16,464$ N, individually. Therefore, we can take the initial selections for the material, the section size, and the connecting forms.

Hereat, another suggestion might be arisen – why not add redundant supports from the ground to reduce the largest tensile forces? Next section will focus on the inner force analysis for such a scheme.

## 6.2 Inner Force Analysis of the Foldable Stair Mechanism

For $k \leq n-1$, the inner force status for each horizontal link is similar which can be illustrated by Fig. 6.6.

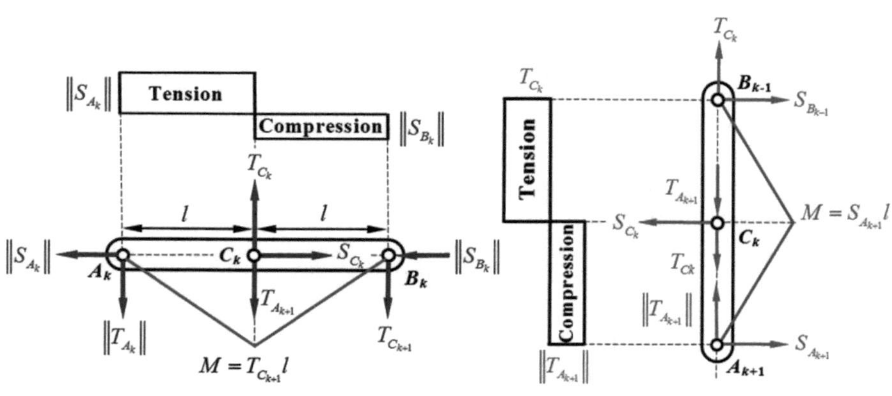


**Fig. 6.6** Inner forces of horizontal and vertical links

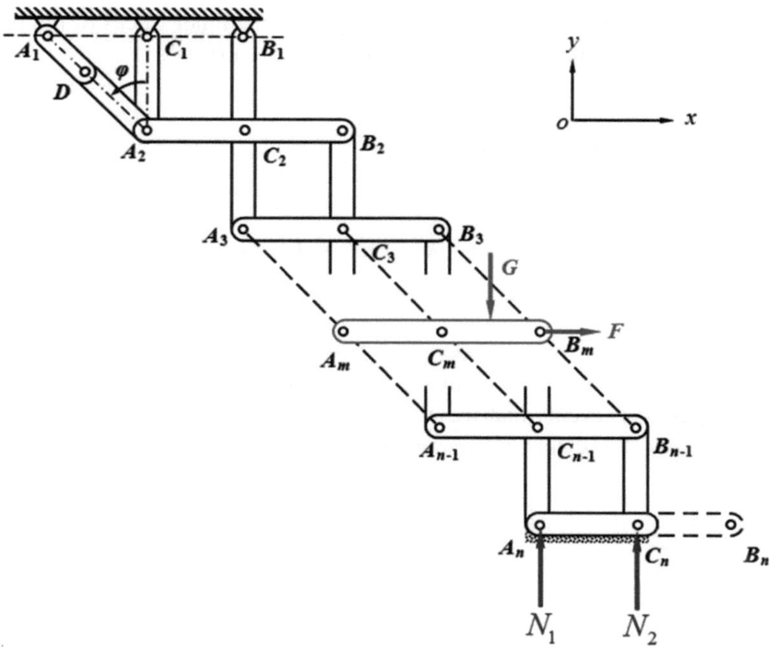


**Fig. 6.7** Redundant statics of the deployed stair mechanism

According to the above analysis, both horizontal and vertical links might be particularly designed to be beams of uniform strength. By this means, we can save a great deal of mass and decrease the self-weight of the stair.

Now, we can analyze the redundant statics of the foldable stair mechanism. Suppose the last stair, namely, the $n$th stair is also supported by the floor which is shown in Fig. 6.6. For simplification and without loss of generality, the redundant forces exerted on the $n$th stair by the floor are assumed to be two vertical forces at joints $B_n$ and $C_n$, which are denoted by $N_1$ and $N_2$, respectively.

In what follows, we can also investigate the equilibrium equations for the deployed stair mechanism. Because the determinate statics has been established in Sect. 6.1, we only need to establish the deformation compatibility equations for the indeterminate structure shown in Fig. 6.7. Suppose that a person weighed $G$ stands on the $m$th ($m \leq n$) stair and the stair also subjects to an additional disturbing force, $F$, which are shown in Fig. 6.7. This problem can be solved in accordance with the second Castigliano's theorem [6] by supposing that the redundant joint forces provided by the ground are $N_1$ and $N_2$, which is shown in Fig. 6.7.

Firstly, the strain energy of the stair structure with the redundant forces $N_1$ and $N_2$ should be established.

When $G$ and $F$ are not considered in the system, the inner forces of the links of every stair can be obtained in accordance to equation set (6.14):

$$\begin{cases} S_{Ak1} = 0 \\ S_{Bk1} = 0 \\ S_{Ck1} = 0 \\ T_{Ak1} = (n-k-1)N_1 + (n-k)N_2 \\ T_{Ck1} = -(n-k)N_1 - (n-k+1)N_2 \end{cases} \tag{6.26}$$

When only $G$ and $F$ are considered in the system, the inner forces of the links of every stair can also be obtained in accordance to equation set (6.14). Assume the weight of a person is evenly distributed to joints $C_m$ and $B_m$. So, the inner forces of the links of every stair can be expressed as

$$\begin{cases} S_{Ak2} = -(m-k+1)F \\ S_{Bk2} = -(m-k-1)F \\ S_{Ck2} = 2(m-k)F \\ T_{Ak2} = -\left(m-k+\frac{1}{2}\right)G \\ T_{Ck2} = \left(m-k+\frac{3}{2}\right)G \end{cases} \tag{6.27}$$

where $k \le m \le n$.

From Eqs. (6.26) and (6.27), the inner forces of the stair structure can be expressed as follows:

1. When $k \le m < n$,

$$\begin{cases} S_{Ak} = -(m-k+1)F \\ S_{Bk} = -(m-k-1)F \\ S_{Ck} = 2(m-k)F \\ T_{Ak} = -\left(m-k+\frac{1}{2}\right)G + (n-k-1)N_1 + (n-k)N_2 \\ T_{Ck} = \left(m-k+\frac{3}{2}\right)G - (n-k)N_1 - (n-k+1)N_2 \end{cases} \tag{6.28}$$

2. When $2 \le m < k \le n$,

$$\begin{cases} S_{Ak} = 0 \\ S_{Bk} = 0 \\ S_{Ck} = 0 \\ T_{Ak} = (n-k-1)N_1 + (n-k)N_2 \\ T_{Ck} = -(n-k)N_1 - (n-k+1)N_2 \end{cases} \tag{6.29}$$

From Fig. 6.7, we can find that the strain energy of every link in a stair mainly result from the bending forces and the tension or compression forces. So the strain energy of the stair structure is obtained:

$$V_{\varepsilon}=\frac{1}{2}\sum_{k=2}^{n}\left(\int_{0}^{\frac{L}{2}}\frac{S_{Ak}^{2}+S_{Bk}^{2}}{Ebh}\mathrm{d}x+\int_{0}^{\frac{L}{2}}\frac{T_{Ck}^{2}+T_{A(k+1)}^{2}}{Ebh}\mathrm{d}x\right.$$
$$\left.+2\int_{0}^{\frac{L}{2}}\frac{T_{Ak}^{2}x^{2}}{\frac{1}{12}Ebh^{3}}\mathrm{d}x+2\int_{0}^{\frac{L}{2}}\frac{S_{A(k+1)}^{2}y^{2}}{\frac{1}{12}Ebh^{3}}\mathrm{d}y\right) \tag{6.30}$$

where $E$ represents the Young's elastic modulus and $b$ and $h$ represent the width and height of the section of each link, individually.

Hereat, the deformation compatibility equations can be expressed in line with the second Castigliano's theorem:

$$\begin{cases}\dfrac{\partial V_{\varepsilon}}{\partial N_1}=0\\[2mm]\dfrac{\partial V_{\varepsilon}}{\partial N_2}=0\end{cases} \tag{6.31}$$

When the foldable stair mechanism is deployed, the constraint forces of joints $A_1$, $B_1$, and $C_1$ can be similarly obtained with equation sets (6.28) and (6.19):

$$\begin{cases}S_{A_2}=-(m-1)F\\ S_{B_2}=-(m-3)F\\ S_{A_3}=-(m-2)F\\ S_{C_2}=2(m-2)F\\ T_{A_2}=-\left(m-\frac{3}{2}\right)G+(n-3)N_1+(n-2)N_2\\ T_{C_2}=\left(m-\frac{1}{2}\right)G-(n-2)N_1-(n-1)N_2\end{cases} \tag{6.32}$$

For joint $A_2$ shown in Fig. 6.5, we have the following equilibrium conditions:

$$\begin{cases}\sum F_x=0, & -R\sin\varphi-S_{A_2}=0\\ \sum F_y=0, & R\cos\varphi+F_{C_1}-T_{A_2}=0\end{cases} \tag{6.33}$$

With Eq. (6.32), we obtain

$$\begin{cases}R=(m-1)F\csc\varphi\\ F_{C_1}=-\left(m-\dfrac{3}{2}\right)G+(n-3)N_1+(n-2)N_2-(m-1)F\cot\varphi\end{cases} \tag{6.34}$$

For vertical link $B_1C_2A_3$ shown in Fig. 6.5, there is

$$\sum F_x=0,\quad S_{B_1}+S_{C_2}+S_{A_3}=0 \tag{6.35}$$

According to Eq. (6.32), we can obtain from Eq. (6.35)

$$S_{B_1} = -S_{C_2} - S_{A_3} = -2(m-2)F + (m-2)F = -(m-2)F \tag{6.36}$$

Therefore, the forces of the fixed joint $B_1$ can be obtained from Eqs. (6.32) and (6.36):

$$\begin{cases} F_{xB_1} = -S_{B_1} = (m-2)F \\ F_{yB_1} = T_{C_2} = \left(m - \dfrac{1}{2}\right)G - (n-2)N_1 - (n-1)N_2 \end{cases} \tag{6.37}$$

where the minus indicates that the force $F_{xB_1}$ is opposite to that of Fig. 6.5 in direction but identical in magnitude.

Therefore, considering $\varphi = \frac{\pi}{4}$, the three joint forces on the supporting plate can be obtained from Eqs. (6.34) and (6.37):

$$\begin{cases} F_{xB_1} = (m-2)F \\ F_{yB_1} = \left(m - \dfrac{1}{2}\right)G - (n-2)N_1 - (n-1)N_2 \\ F_{C_1} = -\left(m - \dfrac{3}{2}\right)G + (n-3)N_1 + (n-2)N_2 - (m-1)F \\ R = \sqrt{2}(m-1)F \end{cases} \tag{6.38}$$

Associating Eqs. (6.31) and (6.38), we obtain the forces of $F_{xB_1}$, $F_{yB_1}$, $F_{C_1}$, $R$, $N_1$, and $N_2$.

## 6.3 Numerical Simulations on the Foldable Stair Mechanism

Based on the analysis of Sect. 6.2, computer simulations can be executed. To get a comprehensive calculation, the algorithm discussed above is programmed at MATLAB, and some numerical simulations are executed. The primary mechanical parameters are listed in Table 6.1.

Equation (6.38) indicates that the forces $N_1$, $N_2$, $F_{xB_1}$, $F_{yB_1}$, $F_{C_1}$, and $R$ are all changing with the shift of standing stair of a person. Computing results based on the analysis of Sect. 6.2 are listed in Table 6.2 where the negative values indicate that

**Table 6.1** Primary parameters of the foldable stair mechanism

| Length of link (m) | $G$ (kg) | $F$ (N) | $E$ (GPa) | $b$ (m) | $h$ (m) |
|---|---|---|---|---|---|
| 0.4 | 100 | $0.2 \times 9.8 \times G$ | 71 | 0.02 | 0.07 |

**Table 6.2** Forces of $N_1$, $N_2$, $F_{xB_1}$, $F_{yB_1}$, $F_{C_1}$, and $R$

| Stair sequence | $N_1$ (N) | $N_2$ (N) | $F_{xB_1}$ (N) | $F_{yB_1}$ (N) | $F_{C_1}$ (N) | $R$ (N) |
|---|---|---|---|---|---|---|
| 1 | 54.2287 | −41.8151 | 0.0000 | 1,338.0244 | −566.4380 | 277.1859 |
| 2 | 201.8832 | −154.3761 | 196.0000 | 1,939.2764 | −1,398.7835 | 554.3717 |
| 3 | 418.5390 | −316.7477 | 392.0000 | 2,321.6697 | −2,031.4610 | 831.5576 |
| 4 | 679.7716 | −507.9947 | 588.0000 | 2,513.1182 | −2,488.8951 | 1,108.7434 |
| 5 | 961.1566 | −707.1819 | 784.0000 | 2,541.5353 | −2,795.5101 | 1,385.9293 |
| 6 | 1,238.2695 | −893.3738 | 980.0000 | 2,434.8349 | −2,975.7305 | 1,663.1151 |
| 7 | 1,486.6857 | −1,045.6354 | 1,176.0000 | 2,220.9306 | −3,053.9809 | 1,940.3010 |
| 8 | 1,681.9809 | −1,143.0313 | 1,372.0000 | 1,927.7361 | −3,054.6857 | 2,217.4869 |
| 9 | 1,799.7305 | −1,164.6262 | 1,568.0000 | 1,583.1651 | −3,002.2695 | 2,494.6727 |
| 10 | 1,815.5101 | −1,089.4848 | 1,764.0000 | 1,215.1314 | −2,921.1566 | 2,771.8586 |
| 11 | 1,704.8951 | −896.6720 | 1,960.0000 | 851.5485 | −2,835.7716 | 3,049.0444 |
| 12 | 1,443.461 | −565.25231 | 2,156 | 520.33025 | −2,770.539 | 3,326.2303 |
| 13 | 1,006.7835 | −74.2906 | 2,352.0000 | 249.3903 | −2,749.8832 | 3,603.4162 |
| 14 | 370.4380 | 597.1484 | 2,548.0000 | 66.6423 | −2,798.2287 | 3,880.602 |
| 15 | −490.0000 | 1,470.0000 | 2,744.0000 | 0.0000 | −2,940.0000 | 4,157.7879 |

the directions of real forces are reverse to those illustrated in Figs. 6.5 and 6.7. The numbers in the left column of Table 6.2 indicate the stair sequence on which stair a person stands. The sequence is numbered from the top stair to the bottom one.

With Table 6.2, we can draw the following conclusions. On the one hand, the magnitudes of both $N_1$ and $N_2$ will increase till when the person goes down to the 10th stair and then decrease to their minimum values; $F_{xB_1}$ and $R$ will monotonously increase to their maximum values as the person goes down; $F_{yB_1}$ and $F_{C_1}$ will first increase and then decrease after reaching to their maximum values when the person goes down to the 5th and 8th stairs, respectively. On the other hand, we also find from Table 6.2 that the force direction of $N_1$ in any of the first 14 stairs is different from that shown in Fig. 6.7. As a matter of fact, the ground can only provide the upward support forces. So a clamp device on the ground should be made to offer the last stair with such forces. In addition, the maximum forces among the components occur within the link $A_1DA_2$ from Table 6.2. In fact, this only results from the assumption that the disturbing force $F = 0.2G$. In real applications, the value of $F$ should be very little which might be supposed as $F = 0$. From Eq. (6.38), we immediately obtain

$$\begin{cases} F_{xB_1} = 0 \\ F_{yB_1} = \left(m - \dfrac{1}{2}\right) G - (n-2)N_1 - (n-1)N_2 \\ F_{C_1} = -\left(m - \dfrac{3}{2}\right) G + (n-3)N_1 + (n-2)N_2 \\ R = 0 \end{cases} \tag{6.39}$$

**Table 6.3** Forces of $N_1$, $N_2$, $F_{xB_1}$, $F_{yB_1}$, $F_{C_1}$, and $R$

| Stair sequence | $N_1$ (N) | $N_2$ (N) | $F_{xB_1}$ (N) | $F_{yB_1}$ (N) | $F_{C_1}$ (N) | $R$ (N) |
|---|---|---|---|---|---|---|
| 1 | 54.2287 | −41.8151 | 0 | 1,338.0244 | −370.4380 | 0 |
| 2 | 201.8832 | −154.3761 | 0 | 1,939.2764 | −1,006.7835 | 0 |
| 3 | 418.5390 | −316.7477 | 0 | 2,321.6697 | −1,443.4610 | 0 |
| 4 | 679.7716 | −507.9947 | 0 | 2,513.1182 | −1,704.8951 | 0 |
| 5 | 961.1566 | −707.1819 | 0 | 2,541.5353 | −1,815.5101 | 0 |
| 6 | 1,238.2695 | −893.3738 | 0 | 2,434.8349 | −1,799.7305 | 0 |
| 7 | 1,486.6857 | −1,045.6354 | 0 | 2,220.9306 | −1,681.9809 | 0 |
| 8 | 1,681.9809 | −1,143.0313 | 0 | 1,927.7361 | −1,486.6857 | 0 |
| 9 | 1,799.7305 | −1,164.6262 | 0 | 1,583.1651 | −1,238.2695 | 0 |
| 10 | 1,815.5101 | −1,089.4848 | 0 | 1,215.1314 | −961.15661 | 0 |
| 11 | 1,704.8951 | −896.67195 | 0 | 851.54851 | −679.77161 | 0 |
| 12 | 1,443.461 | −565.25231 | 0 | 520.33025 | −418.53897 | 0 |
| 13 | 1,006.7835 | −74.290621 | 0 | 249.39028 | −201.88316 | 0 |
| 14 | 370.43799 | 597.14839 | 0 | 66.642301 | −54.22868 | 0 |
| 15 | −490.0000 | 1,470.0000 | 0 | 0.0000 | 0.0000 | 0 |

**Table 6.4** Forces $N_1$, $N_2$, $F_{xB_1}$, $F_{yB_1}$, $F_{C_1}$, and $R$

| $N_1$ (N) | $N_2$ (N) | $F_{xB_1}$ (N) | $F_{yB_1}$ (N) | $F_{C_1}$ (N) | $R$ (N) |
|---|---|---|---|---|---|
| 14,373.3333 | −6,533.3334 | 20,580.0000 | 21,723.3334 | −38,383.3334 | 33,262.3030 |

The computing results based on Eqs. (6.39) and (6.31) are listed in Table 6.3.

Table 6.3 indicates that the forces $N_1$, $N_2$, $F_{yB_1}$, and $F_{C_1}$ are all changing with the shift of standing stair of a person. From Table 6.3, we can find that $F_{xB_1}$ and $R$ are all equal to zeros and the magnitudes of $N_1$ and $N_2$ first increase and then decrease after reaching their maximum values when the person goes down to the 10th stair. The values of both $F_{yB_1}$ and $F_{C_1}$ will increase till when the person goes down to the 5th stair and then decrease.

Suppose that there are 15 persons walking down the stairs one by one. Table 6.4 shows the values of the forces $N_1$, $N_2$, $F_{xB_1}$, $F_{yB_1}$, $F_{C_1}$, and $R$.

To prove the above theoretical results, an 8-stair test rig is built up. Based on the theoretical analysis, the components of the foldable stair are manufactured and assembled. What is more, the structure is designed in line with the form of steel strengthened concrete stairs which brings a lot of convenience to the users. The sectional view of the assembly of one foldable stair is shown in Fig. 6.8. And the assembled 8-stair mechanism is shown in Fig. 1.7. Walking experiments are executed by a group of volunteers with different weights. The experiments mainly cover the qualitative test on the effect of walking habits, the safety and comfort feelings of a person, and the stiffness and strength of the deployed stair.

Walking experiments show that both the stiffness and strength are high enough for a person, whose weight is not more than 100 kg, and the deployed stair has no difference from the steel strengthened concrete stairs, and therefore it is fit well

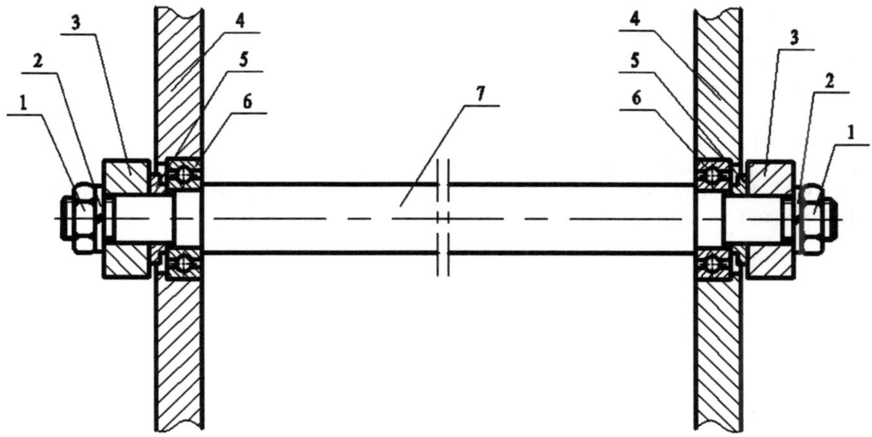


1-nut, 2-spring lock washer, 3-horizontal beam, 4-vertical beam, 5-sleeve, 6-deep groove ball bearing, 7-shaft

**Fig. 6.8** Front view of the assembly unit of the staircase

to the general habits of the residents. Consequently, all the volunteers feel very comfortable and safe when walking up and down the foldable stair.

## 6.4 Conclusions

This chapter has presented a static method for the planar foldable linkage with an example of foldable stair that has advantages of both motion flexibility and high structure stiffness. It examines the geometric property of the stair and investigates the statics of its deployable scissorlike element. The internal forces are examined by using the second Castigliano's theorem. Statics of the foldable stair is then developed with redundant constraints. Experiments verified the study and indicated that the foldable stair not only has the required stiffness and strength but also has the free mobility in folding and unfolding. It is very convenient in use and therefore has a wide application in both indoors and outdoors particularly when the fixed stairs are impossible.

## References

1. Mirats Tur JM, Juan SH (2008) Tensegrity frameworks: static analysis review. Mech Mach Theory 43(7):859–881
2. Seffen KA, You Z, Pellegrino S (2000) Folding and deployment of curved tape springs. Int J Mech Sci 42(10):2055–2073

3. Gan WW, Pellegrino S (2006) Numerical approach to the kinematic analysis of deployable structures forming a closed loop. Proc Inst Mech Eng Part C J Mech Eng Sci 220(7):1045–1056
4. Xu LJ, Tian GY, Duan Y, Yang SX (2001) Inverse kinematic analysis for triple-octahedron variable-geometry truss manipulators. Proc Inst Mech Eng Part C J Mech Eng Sci 215(2): 247–251
5. Nagaraj BP, Pandiyan R, Ghosal A (2010) A constraint Jacobian based approach for static analysis of pantograph masts. Comput Struct 88(1–2):95–104
6. Lai WM, Rubin D, Krempl E (2010) Introduction to continuum mechanics. Butterworth-Heinemann/Elsevier, Boston

# Chapter 7
# Kinetostatics of Double Planar Linkages

This chapter investigates the kinetostatics of overconstraint spatial mechanisms which are all made of double planar linkages. The mechanism is redundantly constrained in structure and therefore has both merits of high structural stiffness and strength of a truss structure and motion flexibility of a mechanism. The kinetostatics is discussed with the structure synthesis of a deployable wing frame and the lift mechanism. It first discusses the kinematics of double planar linkages with the example of a deployable frame for a morphing wing and then presents a stiffness analysis method of the mechanism with an example of lift mechanism.

The major concept structures of previous deployable wings are mostly based on prismatic joints. However, the deflections of the cantilevered links might not satisfy the primary geometry requirements of the prismatic joint. Therefore, revolute joints are used in our deployable frame to avoid violating the geometry conditions for prismatic joints resulting from different deflections of its contacting two parts. The primary element of the foldable frame is synthesized from the viewpoint of identical strength principle. The deflection and slope of every joint node of the foldable frame is investigated within the deploying/folding process. For the kinetostatics of lift mechanism, the focus is the stiffness of the mechanism whose structural parameters might be uncertain. Therefore, the interval analysis method is adopted to establish the stiffness. The forces and moments exerted on the fork frame and connection rods are firstly discussed via coordinate transformations between different Cartesian coordinate systems. The elastic deformation of rods and compatibility equations and the rigid displacement of frame are then analyzed. An equation system is ultimately established to describe the dynamic stiffness of the mechanism. Considering the uncertainty of some parameters in engineering, the trend and interval area of the equivalent stiffness during work is investigated with interval analysis method. Numerical simulations show that this kind of mechanism can provide very good stiffness intervals in both vertical and torsional directions for engineering applications.

J. Zhao et al., *Design of Special Planar Linkages*, Springer Tracts in Mechanical Engineering, DOI 10.1007/978-3-642-38448-6_7,

## 7.1 Introduction

Morphing concept for unmanned aerial vehicles is a topic of current research interest in aerospace engineering [1]. The target is to create a morphing wing that could reconfigure itself to present the optimal performance at all stages of flight [2]. Morphing technologies enable new flight capabilities, such as perching, urban navigation, and indoor flight [3]. Therefore, morphing wing is particularly designed to alter its shape and span for various flight conditions so that the lift-drag ratio could be kept within an ideal range for different fly requirements [1, 4]. The ability to continuously change vehicle geometry provides a single air vehicle with a multiple-mission capacity, as it allows the vehicle to maintain the optimum or near-optimum flight performance from one mission to another [5].

Generally, owing to the increased requirements for weight reduction, aircraft structures are becoming more flexible which renders their performance requirements to be much more stringent [6]. For current engineering, the major means of morphing include camber change, wing twist, wing sweep change, and wing span change while the last two are the widely discussed methods to vary the configuration of a wing for different flight conditions [1]. One of such morphing wing structures is the folding concept [7]. Folding structure concept was initially proposed for moveable theater by a Spanish architect, Pinero, in the early 1960s [8, 9]. Structures whose shapes are formed of folding/deploying processes are subsequently called foldable/deployable structures. Research and development of such structures mainly took place from the 1960s to 1980s [10].

Currently, folding polygonal frames are widely used in aerospace structures to support antenna reflectors and solar blankets the frame of which often consists of a closed loop of bars connected by revolute joints. They can have various forms, from squares and rectangles to hexagons [11]. The foldable systems for structural applications are often composed of a number of primary units such as truss module, which might include other additional elements to increase its stability [12]. Recent needs for space structures and nonconventional structures have extensively stimulated the research on mechanics and design of large flexible space structures. Space structures usually have to be made light and flexible [13]. Foldable plate structures consist of units of plate jointed together along their edges by continuous hinges. These hinges allow the structure to be folded according to an origami pattern formed by intersecting mountain folds and valley folds [14]. Foldable linkages consisting of rigid bars and revolute joints exhibit intriguing motion that is also aesthetically beautiful. Design of such linkages is a complex process by one simple fact that they are kinematically overconstrained. It is their particular arrangement of specially designed and suitably proportioned rigid links that often allows them to move with a single degree of freedom [15]. The elemental module that creates the structural typology is made up of two bars connected with each other by an inner joint. In order to form a foldable framework, these modules are juxtaposed by connecting the ends of every bar. When the truss has a flat covering, all bars prove to be absolutely equal, in a way that the inner connection in each bar is situated exactly at the midpoint of its length [12]. Because substantial geometry variations could provide efficient performance for disparate tasks which might not be possible with a fixed-geometry aircraft [16], various wing geometries can be achieved in flight, allowing for multirole missions with the same aircraft [7]. Obviously, the skin system that can accommodate large shape variations while carrying and transferring aerodynamic loads is an important element of aircraft morphing technology. However, this chapter only focuses on the foldable frame of the wing which provides larger deployment ratio while supporting the loads transferred from the skin. The scissor-hinged structure that is overconstraint in structure but has mobile flexibility of a mechanism provides a possible frame for the morphing wings. So the structural theory of overconstraint mechanisms [17] can be used to the innovative design of redundantly constraint deployable/foldable frame of the wing. What follows will focus on the synthesis of foldable frame of a morphing wing and investigate aeroelastics in folding and unfolding.

With multiple individually articulated sections, various wing geometries can be achieved in flight, allowing for multirole missions with the same aircraft [7]. Unfortunately, the wing root bending moment can be increased considerably due to the increase of the wingspan [1]. Therefore, an investigation of the elastic characteristics of the deployable frame is very important for the variable-span wing. The preliminary design of morphing aircraft requires aerodynamic analysis at widely varying flight conditions and geometric configurations [16]. Design optimization approaches for sizing a morphing aircraft for which the wing can make significant shape changes in flight are discussed in [18]. Optimal design of morphing aircraft wings employing a wing structure composed of an internal layout of cables and struts to provide actuation facility and high stiffness is investigated [19]. The shape and planform of the morphing wing could balance its performance to optimize the missions within the fly speed interval [18]. The effectiveness of a morphing wing can be measured in terms of required stiffness of structure to obtain maneuverability and performance.

The stiffness of the structure is usually analyzed as a measurement to evaluate mechanisms. As a matter of fact, one of the most important characteristics of parallel manipulators is recognized with high stiffness and high strength [20]. Szuminski [21] presented a method of determining the radial and axial stiffness of rolling bearings and rolling kinematic pairs as a function of the external load and the kinematics of motion. Pashkevich et al. [22, 23] presented a methodology to enhance the stiffness analysis of serial and parallel manipulators with passive joints and analyzed the stiffness of overconstrained parallel manipulators. Yue et al. [24] described the relationship among input force, payload, stiffness, and displacement. Furthermore, Li et al. [25] and Carbone et al. [26] also focused on the stiffness analysis of mechanisms, while Jung et al. [27] and Pigoski et al. [28] used stiffness mapping. It should be noticed that stiffness of a mechanism is probably variable [29]. Majou et al. [30] presented a parametric method to analyze stiffness and conduct a compliant modeling and a symbolic expression of the stiffness matrix. In many cases, however, the parameters that stiffness depends on are even uncertain. In order to obtain reasonable results, interval analysis method [31] is often used under the conditions that it is impossible to collect enough test data. The dynamic response problems can be solved and the scope of results could be described by interval analysis [32]. This chapter first discusses the stiffness of double planar linkages used for morphing wing frame and then investigates the dynamic stiffness of a novel lift mechanism used in the implementation for forklift trucks.

## 7.2 Foldable Frame of a Morphing Wing

The primary requirements for the foldable frame of a morphing wing are determined by the aerodynamics of a plane. For the wing with a larger deployment ratio in current researches, the key element is a pressurized telescopic cantilever which is able to undergo large-scale spanwise changes while supporting wing loads [33, 34]. However, such a cantilevered structure easily violates the geometry conditions of the telescopic motions which are the primary requirements for a morphing wing because of the different deflections of the contacting two parts. The revolute joints accompanied by thrust bearings can well avoid this problem for it has less touching area of the joints than the telescopic one. Therefore, this chapter proposes a foldable frame based on revolute-jointed structure for the morphing wing. What follows will first discuss the lift force distribution along the wingspan and then synthesize the foldable frame from the viewpoint of the principle of identical strength.

### *7.2.1 Primary Requirements for the Deployable Frame of a Morphing Wing*

Statistically, the wing circulations of an aircraft with canard wings can be expressed as [35]

$$\Gamma(x) = \Gamma_0\sqrt{1-(x/l)^2} \tag{7.1}$$

where $x$ is the coordinate in the spanwise direction, $\Gamma_0 = \frac{1}{2}C_L b V_\infty$ which represents the maximum value of the wing circulations of an aircraft, and $l$ denotes the wingspan and the deployment ratio is defined as $\varsigma = \frac{l_{max}}{l_{min}}$, $b$ stands for chord of the wing, $V_\infty$ is the wind velocity, $C_L = C_{L\alpha}(\alpha-\alpha_0)$ which indicates the coefficient of lift force produced by the airfoil and $\alpha$ indicates the angle of attack, and $C_{L\alpha}$ and $\alpha_0$ are parameters calculated based on the airfoil shape.

The distribution function of the lift force $L(x)$ with respect to $x$ is [35]

$$L(x) = \rho V_\infty \Gamma(x) \tag{7.2}$$

**Table 7.1** Parameters of a morphing wing

| $C_{L\alpha}(^{\circ -1})$ | $\alpha(^{\circ})$ | $\alpha_0(^{\circ})$ | $b$(m) | $V_\infty$(m/s) | $l_{min}$(m) | $\varsigma$ |
|---|---|---|---|---|---|---|
| 0.106 | 15 | 0 | 2 | 115–200 | 3 | 3 |

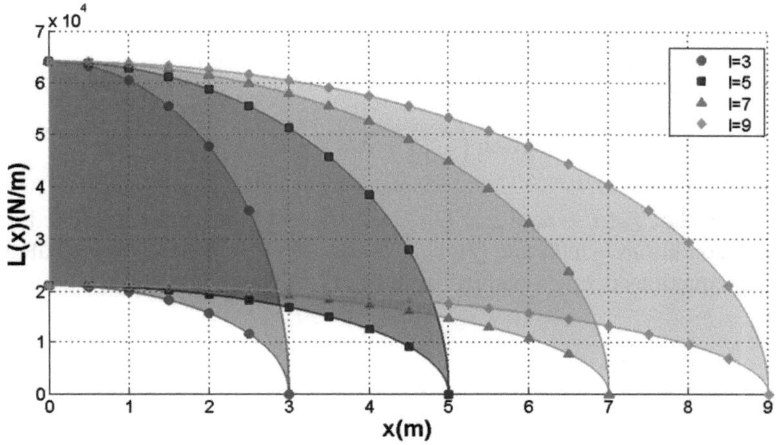


**Fig. 7.1** Distribution of lift force on the wingspan

where $\rho$ denotes the air density. Table 7.1 lists some of parameters of a morphing wing. Equations (7.1) and (7.2) can be found in [35].

With the parameters prescribed in Table 7.1, the distribution of lift force $L(x)$ with respect to the wing span, $x$, is calculated with Eq. (7.2) and which is shown in Fig. 7.1 where the legends $l$s represent different wingspans.

Figure 7.1 shows the changes of intervals of the lift force distribution when $V_\infty$ changes from 115 to 200 m/s. Both the upper limit curve and lower limit curve of the lift force interval are changing to be more and more horizontal when the wingspan extends from 3 to 9 m. From Eq. (7.2), the total lift force gained by the wing can be expressed as

$$L = \int_0^l L(x)\mathrm{d}x = \frac{\pi}{4}\rho l V_\infty \Gamma_0 \tag{7.3}$$

The distributions of shearing force and the moment along the wingspan satisfy that

$$\begin{cases} \dfrac{\mathrm{d}Q}{\mathrm{d}x} = -L(x) \\ \dfrac{\mathrm{d}M}{\mathrm{d}x} = Q(x) \end{cases} \tag{7.4}$$

where $Q$ denotes the shearing force in the cross section of the wing and $M$ represents the bending moment at $x$. From Eq. (7.4), we obtain

$$\begin{cases} Q(x) = -\int L(x)\mathrm{d}x = -\dfrac{Cb}{2}\left[x\sqrt{1-\dfrac{x^2}{l^2}} + l\arcsin\left(\dfrac{x}{l}\right)\right] + Q_C \\ M(x) = \int Q(x)\mathrm{d}x = -\dfrac{Cb}{2}\left[-\dfrac{1}{3}l^2\left(1-\dfrac{x^2}{l^2}\right)^{3/2} + l^2\sqrt{1-\dfrac{x^2}{l^2}}\right. \\ \qquad \left. + Q_C x + lx\arcsin\left(\dfrac{x}{l}\right)\right] + M_C \end{cases} \tag{7.5}$$

where $C = \frac{1}{2}\rho C_{L\alpha}(\alpha - \alpha_0)V_\infty^2$, $Q_C$, and $M_C$ are constants determined by the boundary conditions which are $Q(l) = 0$ and $M(l) = 0$ here. Immediately, the following equations can be obtained from Eq. (7.5) where boundary conditions are applied.

$$\begin{cases} Q(x) = -\dfrac{Cb}{2}\left[x\sqrt{1-\dfrac{x^2}{l^2}} + l\arcsin\left(\dfrac{x}{l}\right) - \dfrac{\pi l}{2}\right] \\ M(x) = -\dfrac{Cb}{2}\left[-\dfrac{1}{3}l^2\left(1-\dfrac{x^2}{l^2}\right)^{3/2} + l^2\sqrt{1-\dfrac{x^2}{l^2}} - \dfrac{\pi lx}{2} + lx\arcsin\left(\dfrac{x}{l}\right)\right] \end{cases} \tag{7.6}$$

For the frame of a morphing wing with large aspect ratio, the major failures often result from the maximum normal stress. Therefore, the shape of the frame should be investigated by using the principle of identical strength. There will be the following rule for the shape of frame if the allowed normal stress of the material of the frame is $[\sigma_A]$ [36]:

$$\begin{aligned} \sigma_{\max}(x) = -\frac{M(x)y(x)}{I_z} = \frac{Cb}{\frac{2I_z}{y(x)}}\left(-\frac{1}{3}l^2\left(1-\frac{x^2}{l^2}\right)^{3/2}\right. \\ \left. + l^2\sqrt{1-\frac{x^2}{l^2}} - \frac{\pi lx}{2} + lx\arcsin\left(\frac{x}{l}\right)\right) \le [\sigma_A] \end{aligned} \tag{7.7}$$

where $I_z = \frac{1}{12}\mathrm{by}^3(x)$ which indicates the moment of inertia of the cross section about the $z$-axis. The rectangular section of $by(x)$ is considered where the side $y$ is parallel to the $z$-axis and the length is $b$. From Eq. (7.7), $y(x)$ is calculated:

$$y(x) \ge \sqrt{\frac{6C}{[\sigma_A]}\left(-\frac{1}{3}l^2\left(1-\frac{x^2}{l^2}\right)^{3/2} + l^2\sqrt{1-\frac{x^2}{l^2}} - \frac{\pi lx}{2} + lx\arcsin\left(\frac{x}{l}\right)\right)}. \tag{7.8}$$

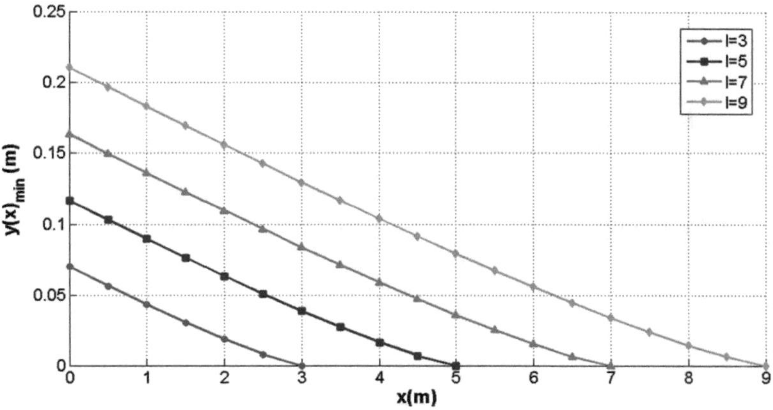


**Fig. 7.2** Ideal surface shape of an unfolded frame

Equation (7.8) indicates that the height of the frame at $x$ should abide by the inequality so that the largest strain at the position of $x$ on the surface of the frame does not exceed the allowed strain by the material. Figure 7.2 illustrates the shape of the surface of an unfolded frame.

The profiles of the wing with respect to span obtained from the principle of identical strength can be all approximated as skew lines. With this conclusion, what follows will focus on the motion analysis and structure synthesis of the deployable frame of a morphing wing.

### *7.2.2 Synthesis of the Deployable Frame of the Morphing Wing*

Based on the principle of identical strength, a terraced rectangular frame of the morphing wing is proposed to approach the identical strength frame shape shown in Fig. 7.2. As shown in Fig. 7.3, the two terraced structures paralleling the horizontal plane on both sides only allow the elements they connected to move in the direction that parallels the horizontal plane, while the zigzag structures set in the middle are parallel to the vertical plane and only permit the elements hinged by them to move along the direction that is parallel to the vertical plane. Therefore, the terraced rectangular frames revolute jointed via two sets of such structures can only move along the intersecting line of the horizontal plane and vertical plane, namely, $x$-axis in Fig. 7.3.

To illustrate the primary requirements for the motions of the deployable frame, a simplified structure is shown in Fig. 7.4. Every level of the terraced structure is made up of a number of quadrilateral elements parallel to the horizontal plane, $xoz$. The zigzag structure is composed of RRR-link groups which are all parallel to the vertical plane, $xoy$. Every RRR-link group in the vertical plane and the corresponding quadrangle in the horizontal plane are connected by two beams paralleling the

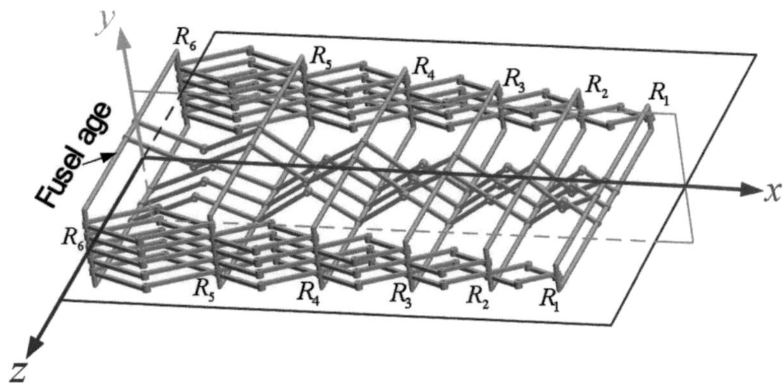


**Fig. 7.3** Architecture of the foldable frame for a morphing wing

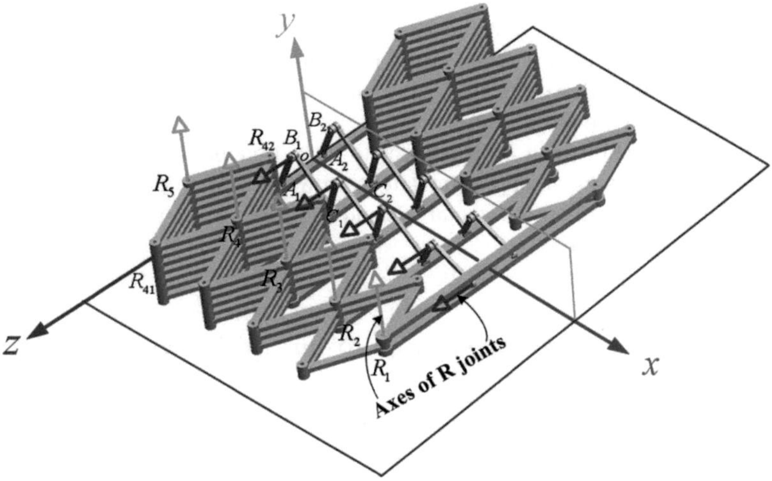


**Fig. 7.4** Coordinate system of the deployable frame for the morphing wing

$z$-axis at both ends of them. In Fig. 7.4, beam $A_1A_2$ is fixed with the fuselage, and therefore the beam $C_1C_2$ forms an end-effector simultaneously connected by three RRR chains, $R_5R_{41}R_4$, $R_5R_{42}R_4$, and $A_1B_1C_1$, where $R_5R_{41}R_4$ and $R_5R_{42}R_4$ have symmetric sizes about a vertical plane. Therefore, they provide the same planar constraints and subsequently increase the strength of the whole frame.

Suppose the coordinates of the revolute joints $R_5$, $R_{41}$, $R_{42}$, and $R_4$ at the instant are $(x_{R_5}\ y_{R_5}\ z_{R_5})$, $(x_{R_{41}}\ y_{R_{41}}\ z_{R_{41}})$, $(x_{R_{42}}\ y_{R_{42}}\ z_{R_{42}})$, and $(x_{R_4}\ y_{R_4}\ z_{R_4})$, individually. Also suppose that the coordinates of the revolute joints $A_1$, $B_1$, and

$C_1$ at the instant are $\left(x_{A_1}\ y_{A_1}\ z_{A_1}\right)$, $\left(x_{B_1}\ y_{B_1}\ z_{B_1}\right)$, and $\left(x_{C_1}\ y_{C_1}\ z_{C_1}\right)$, respectively. The kinematic screws of the two chains can be expressed in the coordinate system shown in Fig. 7.4.

For the kinematic chain $R_5 R_{41} R_4$, the screws can be expressed as

$$\mathbf{\$}_{R_5 R_{41} R_4} = \begin{bmatrix} 0 & 1 & 0 & -z_{R_5} & 0 & x_{R_5} \\ 0 & 1 & 0 & -z_{R_{41}} & 0 & x_{R_{41}} \\ 0 & 1 & 0 & -z_{R_4} & 0 & x_{R_4} \end{bmatrix}^T \tag{7.9}$$

where $\mathbf{\$}_{R_5 R_{41} R_4}$ represents the screw matrix of the chain $R_5 R_{41} R_4$.

Equation (7.9) shows that the $y$-coordinates of the joints never appear in the screw matrix, and this proves that the motion of a planar kinematic chain will not be affected by the changed size in the direction perpendicular to the chain plane, namely, the $y$-direction.

The reciprocal screws of $\mathbf{\$}_{R_5 R_{41} R_4}$ can be obtained by substituting Eq. (7.9) in the following equation:

$$\mathbf{\$}^T \mathbf{\Delta} \mathbf{\$}^r = 0 \tag{7.10}$$

where $\mathbf{\$}^r$ denotes the reciprocal screws to the screw matrix $\mathbf{\$}$, $\mathbf{\Delta} = \begin{bmatrix} 0 & \mathbf{I}_3 \\ \mathbf{I}_3 & 0 \end{bmatrix}$ and $\mathbf{I}_3$ is a $3 \times 3$ identity matrix.

Solving Eq. (7.10) presents the constraint screw matrix

$$\mathbf{\$}^r_{R_5 R_{41} R_4} = \begin{bmatrix} \mathbf{\$}^r_1 & \mathbf{\$}^r_2 & \mathbf{\$}^r_3 \end{bmatrix} \tag{7.11}$$

where $\mathbf{\$}^r_{R_5 R_{41} R_4}$ represents the matrix of reciprocal screws to screw matrix $\mathbf{\$}$, $\mathbf{\$}^r_1 = \begin{pmatrix} 0 & 1 & 0 & 0 & 0 & 0 \end{pmatrix}^T$ which indicates a force constraint along the $y$-axis, $\mathbf{\$}^r_2 = \begin{pmatrix} 0 & 0 & 0 & 1 & 0 & 0 \end{pmatrix}^T$ which represents a moment constraint about the $x$-direction, and $\mathbf{\$}^r_3 = \begin{pmatrix} 0 & 0 & 0 & 0 & 0 & 1 \end{pmatrix}^T$ which represents a moment constraint about the $z$-direction.

For the kinematic chain $R_5 R_{42} R_4$, the screws can be expressed as

$$\mathbf{\$}_{R_5 R_{42} R_4} = \begin{bmatrix} 0 & 1 & 0 & -z_{R_5} & 0 & x_{R_5} \\ 0 & 1 & 0 & -z_{R_{42}} & 0 & x_{R_{42}} \\ 0 & 1 & 0 & -z_{R_4} & 0 & x_{R_4} \end{bmatrix}^T \tag{7.12}$$

where $\mathbf{\$}_{R_5 R_{42} R_4}$ represents the screw matrix of the chain $R_5 R_{42} R_4$.

The reciprocal screws of $\mathbf{\$}_{R_5R_{42}R_4}$ can be obtained by substituting Eq. (7.12) into (7.10):

$$\mathbf{\$}^r_{R_5R_{42}R_4} = \begin{bmatrix} \mathbf{\$}^r_4 & \mathbf{\$}^r_5 & \mathbf{\$}^r_6 \end{bmatrix} \tag{7.13}$$

where $\mathbf{\$}^r_4 = \begin{pmatrix} 0 & 1 & 0 & 0 & 0 & 0 \end{pmatrix}^T$ which indicates a force constraint along the $y$-axis, $\mathbf{\$}^r_5 = \begin{pmatrix} 0 & 0 & 0 & 1 & 0 & 0 \end{pmatrix}^T$ which represents a moment constraint about the $x$-direction, and $\mathbf{\$}^r_6 = \begin{pmatrix} 0 & 0 & 0 & 0 & 0 & 1 \end{pmatrix}^T$ which represents a moment constraint about the $z$-direction.

As a matter of fact, the terminal constraints of kinematic chain $R_5R_{42}R_4$ are the same as those of $R_5R_{41}R_4$ which provide the redundant constraints and which are most useful to increase the strength and stiffness of the structure.

For the chain $A_1B_1C_1$, the kinematic screw matrix can be similarly obtained:

$$\mathbf{\$}_{A_1B_1C_1} = \begin{bmatrix} 0 & 0 & 1 & y_{A_1} & -x_{A_1} & 0 \\ 0 & 0 & 1 & y_{B_1} & -x_{B_1} & 0 \\ 0 & 0 & 1 & y_{C_1} & -x_{C_1} & 0 \end{bmatrix}^T \tag{7.14}$$

Substituting Eq. (7.14) into (7.10) presents the constraint screw matrix:

$$\mathbf{\$}^r_{A_1B_1C_1} = \begin{bmatrix} 0 & 0 & 1 & 0 & 0 & 0 \\ 0 & 0 & 0 & 1 & 0 & 0 \\ 0 & 0 & 0 & 0 & 1 & 0 \end{bmatrix}^T \tag{7.15}$$

The three column screws of matrix $\mathbf{\$}^r_{A_1B_1C_1}$ represent one force constraint along the $z$-axis, one moment constraint about the $x$-direction, and one moment constraint about the $y$-direction, respectively.

The terminal constraint screw matrix of the kinematic chain $A_2B_2C_2$ can be similarly proved to be the same as those of $\mathbf{\$}^r_{A_1B_1C_1}$. So the deployable frame of the morphing wing proposed in this chapter is a highly redundant constraint mechanism which should have higher stiffness and strength than the general ones.

Therefore, the constraint screws exerted to the beam $C_1C_2$ by the kinematic chains $R_5R_{41}R_4$, $R_5R_{42}R_4$, $A_1B_1C_1$, and $A_2B_2C_2$ can be expressed as the union of Eqs. (7.11) and (7.15):

$$\mathbf{\$}^r_{C_1C_2} = \begin{bmatrix} 0 & 1 & 0 & 0 & 0 & 0 \\ 0 & 0 & 1 & 0 & 0 & 0 \\ 0 & 0 & 0 & 1 & 0 & 0 \\ 0 & 0 & 0 & 0 & 1 & 0 \\ 0 & 0 & 0 & 0 & 0 & 1 \end{bmatrix}^T \tag{7.16}$$

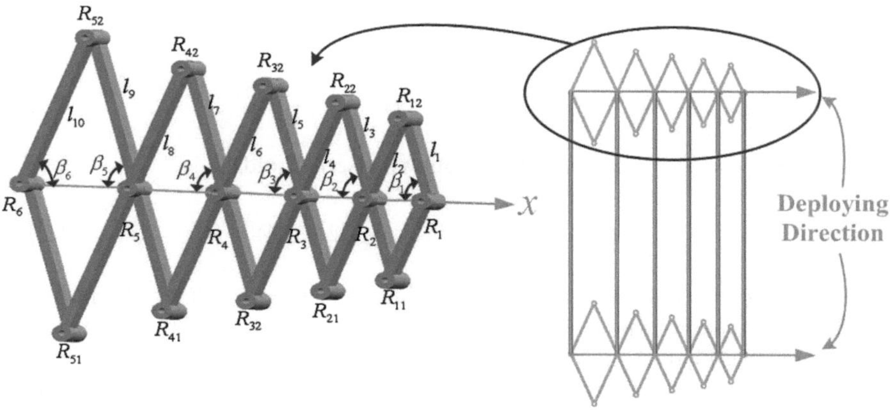


**Fig. 7.5** Geometry parameters of the tapered structure

Substituting Eq. (7.16) into (7.10) again yields the free relative motion of beam $C_1C_2$ with respect to beam $A_1A_2$ under the constraints of these kinematic chains:

$$\$_{C_1C_2} = \begin{bmatrix} 0 & 0 & 0 & 1 & 0 & 0 \end{bmatrix}^T \tag{7.17}$$

Equation (7.17) indicates that the beam $C_1C_2$ only makes a translation with respect to beam $A_1A_2$. The same conclusion will be drawn when the above processes are applied to the next set of elements. Therefore, the frame has and only has one translational degree of freedom in the direction of the $x$-axis. As a result, the primary requirements for stretching and compacting motions of the foldable frame are that there are at least two sets of planar kinematic chains locating in two planes that are not parallel to each other. This kind of mechanism was initially designed by Sarrus in 1853 [37]. But for the use of foldable/deployable frame of a morphing wing, the more overconstraints induced into the structure, the higher the stiffness and strength of the structure will be. The four sets of overconstraint planar linkages in the horizontal plane and vertical plane for the deployable frame shown in Fig. 7.3 are much better than the initial Sarrus mechanism for a morphing wing.

As the bearing frame of a wing, it should keep a proper aspect ratio. Therefore, a deployable structure with a certain slope is particularly designed. Each quadrangle shown in Fig. 7.5 is symmetric about its horizontal diagonal. For example, in triangle $R_1R_{12}R_2$, there is the following relation in line with the sine theorem:

$$\frac{\sin\beta_2}{\sin\beta_1} = \frac{l_1}{l_2} \tag{7.18}$$

Similarly, there is the following equation in triangle $R_2R_{22}R_3$:

$$\frac{\sin\beta_3}{\sin\beta_2}=\frac{l_3}{l_4}$$

If $\frac{l_9}{l_{10}}=\cdots=\frac{l_3}{l_4}=\frac{l_1}{l_2}=\eta$, then the following series holds:

$$\sin\beta_i=\eta^{i-1}\sin\beta_1(i=1,2,\ldots 6) \quad (7.19)$$

Equation (7.19) indicates that the deployable frame proposed in this chapter only needs one actuation to propel the folding or unfolding motion of the whole wing. Therefore, this kind of morphing wing could perfect the actuator setting on the one hand and save the energy consumption comparing with those based on shape-memory alloy on the other hand.

## 7.3 Aeroelastics of the Deployable Frame for the Morphing Wing

Since the frame can be deployed or folded in a fixed direction, the side structure can only stretch or compact in a line which is represented by the $x$-axis shown in Fig. 7.6 where $\alpha$ indicates the deploying angle of the structure.

With the distribution function of lift force specified by Eq. (7.2), the equivalent central forces exerted to every joint node can be shown in Fig. 7.6 by integrating the distribution function within the neighborhood between every two-joint node.

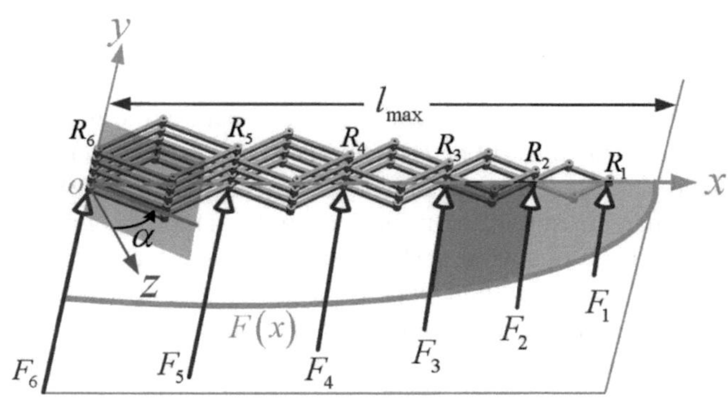


**Fig. 7.6** The bearing structure

$$\begin{cases} F_{R_i(L)} + F_{R_{i+1}(R)} = \int_{x_{i+1}}^{x_i} L(x)\mathrm{d}x \\ x_i F_{R_i(L)} + x_{i+1} F_{R_{i+1}(R)} = \int_{x_{i+1}}^{x_i} xL(x)\mathrm{d}x \end{cases} \tag{7.20}$$

where $i = 1, 2, \ldots, 5$, $x_i$ represents the $x$-coordinate of joint $R_i$, $F_{R_i(L)}$ indicates the left equivalent force for joint $R_i$, and $F_{R_i(R)}$ indicates the right equivalent force for joint $R_i$. Expanding Eq. (7.20) presents

$$\begin{cases} F_{R_i(L)} + F_{R_{i+1}(R)} = \frac{Cbl}{2}\left[\arcsin\left(\frac{x_i}{l}\right) - \arcsin\left(\frac{x_{i+1}}{l}\right) + \frac{x_i}{l}\sqrt{1-\left(\frac{x_i}{l}\right)^2} \right. \\ \qquad \left. - \frac{x_{i+1}}{l}\sqrt{1-\left(\frac{x_{i+1}}{l}\right)^2}\right] \\ x_i F_{R_i(L)} + x_{i+1} F_{R_{i+1}(R)} = \frac{Cbl^2}{3}\left\{\cos^3\left[\arcsin\left(\frac{x_{i+1}}{l}\right)\right] - \cos^2\left[\arcsin\left(\frac{x_i}{l}\right)\right]\right\} \end{cases} \tag{7.21}$$

The solutions of Eq. (7.21) are

$$\begin{cases} F_{R_i(L)} = \frac{Cbl^2}{3(x_i - x_{i+1})}\left\{\cos^3\left[\arcsin\left(\frac{x_{i+1}}{l}\right)\right] - \cos^2\left[\arcsin\left(\frac{x_i}{l}\right)\right]\right\} \\ \qquad - \frac{Cblx_{i+1}}{2(x_i - x_{i+1})}\left[\arcsin\left(\frac{x_i}{l}\right) - \arcsin\left(\frac{x_{i+1}}{l}\right) + \frac{x_i}{l}\sqrt{1-\left(\frac{x_i}{l}\right)^2} \right. \\ \qquad \left. - \frac{x_{i+1}}{l}\sqrt{1-\left(\frac{x_{i+1}}{l}\right)^2}\right] \\ F_{R_{i+1}(R)} = \frac{Cblx_i}{2(x_i - x_{i+1})}\left[\arcsin\left(\frac{x_i}{l}\right) - \arcsin\left(\frac{x_{i+1}}{l}\right) + \frac{x_i}{l}\sqrt{1-\left(\frac{x_i}{l}\right)^2} \right. \\ \qquad \left. - \frac{x_{i+1}}{l}\sqrt{1-\left(\frac{x_{i+1}}{l}\right)^2}\right] \\ \qquad - \frac{Cbl^2}{3(x_i - x_{i+1})}\left\{\cos^3\left[\arcsin\left(\frac{x_{i+1}}{l}\right)\right] - \cos^2\left[\arcsin\left(\frac{x_i}{l}\right)\right]\right\} \end{cases} \tag{7.22}$$

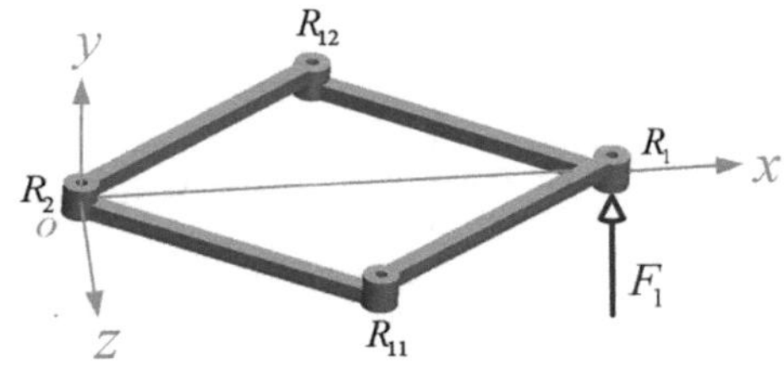


**Fig. 7.7** The distal quadrilateral

$F_{R_1(R)}$ can be directly calculated from Eq. (7.23):

$$F_{R_1(R)} = \int_{x_1}^{l} L(x)\mathrm{d}x = \frac{Cbl}{2}\left[\frac{\pi}{2} - \arcsin\left(\frac{x_1}{l}\right) - \frac{x_1}{l}\sqrt{1-\left(\frac{x_1}{l}\right)^2}\right] \quad (7.23)$$

Without loss of generality, $F_{R_6(L)}$ can be assumed as

$$F_{R_6(L)} = 0$$

So, the equivalent concentrated force for each node shown in Fig. 7.6 can be expressed as

$$F_i = F_{R_i(L)} + F_{R_i(R)} \quad (7.24)$$

where $i = 1, 2, \ldots, 6$.

### 7.3.1 Statics of the Deployable Frame

To analyze statics of the structure, a quadrilateral element shown in Fig. 7.7 is investigated firstly.

There is the following equilibrium equation:

$$F_{11} + F_{12} = F_1 \quad (7.25)$$

where $F_{11}$ and $F_{12}$ represent the forces exerted to the distal joints of kinematic chains $R_2R_{11}R_1$ and $R_2R_{12}R_1$, individually. Figure 7.8 illustrates the force exerted to the left half of the quadrilateral.

As shown in Fig. 7.8, $R_i$ is the $i$th($i = 1, 2, \ldots$) revolute joint at intersections, numbered from the outer end to the inner end. Assume that $d_{mn}$ denotes the distance from $R_m$ to $R_n$, there is

$$d_{mn} = \sum_{j=m}^{n-1} (l_{j1}\cos\alpha_j + l_{j2}\cos\alpha_{j+1}) \quad (7.26)$$

**Fig. 7.8** Deflection analysis of the first quadrilateral

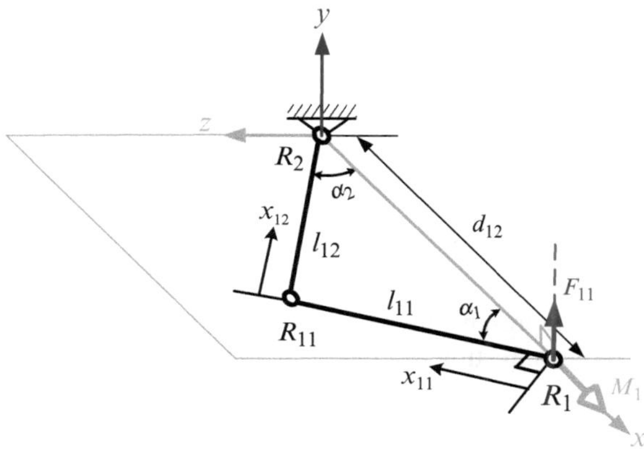


where $m, n = 1, 2, \ldots, 6$ and $n > m$.

The bending moment and the torque of every link can be expressed with

$$\begin{cases} M_{11}(x_{11}) = F_{11}x_{11} + M_1 \sin\alpha_1 \\ T_{11}(x_{11}) = M_1 \cos\alpha_1 \\ M_{12}(x_{12}) = F_{11}(x_{12} - l_{12} + d_{12}\cos\alpha_2) - M_1 \sin\alpha_2 \\ T_{12}(x_{12}) = F_{11}d_{12}\sin\alpha_2 + M_1 \cos\alpha_2 \end{cases} \tag{7.27}$$

where $F_{11}$ is the external force applied to joint $R_1$ and $M_1$ stands for the additional moment of torque because of the action of $F_{11}$ on $R_1$ and $x_{1i} \in [0, l_{1i}], i = 1, 2$ which are shown in Fig. 7.8.

In line with Castigliano's theorem and considering the symmetry of the structure [37], the relative slope of the joint $R_1$ is

$$\begin{aligned} \theta(M_1) = & \int_0^{l_{11}} \left[ \frac{M_{11}(x)}{EI}\frac{\partial M_{11}(x)}{\partial M_1} + \frac{T_{11}(x)}{GI_\mathrm{p}}\frac{\partial T_{11}(x)}{\partial M_1} \right] \mathrm{d}x \\ & + \int_0^{l_{12}} \left[ \frac{M_{12}(x)}{EI}\frac{\partial M_{12}(x)}{\partial M_1} + \frac{T_{12}(x)}{GI_\mathrm{p}}\frac{\partial T_{12}(x)}{\partial M_1} \right] \mathrm{d}x = 0 \end{aligned} \tag{7.28}$$

where $E$ is the Young's elastic modulus, $G$ represents the shear modulus of elasticity, $I$ is the bending moment of inertia of the beam, and $I_\mathrm{p}$ stands for the rotational moment of inertia of the beam.

Substituting Eq. (7.27) into (7.28) and rearranging yields

$$M_1 = r_1 F_{11} \tag{7.29}$$

where

$$r_1 = \frac{\dfrac{l_{12}\left(d_{12}\cos\alpha_2 - l_{12}\right)\sin\alpha_2 - \dfrac{1}{2}\left(l_{11}^2\sin\alpha_1 - l_{12}^2\sin\alpha_2\right)}{EI} - \dfrac{l_{12}d_{12}\sin\alpha_2\cos\alpha_2}{GI_{\mathrm{p}}}}{\dfrac{l_{11}\sin^2\alpha_1 + l_{12}\sin^2\alpha_2}{EI} + \dfrac{l_{11}\cos^2\alpha_1 + l_{12}\cos^2\alpha_2}{GI_{\mathrm{p}}}} \tag{7.30}$$

which represents the virtual distance from force $F_{11}$ to joint $R_1$.

The deflection of joint $R_1$ with respect to joint $R_2$ is

$$\begin{aligned}\delta_{R_1R_2} &= \int_0^{l_{11}}\left[\frac{M_{11}(x)}{EI}\frac{\partial M_{11}(x)}{\partial F_{11}} + \frac{T_{11}(x)}{GI_{\mathrm{p}}}\frac{\partial T_{11}(x)}{\partial F_{11}}\right]\mathrm{d}x \\ &+ \int_0^{l_{12}}\left[\frac{M_{12}(x)}{EI}\frac{\partial M_{12}(x)}{\partial F_{11}} + \frac{T_{12}(x)}{GI_{\mathrm{p}}}\frac{\partial T_{12}(x)}{\partial F_{11}}\right]\mathrm{d}x\end{aligned} \tag{7.31}$$

Substituting Eqs. (7.27), (7.29), and (7.30) into (7.31) gives

$$\begin{aligned}\delta_{R_1R_2} &= \int_0^{l_{11}} \frac{F_{11}\left(x_{11} + r_1\sin\alpha_1\right)x_{11}}{EI}\mathrm{d}x_{11} \\ &+ \int_0^{l_{12}}\left\{\begin{aligned}&\frac{F_{11}\left[\left(x_{12} - l_{12} + L_{12}\cos\alpha_2\right) - r_1\sin\alpha_2\right]\left(x_{12} - l_{12} + L_{12}\cos\alpha_2\right)}{EI} \\ &+ \frac{F_{11}\left(L_{12}\sin\alpha_2 + r_1\cos\alpha_2\right)L_{12}\sin\alpha_2}{GI_p}\end{aligned}\right\}\mathrm{d}x_{12} \\ &= \left\{\begin{aligned}&\frac{r_1\sin\alpha_1 l_{11}^2 - r_1\sin\alpha_2\left[\left(L_{12}\cos\alpha_2\right)^2 - \left(L_{12}\cos\alpha_2 - l_{12}\right)^2\right]}{2EI} \\ &+ \frac{l_{11}^3 + \left(L_{12}\cos\alpha_2\right)^3 - \left(L_{12}\cos\alpha_2 - l_{12}\right)^3}{3EI} \\ &+ \frac{l_{12}L_{12}\sin\alpha_2\left(L_{12}\sin\alpha_2 + r_1\cos\alpha_2\right)}{GI_{\mathrm{p}}}\end{aligned}\right\}F_{11}\end{aligned} \tag{7.32}$$

When the right kinematic chain $R_2R_{12}R_1$ is analyzed with the above process, the deflection should also have the same form of Eq. (7.32) by just replacing $F_{11}$ with $F_{12}$.

$$\delta'_{R_1R_2} = \left\{ \begin{array}{l} \dfrac{r_1 \sin\alpha_1 l_{11}^2 - r_1 \sin\alpha_2 \left[ (L_{12}\cos\alpha_2)^2 - (L_{12}\cos\alpha_2 - l_{12})^2 \right]}{2EI} \\ + \dfrac{l_{11}^3 + (L_{12}\cos\alpha_2)^3 - (L_{12}\cos\alpha_2 - l_{12})^3}{3EI} \\ + \dfrac{l_{12}L_{12}\sin\alpha_2 (L_{12}\sin\alpha_2 + r_1\cos\alpha_2)}{GI_{\mathrm{p}}} \end{array} \right\} F_{12} \tag{7.33}$$

Because the left and the right kinematic chains, $R_2R_{11}R_1$ and $R_2R_{12}R_1$, are jointed at $R_1$ and $R_2$, there must be

$$\delta'_{R_1R_2} = \delta_{R_1R_2} \tag{7.34}$$

Substituting Eqs. (7.32) and (7.33) into (7.34) and taking Eq. (7.25) into account, we gain

$$F_{11} = F_{12} = \frac{1}{2}F_1 \tag{7.35}$$

Substituting Eq. (7.35) into (7.32) or (7.33) produces

$$\delta_{R_1R_2} = \frac{1}{2} \left\{ \begin{array}{l} \dfrac{r_1 \sin\alpha_1 l_{11}^2 - r_1 \sin\alpha_2 \left[ (L_{12}\cos\alpha_2)^2 - (L_{12}\cos\alpha_2 - l_{12})^2 \right]}{2EI} \\ + \dfrac{l_{11}^3 + (L_{12}\cos\alpha_2)^3 - (L_{12}\cos\alpha_2 - l_{12})^3}{3EI} \\ + \dfrac{l_{12}L_{12}\sin\alpha_2 (L_{12}\sin\alpha_2 + r_1\cos\alpha_2)}{GI_{\mathrm{p}}} \end{array} \right\} F_1 \tag{7.36}$$

The deflections of the structure when two quadrilaterals are considered will be investigated next. With a similar process above, we can find that the force applied to every node will be equally undertaken by its two symmetry chains. Therefore, the node force exerted to the $R_n (n = 1, 2, \ldots)$ in what follows is directly represented by $\frac{1}{2}F_n$ which is shown in Fig. 7.9.

The bending moment and the torque of every link can also be similarly obtained in line with Eq. (7.27)

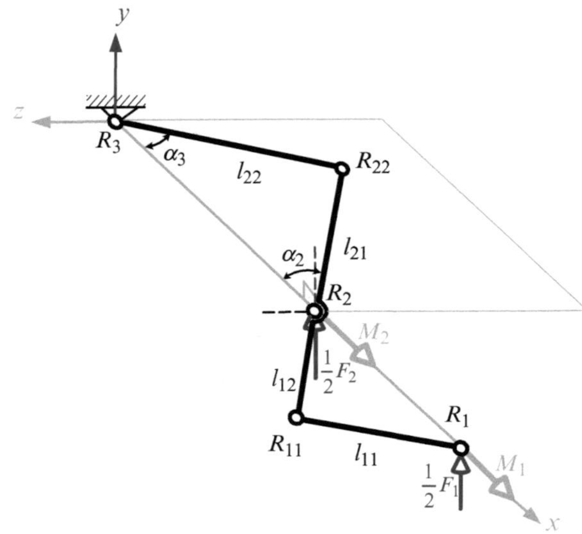


**Fig. 7.9** Deflection analysis of the first two quadrilaterals

$$\begin{cases} M_{11}(x_{11}) = \frac{1}{2}F_1 x_{11} + M_1 \sin\alpha_1 \\ T_{11}(x_{11}) = M_1 \cos\alpha_1 \\ M_{12}(x_{12}) = \frac{1}{2}F_1(x_{12} - l_{12} + d_{12}\cos\alpha_2) - M_1 \sin\alpha_2 \\ T_{12}(x_{12}) = \frac{1}{2}F_1 d_{12}\sin\alpha_2 + M_1 \cos\alpha_2 \\ M_{21}(x_{21}) = \frac{1}{2}F_1(x_{21} + d_{12}\cos\alpha_2) + \frac{1}{2}F_2 x_{21} - (M_1 + M_2)\sin\alpha_2 \\ T_{21}(x_{21}) = \frac{1}{2}F_1 d_{12}\sin\alpha_2 + (M_1 + M_2)\cos\alpha_2 \\ M_{22}(x_{22}) = \frac{1}{2}F_1(x_{22} - l_{22} + d_{13}\cos\alpha_3) + \frac{1}{2}F_2(x_{22} - l_{22} + d_{23}\cos\alpha_3) \\ \qquad + (M_1 + M_2)\sin\alpha_3 \\ T_{22}(x_{22}) = -\frac{1}{2}F_1 d_{13}\sin\alpha_3 - \frac{1}{2}F_2 d_{23}\sin\alpha_3 + (M_1 + M_2)\cos\alpha_3 \end{cases} \tag{7.37}$$

where $M_2$ represents the additional moment of torque to the external force $F_2$ at joint $R_2$.

In accordance to the Castigliano's theorem and considering the symmetry of the structure [36], the following relationship should hold:

$$\begin{cases} \theta(M_1) = 2\displaystyle\int_0^{l_{11}} \left[ \frac{M_{11}(x)}{EI} \frac{\partial M_{11}(x)}{\partial M_1} + \frac{T_{11}(x)}{GI_\mathrm{p}} \frac{\partial T_{11}(x)}{\partial M_1} \right] \mathrm{d}x \\ \qquad + 2\displaystyle\int_0^{l_{12}} \left[ \frac{M_{12}(x)}{EI} \frac{\partial M_{12}(x)}{\partial M_1} + \frac{T_{12}(x)}{GI_\mathrm{p}} \frac{\partial T_{12}(x)}{\partial M_1} \right] \mathrm{d}x \\ \qquad + 2\displaystyle\int_0^{l_{21}} \left[ \frac{M_{21}(x)}{EI} \frac{\partial M_{21}(x)}{\partial M_1} + \frac{T_{21}(x)}{GI_\mathrm{p}} \frac{\partial T_{21}(x)}{\partial M_1} \right] \mathrm{d}x \\ \qquad + 2\displaystyle\int_0^{l_{22}} \left[ \frac{M_{22}(x)}{EI} \frac{\partial M_{22}(x)}{\partial M_1} + \frac{T_{22}(x)}{GI_\mathrm{p}} \frac{\partial T_{22}(x)}{\partial M_1} \right] \mathrm{d}x = 0 \\ \theta(M_2) = 2\displaystyle\int_0^{l_{11}} \left[ \frac{M_{11}(x)}{EI} \frac{\partial M_{11}(x)}{\partial M_2} + \frac{T_{11}(x)}{GI_\mathrm{p}} \frac{\partial T_{11}(x)}{\partial M_2} \right] \mathrm{d}x \\ \qquad + 2\displaystyle\int_0^{l_{12}} \left[ \frac{M_{12}(x)}{EI} \frac{\partial M_{12}(x)}{\partial M_2} + \frac{T_{12}(x)}{GI_\mathrm{p}} \frac{\partial T_{12}(x)}{\partial M_2} \right] \mathrm{d}x \\ \qquad + 2\displaystyle\int_0^{l_{21}} \left[ \frac{M_{21}(x)}{EI} \frac{\partial M_{21}(x)}{\partial M_2} + \frac{T_{21}(x)}{GI_\mathrm{p}} \frac{\partial T_{21}(x)}{\partial M_2} \right] \mathrm{d}x \\ \qquad + 2\displaystyle\int_0^{l_{22}} \left[ \frac{M_{22}(x)}{EI} \frac{\partial M_{22}(x)}{\partial M_2} + \frac{T_{22}(x)}{GI_\mathrm{p}} \frac{\partial T_{22}(x)}{\partial M_2} \right] \mathrm{d}x = 0 \end{cases} \tag{7.38}$$

And the deflections of joints $R_1$ and $R_2$ are expressed as

$$\begin{cases} \delta_{R_1} = 2\displaystyle\int_0^{l_{11}} \left[ \frac{M_{11}(x)}{EI} \frac{\partial M_{11}(x)}{\partial F_1} + \frac{T_{11}(x)}{GI_\mathrm{p}} \frac{\partial T_{11}(x)}{\partial F_1} \right] \mathrm{d}x \\ \qquad + 2\displaystyle\int_0^{l_{12}} \left[ \frac{M_{12}(x)}{EI} \frac{\partial M_{12}(x)}{\partial F_1} + \frac{T_{12}(x)}{GI_\mathrm{p}} \frac{\partial T_{12}(x)}{\partial F_1} \right] \mathrm{d}x \\ \qquad + 2\displaystyle\int_0^{l_{21}} \left[ \frac{M_{21}(x)}{EI} \frac{\partial M_{21}(x)}{\partial F_1} + \frac{T_{21}(x)}{GI_\mathrm{p}} \frac{\partial T_{21}(x)}{\partial F_1} \right] \mathrm{d}x \\ \qquad + 2\displaystyle\int_0^{l_{22}} \left[ \frac{M_{22}(x)}{EI} \frac{\partial M_{22}(x)}{\partial F_1} + \frac{T_{22}(x)}{GI_\mathrm{p}} \frac{\partial T_{22}(x)}{\partial F_1} \right] \mathrm{d}x \\ \delta_{R_2} = 2\displaystyle\int_0^{l_{11}} \left[ \frac{M_{11}(x)}{EI} \frac{\partial M_{11}(x)}{\partial F_2} + \frac{T_{11}(x)}{GI_\mathrm{p}} \frac{\partial T_{11}(x)}{\partial F_2} \right] \mathrm{d}x \\ \qquad + 2\displaystyle\int_0^{l_{12}} \left[ \frac{M_{12}(x)}{EI} \frac{\partial M_{12}(x)}{\partial F_2} + \frac{T_{12}(x)}{GI_\mathrm{p}} \frac{\partial T_{12}(x)}{\partial F_2} \right] \mathrm{d}x \\ \qquad + 2\displaystyle\int_0^{l_{21}} \left[ \frac{M_{21}(x)}{EI} \frac{\partial M_{21}(x)}{\partial F_2} + \frac{T_{21}(x)}{GI_\mathrm{p}} \frac{\partial T_{21}(x)}{\partial F_2} \right] \mathrm{d}x \\ \qquad + 2\displaystyle\int_0^{l_{22}} \left[ \frac{M_{22}(x)}{EI} \frac{\partial M_{22}(x)}{\partial F_2} + \frac{T_{22}(x)}{GI_\mathrm{p}} \frac{\partial T_{22}(x)}{\partial F_2} \right] \mathrm{d}x \end{cases} \tag{7.39}$$

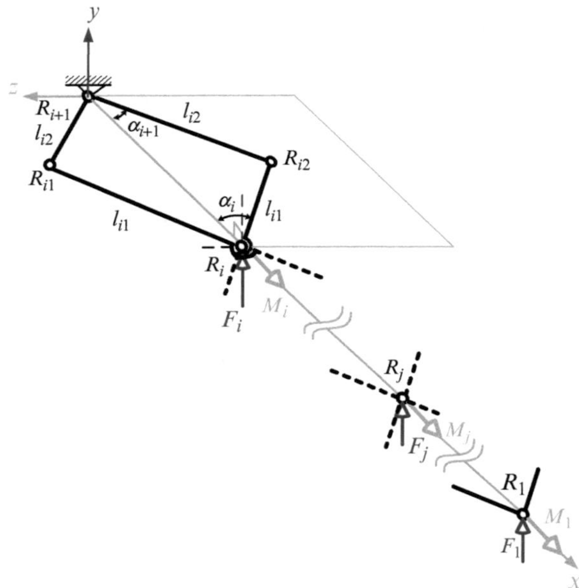


**Fig. 7.10** Deflection analysis of the branch structure

For the $n$th quadrilateral elements shown in Fig. 7.10, $R_n R_{n1} R_{n+1}$ is taken in analysis when $n$ is odd and $R_n R_{n2} R_{n+1}$ is taken when $n$ is even. The bending moment and torque of each segment can be expressed as follows:

$$\begin{cases} M_{i1}(x_{i1}) = \dfrac{1}{2}\sum_{j=1}^{i} F_j(x_{i1} + d_{ji}\cos\alpha_i) + \sum_{j=1}^{i}(-1)^{i+1}M_j \sin\alpha_i \\ T_{i1}(x_{i1}) = \dfrac{1}{2}\sum_{j=1}^{i}(-1)^i F_j d_{ji}\sin\alpha_i + \sum_{j=1}^{i} M_j \cos\alpha_i \\ M_{i2}(x_{i2}) = \dfrac{1}{2}\sum_{j=1}^{i} F_j\left(x_{i2} - l_{i2} + d_{j(i+1)}\cos\alpha_{i+1}\right) + \sum_{j=1}^{i}(-1)^i M_j \sin\alpha_{i+1} \\ T_{i2}(x_{i2}) = \dfrac{1}{2}\sum_{j=1}^{i}(-1)^{i+1} F_j d_{j(i+1)}\sin\alpha_{i+1} + \sum_{j=1}^{i} M_j \cos\alpha_{i+1} \end{cases} \tag{7.40}$$

and

$$\begin{cases} \dfrac{\partial M_{i1}(x)}{\partial F_j} = \dfrac{1}{2}(x_{i1} + d_{ji}\cos\alpha_{i1}), \dfrac{\partial M_{i1}(x)}{\partial M_j} = (-1)^{i+1}\sin\alpha_i \\ \dfrac{\partial T_{i1}(x)}{\partial F_j} = \dfrac{1}{2}(-1)^i d_{ji}\sin\alpha_{i1}, \dfrac{\partial T_{i1}(x)}{\partial M_j} = \cos\alpha_i \\ \dfrac{\partial M_{i2}(x)}{\partial F_j} = \dfrac{1}{2}(x_{i2} - l_{i2} + d_{j(i+1)}\cos\alpha_{i2}), \dfrac{\partial M_{i2}(x)}{\partial M_j} = (-1)^i \sin\alpha_{i+1} \\ \dfrac{\partial T_{i2}(x)}{\partial F_j} = \dfrac{1}{2}(-1)^{i+1} d_{j(i+1)}\sin\alpha_{i+1}, \dfrac{\partial T_{i2}(x)}{\partial M_j} = \cos\alpha_{i+1} \end{cases} \quad (i > j) \tag{7.41}$$

According to the Castigliano's theorem and considering the symmetry of the structure [36], the following relationship holds for each link (also see Eq. (7.27)):

$$\begin{aligned}\theta(M_i) &= 2\sum_{j=0}^{i}\left\{\int_0^{l_{j1}}\left[\frac{M_{j1}(x)}{EI}\frac{\partial M_{j1}(x)}{\partial M_i}+\frac{T_{j1}(x)}{GI_\mathrm{p}}\frac{\partial T_{j1}(x)}{\partial M_i}\right]\mathrm{d}x\right.\\&\quad\left.+\int_0^{l_{j2}}\left[\frac{M_{j1}(x)}{EI}\frac{\partial M_{j2}(x)}{\partial M_i}+\frac{T_{j2}(x)}{GI_\mathrm{p}}\frac{\partial T_{j2}(x)}{\partial M_i}\right]\mathrm{d}x\right\}\\&=0\end{aligned} \tag{7.42}$$

The deflection $\delta_{R_i}$ of joint $R_i$ and the slope $R_i$ with respect to spanwise direction are then obtained as follows:

$$\begin{cases}\delta_{R_i} = 2\sum_{j=0}^{i}\left\{\int_0^{l_{j1}}\left[\frac{M_{j1}(x)}{EI}\frac{\partial M_{j1}(x)}{\partial M_i}+\frac{T_{j1}(x)}{GI_\mathrm{p}}\frac{\partial T_{j1}(x)}{\partial M_i}\right]\mathrm{d}x\right.\\\quad\left.+\int_0^{l_{j2}}\left[\frac{M_{j1}(x)}{EI}\frac{\partial M_{j2}(x)}{\partial M_i}+\frac{T_{j2}(x)}{GI_\mathrm{p}}\frac{\partial T_{j2}(x)}{\partial M_i}\right]\mathrm{d}x\right\}\\\theta_i = 2\sum_{j=0}^{i}\left\{\int_0^{l_{j1}}\left[\frac{M_{j1}(x)}{EI}\frac{\partial M_{j1}(x)}{\partial M_{zi}}+\frac{T_{j1}(x)}{GI_\mathrm{p}}\frac{\partial T_{j1}(x)}{\partial M_{zi}}\right]\mathrm{d}x\right.\\\quad\left.+\int_0^{l_{j2}}\left[\frac{M_{j1}(x)}{EI}\frac{\partial M_{j2}(x)}{\partial M_{zi}}+\frac{T_{j2}(x)}{GI_\mathrm{p}}\frac{\partial T_{j2}(x)}{\partial M_{zi}}\right]\mathrm{d}x\right\}\end{cases} \tag{7.43}$$

where $\theta_i$ denotes the slope of $R_i$ with respect to the spanwise direction.

### 7.3.2 Primary Theory of Interval Analysis

The design parameters in engineering are generally considered as deterministic, but there are many cases that some parameters are uncertain and therefore it is very difficult to roughly understand the product performance. These uncertain parameters come from either design tolerance, manufacture error, or even the error in assembly or working process [31, 32, 38]. When the ranges of uncertain parameters become large or the number of uncertain parameters has increased greatly, the deterministic method could not give a definite solution. To solve these problems, interval analysis [31, 32] is a good method to describe all of the possible results with interval expressions.

An uncertain real can be considered as a real set which contains all of the possible values within a real interval whose lower and upper bounds are all known. At the same time, a real set can be expressed as the real interval which contains the lower bound and upper bound of the real set. When the interval analysis method is utilized to solve the problems which consist of uncertain parameters, the cost will be reduced a lot. Furthermore, the interval method does not need to know the probabilistic distribution or membership function of uncertain parameters, so it can be used widely in engineering.

Let us define a real interval $[x]$ as a connected nonempty subset of real set $R$. It can be expressed as [31, 32]

$$[x] = [\underline{x}, \bar{x}] = \left\{x \in R : \underline{x} \leq x \leq \bar{x}\right\} \tag{7.44}$$

where $\underline{x}$ is the lower bound of interval $[x]$ which can also be noted as $\inf([x])$, the largest number on the left of $[x]$, and $\bar{x}$ is the upper bound of interval $[x]$ which can be noted as $\sup([x])$, the smallest number on the right of $[x]$. The set of all intervals over $R$ is denoted by $IR$ where

$$IR = \left\{[\underline{x}, \bar{x}] : \underline{x}, \bar{x} \in R : \underline{x} \leq \bar{x}\right\} \tag{7.45}$$

The midpoint of $[x]$ can be expressed as

$$x_c = \text{mid}([x]) = \frac{1}{2}(\underline{x} + \bar{x}) \tag{7.46}$$

The radius of $[x]$ is

$$\text{rad}([x]) = \frac{1}{2}(\bar{x} - \underline{x}) \tag{7.47}$$

The width of $[x]$ can be denoted by $w([x]) = 2\text{rad}([x])$. In addition, another real interval related with interval $[x]$ can be defined as

$$[\Delta x] = [-\text{rad}([x]), \quad \text{rad}([x])] \tag{7.48}$$

The real interval $[x]$ can be transformed to the center form

$$[x] = x_c + [\Delta x] \tag{7.49}$$

For example,

$$[x] = [2, 4] = 3 + [-1, 1] \quad \text{where } x_c = 3, \ [\Delta x] = [-1, 1] \tag{7.50}$$

For further discussion, the mathematic operations are briefly introduced for the interval parameters. Interval arithmetic operations are defined on the real set $R$

such that the interval expression covers all possible real results. Given the two real intervals $[x]$ and $[y]$, the operation below holds

$$[x] * [y] = \{x * y : x \in [x], y \in [y]\} \quad \text{for} \quad * \in \{+, -, \times, \div\} \tag{7.51}$$

Then, the four elementary operations are defined by

$$\begin{cases} [x] + [y] = [\underline{x} + \underline{y}, \bar{x} + \bar{y}],\ [x] - [y] = [\underline{x} - \bar{y}, \bar{x} - \underline{y}] \\ [x] \times [y] = \left[\min(\underline{x}\,\underline{y}, \underline{x}\,\bar{y}, \bar{x}\,\underline{y}, \bar{x}\,\bar{y}), \max(\underline{x}\,\underline{y}, \underline{x}\,\bar{y}, \bar{x}\,\underline{y}, \bar{x}\,\bar{y})\right] \\ 1/[x] = [1/\bar{x}, 1/\underline{x}] \quad \text{if } 0 \notin [x] \\ [x] \div [y] = [x] \times 1/[y] \end{cases} \tag{7.52}$$

Particularly

$$[x]^2 = \begin{cases} \left[0, \quad \max\left(\underline{x}^2, \bar{x}^2\right)\right], & \text{if } 0 \in [x] \\ \left[\min\left(\underline{x}^2, \bar{x}^2\right), \quad \max\left(\underline{x}^2, \bar{x}^2\right)\right], & \text{if } 0 \notin [x] \end{cases} \tag{7.53}$$

In most cases, interval arithmetic will enlarge the width of the result. For example, if $[x] = [-1, 1]$, then

$$4[x] - 2[x] = [-4, 4] - [-2, 2] = [-6, 6] \tag{7.54}$$

But the accurate and narrow result can be obtained by altering the form

$$4[x] - 2[x] = (4 - 2)[x] = 2 \times [-1, 1] = [-2, 2] \tag{7.55}$$

The result will be narrower when the number of the occurrences of each interval variable is as little as possible [38].

Consider a function $f$ from $R^n$ to $R^m$. The interval function $[f]$ from $I\,R^n$ to $I\,R^m$ is an inclusion function for $f$ if

$$\forall [x] \in I\,R^n, \quad f([x]) \subset [f]([x]) \tag{7.56}$$

where $\forall [x]$ indicates any interval $[x]$.

One of the most important purposes of interval analysis is to provide, for a large class of functions $f$, inclusion functions $[f]$ that can be evaluated reasonably and quickly such that the result interval is not too wide to engineering applications. Natural inclusion function which just needs to replace each real variable by interval variable is the simplest method to get interval result of a function. Natural inclusion function can obtain the minimal interval result if each of the variables occurs once at most in the formal function of $f$ and the function is continuous [39].

**Table 7.2** Parameters of the scissor-like structure

| $E$(GPa) | $\mu$ | $l_{i1} = l_{i2}$(m) |
|---|---|---|
| 206 | 0.3 | 1 |

**Table 7.3** Bending moment of inertia and rotational moment of inertia of every level structure

| Item | 1 | 2 | 3 | 4 | 5 |
|---|---|---|---|---|---|
| $I$(m$^4$) | $1.25 \times 10^{-7}$ | $1.00 \times 10^{-6}$ | $3.38 \times 10^{-6}$ | $8.00 \times 10^{-6}$ | $1.56 \times 10^{-5}$ |
| $I_p$(m$^4$) | $2.12 \times 10^{-7}$ | $1.19 \times 10^{-6}$ | $3.85 \times 10^{-6}$ | $1.03 \times 10^{-5}$ | $2.52 \times 10^{-5}$ |

Table 7.2 lists the basic structure parameters, and Table 7.3 lists the bending moment of inertia and rotational moment of inertia of every level structure.

With the material parameters shown in Tables 7.2 and 7.3 and Eq. (7.42), the deflections and slopes of every node with respect to the deploying angle, $\alpha$, shown in Fig. 7.3, during unfolding are illustrated in Figs. 7.11 and 7.12, individually.

Figures 7.11 and 7.12 illustrate the deflections and slopes of every joint node on the foldable frame of the morphing wing when deploying from the folded state to the completely unfolded one. Figure 7.11 shows the upper and lower limit curves of the deflection of each joint node, and Fig. 7.12 shows the upper and lower limit curves of the slope of each joint node. Numerical analysis indicates that the changes of both deflections and slopes of every node of the morphing wing during unfolding process are limited into very narrow intervals when $\varsigma = 3$ which is prescribed in Table 7.1. Hence, the deployment ratio, $\varsigma$, of the foldable frame can be designed much higher than the existing morphing wings, which is usually specified as $\varsigma = 3$, even considering the allowable deflection under the completely unfolded situations.

Traditionally, morphing wing skin is made of shape-memory materials [40]. Considering that the shape-memory materials are unavailable to follow the morphing action for a foldable frame with much larger deployment ratios, for example, this chapter proposes a convolute skin consisting of soft but strengthened ductile material to keep the smoothness of the skin during folding and unfolding. The skin shown in Fig. 7.13a is set on the foldable frame through lateral beams attached to the joints of horizontal scissor frame. It allows the skin to be stretched simultaneously with the wing frame. So this kind of morphing wing always has smooth skin which is shown in Fig. 7.13b. Therefore, there is no ill effect on the aerodynamics of the aircraft. Theoretically, one actuator is enough to control the folding and unfolding of the metamorphic wing. The actuator can be mounted on the first revolute joint of middle scissor linkage shown in Fig. 7.13c. Therefore, the number of actuators needed to control the wing should be the least.

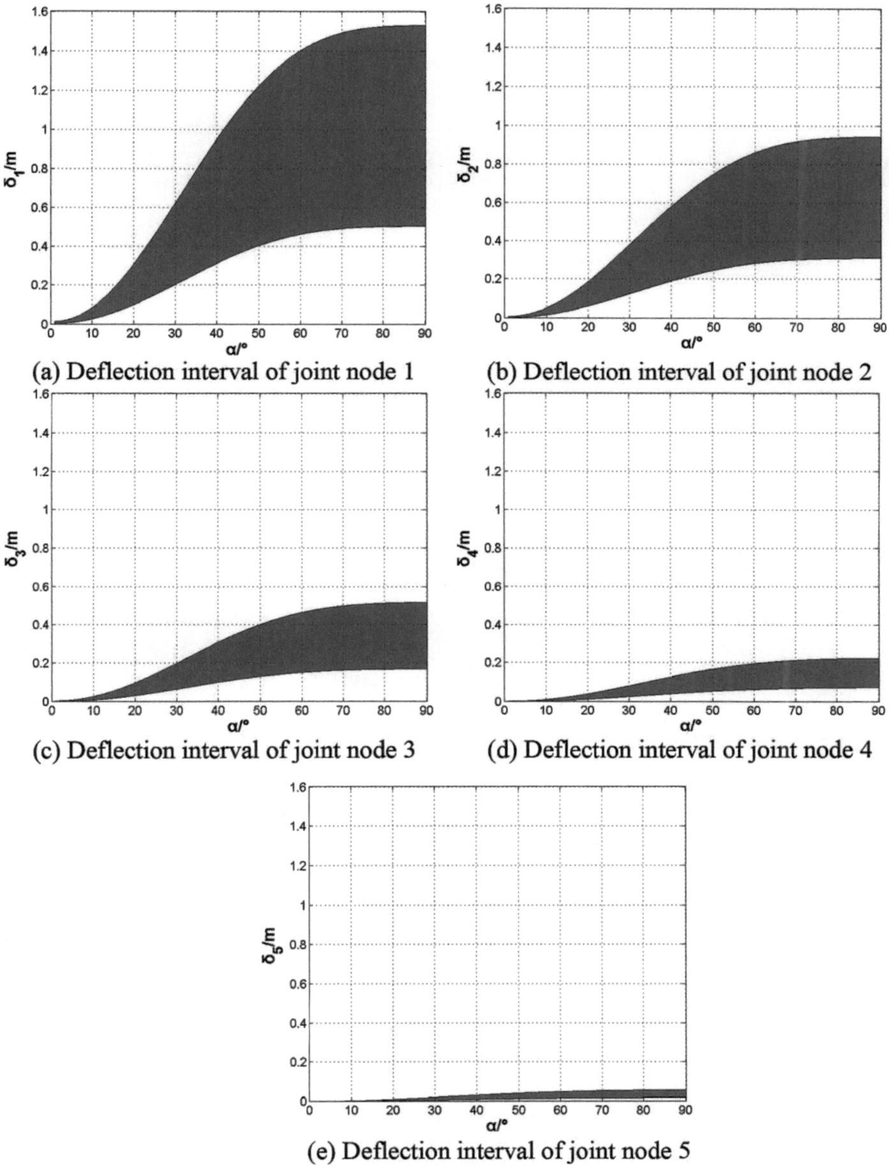


(a) Deflection interval of joint node 1

(b) Deflection interval of joint node 2

(c) Deflection interval of joint node 3

(d) Deflection interval of joint node 4

(e) Deflection interval of joint node 5

**Fig. 7.11** Deflections of the joint nodes with respect to deploying angle

## 7.4 Lifting Guidance Mechanism for Forklift Truck

Forklift trucks are usually used at warehouses, railway station, ports, and factories for loading, unloading, and conveying. However, the general weight-balanced forklift trucks, which often have to contain a mast system, have shown a lot of

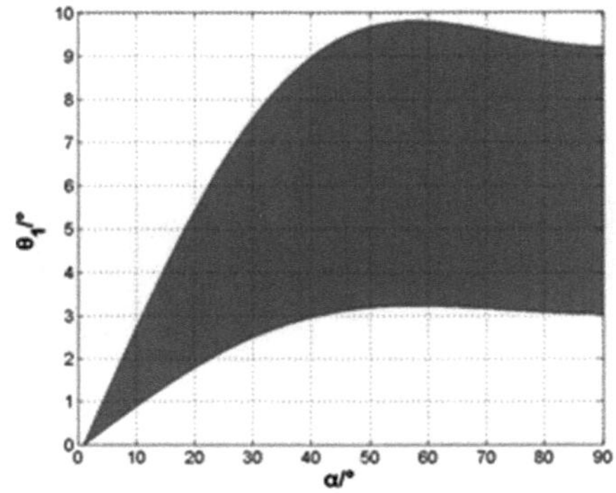


(a) Slope interval of joint node 1

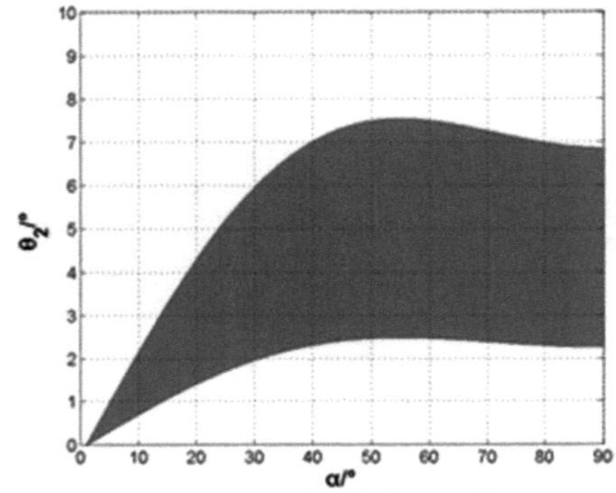


(b) Slope interval of joint node 2

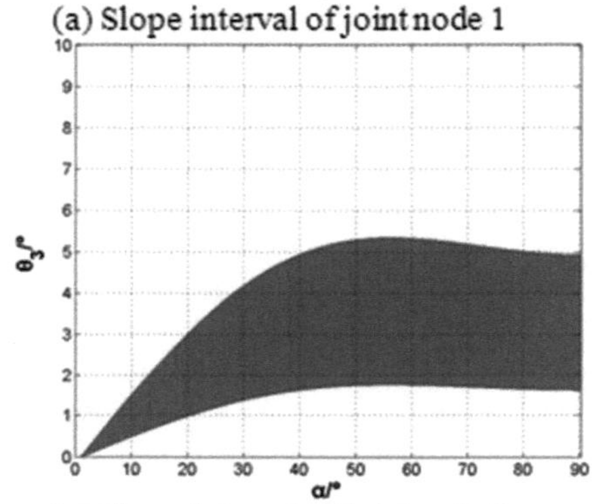


(c) Slope interval of joint node 3

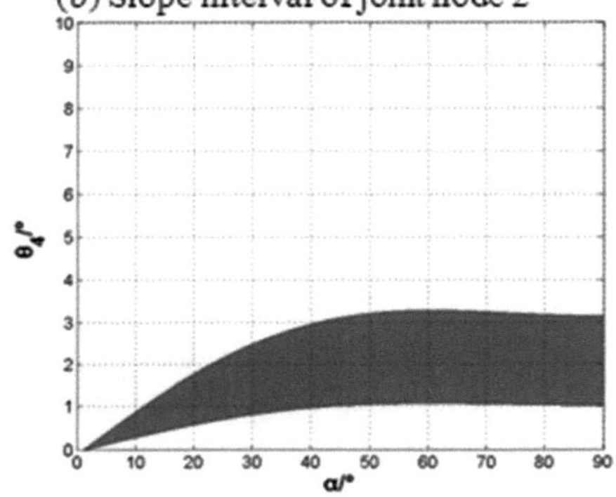


(d) Slope interval of joint node 4

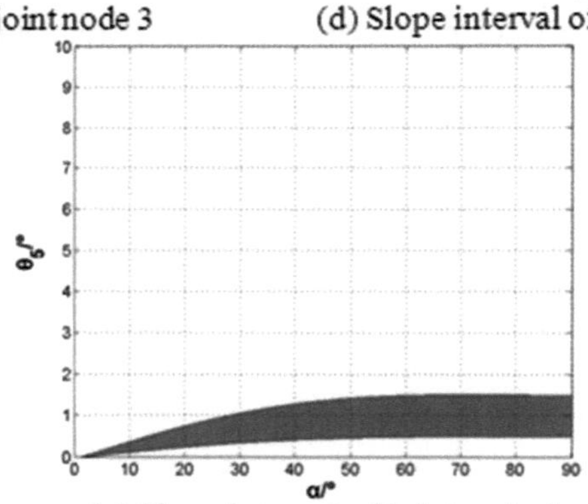


(e) Slope interval of joint node 5

**Fig. 7.12** Slopes of the nodes with respect to deploying angle

disadvantages. For example, because the mast system is composed of several large components and is assembled in front of the driver's cab, it will obstruct the driver's vision seriously. A historical review indicates that many accidents involving collisions between pedestrians and trucks are due to inherently bad visibilities of the forklift trucks [41]. Moreover, the mast system's high self-weight increases the mass of the rear equilibrator and the unnecessary waste of energy.

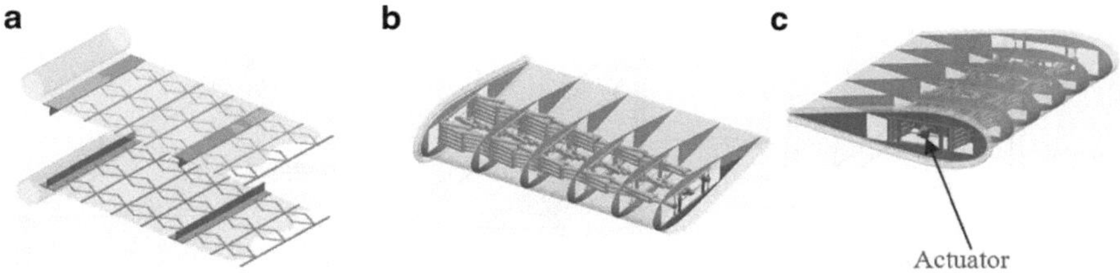


**Fig. 7.13** Actuation scheme and the stretchable skin (**a**) Ductile skin attached to wing frame (**b**) Completely unfolded wing (**c**) Position of the actuator

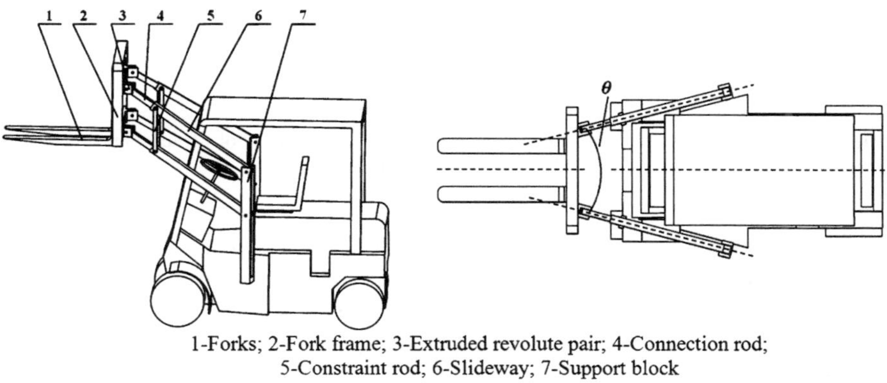


**Fig. 7.14** A new lift-guidance mechanism for forklift trucks

Consequently, an innovative lift mechanism for forklift trucks is invented [42]. It is designed based on the spatial multi-link linear guidance mechanism. The lift mechanism assembled in a forklift truck is shown in Fig. 7.14. Two support blocks are connected on both sides of the forklift truck. The two vertical planes in which both of the support blocks located are symmetrical with respect to the longitudinal symmetrical plane of the vehicle and the subtended angle between the two planes is denoted by $\theta(0° < \theta < 180°)$. Two RPR-kinematic chains locate in the two planes, respectively. Each of the RPR-kinematic chains consists of two revolute joints, a prismatic joint, a connection rod, and a slideway. The connection rod and the slideway are connected through one prismatic joint. The fork frame has two extruded revolute joints that are connected with the ends of the two RPR-kinematic chains. Under the redundant constraints of the lift mechanism, the fork frame with the cargoes can be only lifted up and down along a vertical straight line. Compared with the mast systems in general forklift trucks, this new kind of lift mechanism not only expands the vision field of the driver widely but also reduces the whole self-weight of the vehicle.

For implementation, windlass and soft steel cables are utilized to lift the fork frame and cargoes on it. The structure is shown in Fig. 7.15a. The windlass is set up

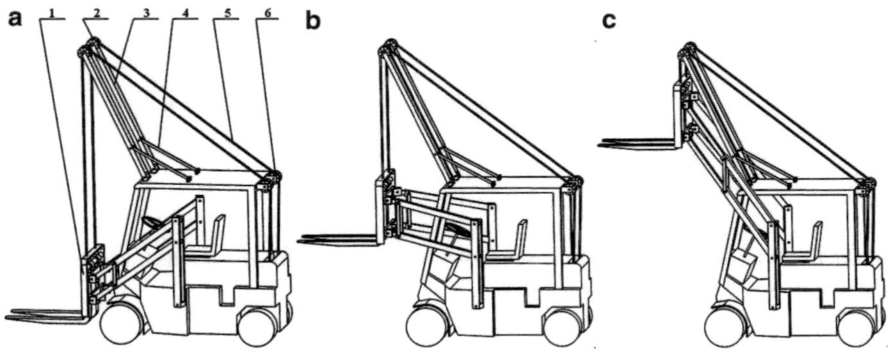


1-Fork frame; 2-Front pulley; 3-Front supporting rods;
4-Back supporting rods; 5-Cables; 6-Back pulley

**Fig. 7.15** Implementation of the new lift mechanism (**a**) Lateral view of lift mechanism (**b**) Lift mechanism in a certain position (**c**) Lift mechanism in another position

at the rear of the forklift truck and some pulleys and supporting rods are placed on the vehicle. One end of each cable is connected with the fork frame and the other end is fixed to the windlass. When the windlass rotates, the cables will lift or drop the fork frame. Figure 7.15a–c shows different positions when the mechanism is lifting the fork frame up.

As is shown above, the fork frame is restrained to move in a straight line due to the redundant constraints of the mechanism. When loading cargoes, however, the fork frame possibly generates deformation or rotation under load and therefore might cause sliding down of the cargoes. Moreover, these changes during lifting up heavily depend on the stiffness of structure and work conditions. Therefore, it is necessary to analyze the dynamic stiffness of this indeterminate structure. And then, the stiffness of this mechanism under different conditions can be analyzed and described.

## 7.5 Dynamics Analysis for Lifting Guidance Mechanism Forklift Truck

### *7.5.1 Dynamics of a Lift Mechanism*

While loading cargoes with this new type of lift mechanism for forklift trucks, the long rods connected to the fork frame through revolute joints are subjected to major forces and moments. The fork frame possibly generates deformation or rotation under the load and may cause slide down of the cargoes. Consequently, dynamics of the lift mechanism should be first analyzed.

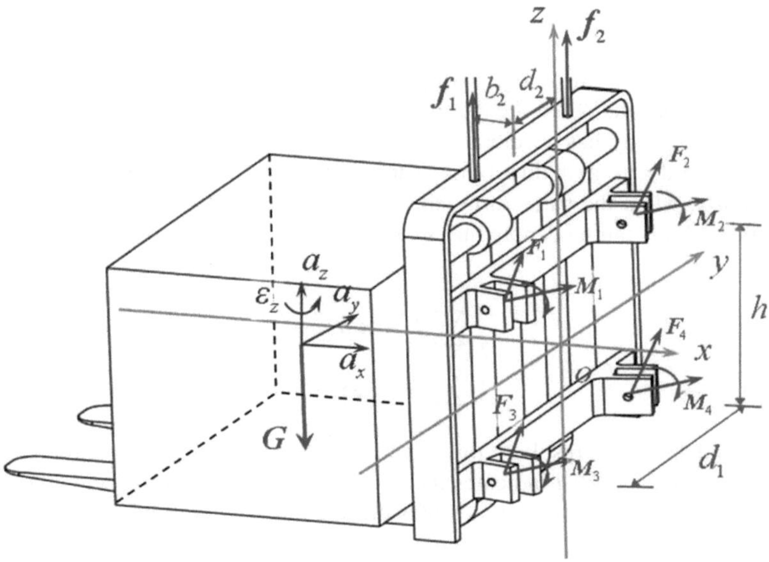


**Fig. 7.16** Cartesian coordinate system on the fork frame

For the actuation described in Sect. 7.4, windlass and cables are utilized to lift the fork frame and cargoes. In order to analyze the forces exerted on the fork frame, we can set a Cartesian coordinate system, where the origin of coordinate system is the center of four extruded joints, $z$-axis is perpendicular to the ground, $x$-axis is parallel to the longitudinal symmetry axis of the vehicle, and $y$-axis is determined in accordance to the right-hand rule which is shown in Fig. 7.16.

Suppose that the dynamic loads of cargoes are denoted by a vector, $\boldsymbol{G}$, which is exerted on the frame and it is divided into three components denoted by $\left(G_x\ G_y\ G_z\right)$ where the subscripts indicate the positive directions of the forces. The coordinates of the position of force are denoted by $(x_G\ y_G\ z_G)$. The self-weight of the fork frame is denoted by $G_0$, the gravity center of which are $(x_{G0}\ y_{G0}\ z_{G0})$. The mass of cargoes and fork frame are $m_G$ and $m_0$, respectively.

As is shown Fig. 7.1, the four long connection rods exert constraints on the fork frame and the constraints are denoted by $\boldsymbol{F}_i$ and $\boldsymbol{M}_i(i = 1, 2, 3, 4)$. In the Cartesian coordinate system, the constraints can be divided into components along the three coordinate-axis directions and be denoted by $F_{ix}$, $F_{iy}$, $F_{iz}$, and $M_{ix}$, $M_{iy}$, $M_{iz}(i = 1, 2, 3, 4)$, individually. The subscripts present the positive directions of the forces or moments.

The distance between the two extruded joints that locate in the same horizontal plane is denoted by $d_1$. The distance between the two extruded joints that locate in the same vertical plane is denoted by $h$. The two vertical forces exerted to the fork frame by the two cables are denoted by $f_1$ and $f_2$, respectively. The distance from them to the $yoz$-plane is denoted by $b_2$, and the distance from each force to the $xoz$-plane is denoted by $d_2$. These geometry relationships are shown in Fig. 7.16.

In the process of lifting or moving ahead, the body is not always in the state of static equilibrium. When the cargoes are being lifted or the truck is accelerating or braking, the body has acceleration and it can be divided into three components along the coordinate-axis directions. The components are denoted by $a_x$, $a_y$, and $a_z$. Similarly, when the truck turns, the body has angular accelerations around the $z$-direction which can be denoted by $\varepsilon_z$. Consequently, we can obtain the equations of motion:

$$\begin{cases} \sum \boldsymbol{F} - m\boldsymbol{a} = 0 \\ \sum \boldsymbol{M} - \mathbf{J}\boldsymbol{\varepsilon} = 0 \end{cases} \tag{7.57}$$

where $\boldsymbol{F}$ and $\boldsymbol{M}$ are the resultant forces and moments, $m$ represents the total mass of cargoes and fork frame, $\boldsymbol{a}$ indicates the acceleration vector, $\mathbf{J}$ represents the matrix of the moment of inertia, and $\boldsymbol{\varepsilon}$ indicates the angular acceleration vector.

Equation (7.57) can be rewritten in scalar form:

$$\begin{cases} R_x = \sum_{i=1}^{4} F_{ix} + G_x - (m_G + m_0)a_x = 0 \\ R_y = \sum_{i=1}^{4} F_{iy} + G_y - (m_G + m_0)a_y = 0 \\ R_z = \sum_{i=1}^{4} F_{iz} + f_1 + f_2 + G_z + G_0 - (m_G + m_0)a_z = 0 \\ M_x = \sum_{i=1}^{4} (y_i F_{iz} - z_i F_{iy}) + y_G G_z + y_0 G_0 - z_G G_y - f_1 d_2 + f_2 d_2 + \sum_{i=1}^{4} M_{ix} = 0 \\ M_y = \sum_{i=1}^{4} (z_i F_{ix} - x_i F_{iz}) + z_G G_x - x_G G_z - x_0 G_0 + (f_1 + f_2) b_2 + \sum_{i=1}^{4} M_{iy} = 0 \\ M_z = \sum_{i=1}^{4} (x_i F_{iy} - y_i F_{ix}) + x G_y - y G_x + \sum_{i=1}^{4} M_{iz} - J_z \varepsilon_z = 0 \end{cases} \tag{7.58}$$

where $m_G$ represents the mass of cargoes and $m_0$ is the mass of the fork frame, $x_i = 0$ for $i = 1, 2, 3, 4,\ y_i = -d_1/2$, for $i = 1, 3,\ y_i = d_1/2$, for $i = 2, 4,\ z_i = h$, for $i = 1, 2$, and $z_i = 0$ for $i = 3, 4$, and $J_z$ indicates the moment of inertia of cargoes and fork frame about the $z$-axis and $J_z$ is assumed to be constant for convenience.

In order to solve Eq. (7.58) and study the reaction forces/moments exerted on the long connection rods, we can constitute a local Cartesian coordinate system fixed with one of the long connection rods, where $x_{i0}$-axis is parallel to it and $y_{i0}$-axis is

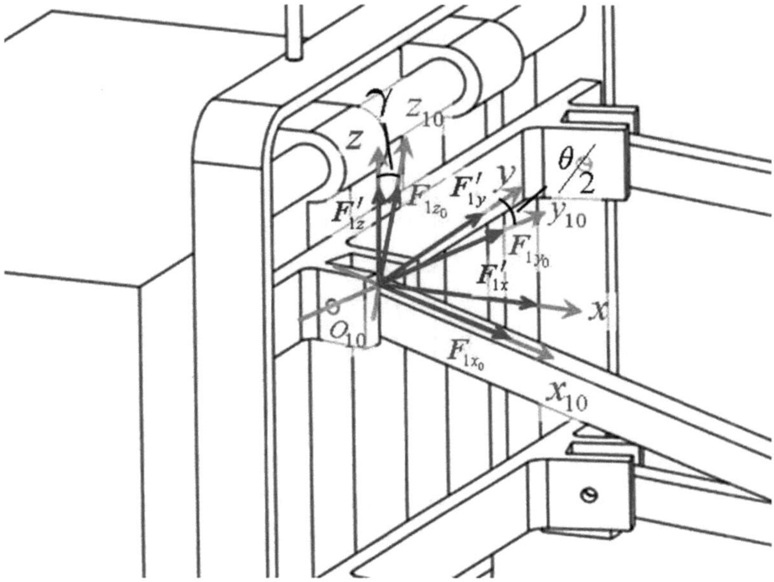


**Fig. 7.17** Local Cartesian coordinate system fixed with one of the long connection rods

superimposed with the axis of the revolute joint and $z_{i0}$-axis is determined in line with the right-hand rule which is shown in Fig. 7.17. The subscript $i$ can be $1, 2, 3, 4$ which denotes the sequence of a connection rod. The rods' subscript numbers are the same to those of the corresponding forces shown in Fig. 7.17. The $x$-, $y$-, and $z$-directions in Fig. 7.17 denote, respectively, the directions that are parallel to the axes in Cartesian coordinate system shown in Fig. 7.16.

According to Sect. 7.4, it is not difficult to find that the angle between $y_{i0}$-direction and $y$-direction is $\frac{\theta}{2}$. When the long connection rod rotates an angle of $\gamma$ around the revolute joint from the horizontal position, the angle between $z_{i0}$-direction and $z$-direction is therefore equal to $\gamma$. In the Cartesian coordinate system $o_{i0}-xyz$, the forces exerted on one of the long connection rods by the fork frame are all divided into three orthogonal components which are denoted by $F'_{ix}$, $F'_{iy}$, $F'_{iz}$, where the subscripts indicate the positive directions of the force components and the superscripts indicate that the forces are reaction forces. According to Newton's third law, we have

$$\begin{cases} F'_{ix} = -F_{ix} \\ F'_{iy} = -F_{iy} \\ F'_{iz} = -F_{iz} \end{cases} \tag{7.59}$$

where $i$ can be $1, 2, 3, 4$ and the minus signs indicate the reverse directions of both sides.

In the local coordinate system $o_{i0} - x_{i0}y_{i0}z_{i0}$, the forces exerted on the long connection rod are all divided into three orthogonal components which are denoted by $F_{ix_0}$, $F_{iy_0}$, $F_{iz_0}$, respectively. Assume that $\mathbf{R}_i$ denotes the coordinate transformation matrix from the coordinate system $o_{i0} - xyz$ to $o_{i0} - x_{i0}y_{i0}z_{i0}$; therefore, we have [43]

$$\begin{cases} \begin{bmatrix} F_{ix_0} \\ F_{iy_0} \\ F_{iz_0} \end{bmatrix} = \mathbf{R}_i \begin{bmatrix} F'_{ix} \\ F'_{iy} \\ F'_{iz} \end{bmatrix} = \mathbf{R}_i \begin{bmatrix} -F_{ix} \\ -F_{iy} \\ -F_{iz} \end{bmatrix} \\ \begin{bmatrix} M_{ix_0} \\ M_{iy_0} \\ M_{iz_0} \end{bmatrix} = \mathbf{R}_i \begin{bmatrix} M'_{ix} \\ M'_{iy} \\ M'_{iz} \end{bmatrix} = \mathbf{R}_i \begin{bmatrix} -M_{ix} \\ -M_{iy} \\ -M_{iz} \end{bmatrix} \end{cases} \tag{7.60}$$

where

$$\begin{aligned} \mathbf{R}_1 = \mathbf{R}_3 &= \begin{bmatrix} \cos\gamma & 0 & -\sin\gamma \\ 0 & 1 & 0 \\ \sin\gamma & 0 & \cos\gamma \end{bmatrix} \begin{bmatrix} \cos\frac{\theta}{2} & -\sin\frac{\theta}{2} & 0 \\ \sin\frac{\theta}{2} & \cos\frac{\theta}{2} & 0 \\ 0 & 0 & 1 \end{bmatrix} \\ &= \begin{bmatrix} \cos\gamma\cos\frac{\theta}{2} & -\cos\gamma\sin\frac{\theta}{2} & -\sin\gamma \\ \sin\frac{\theta}{2} & \cos\frac{\theta}{2} & 0 \\ \sin\gamma\cos\frac{\theta}{2} & -\sin\gamma\sin\frac{\theta}{2} & \cos\gamma \end{bmatrix} \end{aligned} \tag{7.61}$$

$$\begin{aligned} \mathbf{R}_2 = \mathbf{R}_4 &= \begin{bmatrix} \cos\gamma & 0 & -\sin\gamma \\ 0 & 1 & 0 \\ \sin\gamma & 0 & \cos\gamma \end{bmatrix} \begin{bmatrix} \cos\frac{\theta}{2} & \sin\frac{\theta}{2} & 0 \\ -\sin\frac{\theta}{2} & \cos\frac{\theta}{2} & 0 \\ 0 & 0 & 1 \end{bmatrix} \\ &= \begin{bmatrix} \cos\gamma\cos\frac{\theta}{2} & \cos\gamma\sin\frac{\theta}{2} & -\sin\gamma \\ -\sin\frac{\theta}{2} & \cos\frac{\theta}{2} & 0 \\ \sin\gamma\cos\frac{\theta}{2} & \sin\gamma\sin\frac{\theta}{2} & \cos\gamma \end{bmatrix} \end{aligned} \tag{7.62}$$

The forces exerted on the long connection rod are also shown in Fig. 7.18. If the friction in the prismatic joint is ignored, the connection rod will suffer no constraint along the $x_0$-direction. Therefore, we have $F_{ix_0} = 0$. The long connection rod is connected to the fork frame by a revolute joint, so $M_{iy_0} = 0$. In addition, the slideway is connected to the chassis by a revolute joint; therefore $F_{iz_0} = 0$; otherwise the connection rod and slideway will rotate around the revolute joint shown in Fig. 7.16. Associating Eqs. (7.60), (7.61), and (7.62) and substituting $F_{ix_0} = 0$, $M_{iy_0} = 0$, and $F_{iz_0} = 0$ into them present

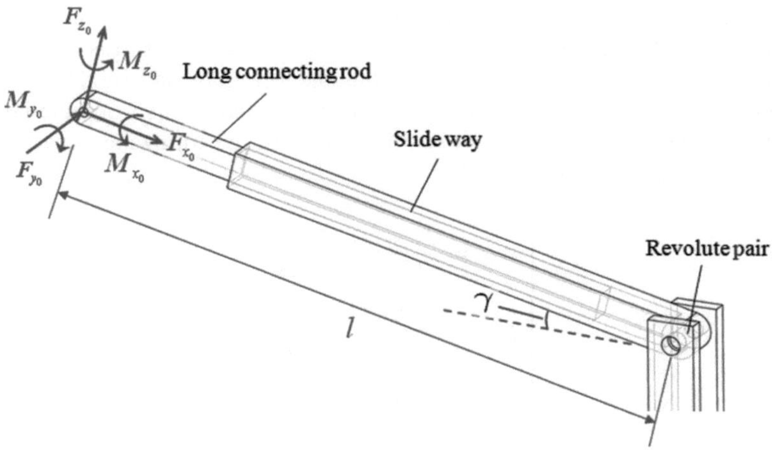


**Fig. 7.18** Forces exerted on the long connection rod

$$\left\{\begin{bmatrix} F_{ix} \\ F_{iy} \\ F_{iz} \\ M_{ix} \\ M_{iy} \\ M_{iz} \end{bmatrix} = \mathbf{S}_i \begin{bmatrix} F_{i\,y0} \\ M_{i\,x0} \\ M_{i\,z0} \end{bmatrix}\right. \tag{7.63}$$

where

$$\mathbf{S}_1 = \mathbf{S}_3 = \begin{bmatrix} -\sin\frac{\theta}{2} & 0 & 0 \\ -\cos\frac{\theta}{2} & 0 & 0 \\ 0 & 0 & 0 \\ 0 & -\cos\gamma\cos\frac{\theta}{2} & -\sin\gamma\cos\frac{\theta}{2} \\ 0 & \cos\gamma\sin\frac{\theta}{2} & \sin\gamma\sin\frac{\theta}{2} \\ 0 & \sin\gamma & -\cos\gamma \end{bmatrix} \tag{7.64}$$

$$\mathbf{S}_2 = \mathbf{S}_4 = \begin{bmatrix} \sin\frac{\theta}{2} & 0 & 0 \\ -\cos\frac{\theta}{2} & 0 & 0 \\ 0 & 0 & 0 \\ 0 & -\cos\gamma\cos\frac{\theta}{2} & -\sin\gamma\cos\frac{\theta}{2} \\ 0 & -\cos\gamma\sin\frac{\theta}{2} & -\sin\gamma\sin\frac{\theta}{2} \\ 0 & \sin\gamma & -\cos\gamma \end{bmatrix} \tag{7.65}$$

Consequently, there are still three unknown forces and moments about the connection rod. In order to obtain all of the forces and moments, it is necessary to establish compatibility equations. Assume that the fork frame is a rigid body, while the long connection rods are elastic ones. According to mechanics of materials and considering the connection rods as cantilever beams, the following equations are immediately obtained from Fig. 7.18:

$$\begin{cases} \varepsilon_{i\,y_0} = \dfrac{F_{i\,y_0} l^3}{3EI} - \dfrac{M_{i\,z_0} l^2}{2EI} \\ \mu_{i\,z_0} = -\dfrac{F_{i\,y_0} l^2}{2EI} + \dfrac{M_{i\,z_0} l}{EI} \\ \alpha_{i\,x_0} = \dfrac{M_{i\,x_0} l}{G I_p} \\ l = \dfrac{l_0}{\cos\gamma} \end{cases} \tag{7.66}$$

where the subscripts $i = 1, 2, 3, 4$ indicate which connection rod the forces and moments are exerted to, $\varepsilon_{i\,y_0}$ denotes the elastic deformation along $y_0$-direction, $\mu_{i\,z_0}$ is the angle of elastic rotation around $z_0$-direction, $\alpha_{i\,x_0}$ represents the torsional angle of elastic deformation around $x_0$-direction, $l_0$ stands for the distance between the two revolute joints when the connection rod is at horizontal position, $E$ denotes the elastic modulus, $G$ is the shearing modulus, and the rotational and torsional moment of inertia, $I$ and $I_{\mathrm{p}}$, are supposed to be constant by ignoring the change of cross section for convenience. The plus or minus signs indicate that the directions of the deformations are the same or reverse to the positive directions of axes in the local Cartesian coordinate system.

The ends of the four connection rods are connected with the fork frame. Therefore, while the ends of the four connection rods generate elastic deformation, the entire fork frame has a micro-displacement and a micro-rotation around the center of the four extruded joints because the whole frame is considered as a rigid body. Suppose that the elastic deformations of the connection rods are denoted by $(\varepsilon_{ix}\ \varepsilon_{iy}\ \varepsilon_{iz})^{\mathrm{T}}$, $i = 1, 2, 3, 4$, while the micro-displacement of the fork frame along the three axial directions is denoted by $(s_x\ s_y\ s_z)^{\mathrm{T}}$ and the rotation angles of the fork frame around the three axial directions are denoted by $(\alpha_x\ \beta_y\ \mu_z)^{\mathrm{T}}$, respectively. According to the coordinate system shown in Fig. 7.16, we can find the coordinate positions of the end of No.1 connection rod to be $\left(0, -\frac{d_1}{2}, \frac{h}{2}\right)$. Because $\alpha_x$, $\beta_y$, and $\mu_z$ are all infinitesimal, the displacements of the end of No.1 connection rod can be approximately expressed as

$$
\begin{aligned}
(\varepsilon_{1x}\ \varepsilon_{1y}\ \varepsilon_{1z})^{\mathrm{T}} &= \begin{bmatrix} 0 & -\mu_z & \beta_y \\ \mu_z & 0 & -\alpha_x \\ -\beta_y & \alpha_x & 0 \end{bmatrix} \begin{bmatrix} 0 \\ -\dfrac{d_1}{2} \\ \dfrac{h}{2} \end{bmatrix} + \begin{bmatrix} s_x \\ s_y \\ s_z \end{bmatrix} \\
&= \left( \frac{\beta_y h}{2} + \frac{\mu_z d_1}{2} + s_x \ -\frac{\alpha_x h}{2} + s_y \ -\frac{\alpha_x d_1}{2} + s_z \right)^{\mathrm{T}}
\end{aligned} \tag{7.67}
$$

Similarly, the following equations hold:

$$
\begin{cases}
(\varepsilon_{1x}\ \varepsilon_{1y}\ \varepsilon_{1z})^{\mathrm{T}} = \left( \dfrac{\beta_y h}{2} + \dfrac{\mu_z d_1}{2} + s_x \ -\dfrac{\alpha_x h}{2} + s_y \ -\dfrac{\alpha_x d_1}{2} + s_z \right)^{\mathrm{T}} \\
(\varepsilon_{2x}\ \varepsilon_{2y}\ \varepsilon_{2z})^{\mathrm{T}} = \left( \dfrac{\beta_y h}{2} - \dfrac{\mu_z d_1}{2} + s_x \ -\dfrac{\alpha_x h}{2} + s_y \ \dfrac{\alpha_x d_1}{2} + s_z \right)^{\mathrm{T}} \\
(\varepsilon_{3x}\ \varepsilon_{3y}\ \varepsilon_{3z})^{\mathrm{T}} = \left( -\dfrac{\beta_y h}{2} + \dfrac{\mu_z d_1}{2} + s_x \ \dfrac{\alpha_x h}{2} + s_y \ -\dfrac{\alpha_x d_1}{2} + s_z \right)^{\mathrm{T}} \\
(\varepsilon_{4x}\ \varepsilon_{4y}\ \varepsilon_{4z})^{\mathrm{T}} = \left( -\dfrac{\beta_y h}{2} - \dfrac{\mu_z d_1}{2} + s_x \ \dfrac{\alpha_x h}{2} + s_y \ \dfrac{\alpha_x d_1}{2} + s_z \right)^{\mathrm{T}}
\end{cases} \tag{7.68}
$$

As is expressed in Eqs. (7.61) and (7.62), $\mathbf{R}_i$ denotes the coordinate transformation matrix from the coordinate system $o_{i0} - xyz$ to $o_{i0} - x_{i0}y_{i0}z_{i0}$; therefore, we obtain

$$
\begin{cases}
\begin{bmatrix} \varepsilon_{ix0} \\ \varepsilon_{iy0} \\ \varepsilon_{iz0} \end{bmatrix} = \mathbf{R}_i \begin{bmatrix} \varepsilon_{ix} \\ \varepsilon_{iy} \\ \varepsilon_{iz} \end{bmatrix} \\
\begin{bmatrix} \alpha_{ix0} \\ \beta_{iy0} \\ \mu_{iz0} \end{bmatrix} = \mathbf{R}_i \begin{bmatrix} \alpha_x \\ \beta_y \\ \mu_z \end{bmatrix}
\end{cases} \tag{7.69}
$$

where $i = 1, 2, 3, 4$. Substituting Eqs. (7.61), (7.62), (7.66), and (7.68) into (7.69), the following equations are gained:

$$\begin{cases}
\dfrac{F_{1y_0}l^3}{3EI} - \dfrac{M_{1z_0}l^2}{2EI} = \left(\dfrac{\beta_y h}{2} + \dfrac{\mu_z d_1}{2} + s_x\right)\sin\dfrac{\theta}{2} + \left(-\dfrac{\alpha_x h}{2} + s_y\right)\cos\dfrac{\theta}{2} \\
\dfrac{F_{2y_0}l^3}{3EI} - \dfrac{M_{2z_0}l^2}{2EI} = -\left(\dfrac{\beta_y h}{2} - \dfrac{\mu_z d_1}{2} + s_x\right)\sin\dfrac{\theta}{2} + \left(-\dfrac{\alpha_x h}{2} + s_y\right)\cos\dfrac{\theta}{2} \\
\dfrac{F_{3y_0}l^3}{3EI} - \dfrac{M_{3z_0}l^2}{2EI} = \left(-\dfrac{\beta_y h}{2} + \dfrac{\mu_z d_1}{2} + s_x\right)\sin\dfrac{\theta}{2} + \left(\dfrac{\alpha_x h}{2} + s_y\right)\cos\dfrac{\theta}{2} \\
\dfrac{F_{4y_0}l^3}{3EI} - \dfrac{M_{4z_0}l^2}{2EI} = -\left(-\dfrac{\beta_y h}{2} - \dfrac{\mu_z d_1}{2} + s_x\right)\sin\dfrac{\theta}{2} + \left(\dfrac{\alpha_x h}{2} + s_y\right)\cos\dfrac{\theta}{2} \\
\dfrac{M_{1x_0}l}{GI_p} = \alpha_x\cos\gamma\cos\dfrac{\theta}{2} - \beta_y\cos\gamma\sin\dfrac{\theta}{2} - \mu_z\sin\gamma \\
\dfrac{M_{2x_0}l}{GI_p} = \alpha_x y\gamma\cos\dfrac{\theta}{2} + \beta_y\cos\gamma\sin\dfrac{\theta}{2} - \mu_z\sin\gamma \\
\dfrac{M_{3x_0}l}{GI_p} = \alpha_x\cos\gamma\cos\dfrac{\theta}{2} - \beta_y\cos\gamma\sin\dfrac{\theta}{2} - \mu_z\sin\gamma \\
\dfrac{M_{4x_0}l}{GI_p} = \alpha_x\cos\gamma\cos\dfrac{\theta}{2} + \beta_y\cos\gamma\sin\dfrac{\theta}{2} - \mu_z\sin\gamma \\
-\dfrac{F_{1y_0}l^2}{2EI} + \dfrac{M_{1z_0}l}{EI} = \alpha_x\sin\gamma\cos\dfrac{\theta}{2} - \beta_y\sin\gamma\sin\dfrac{\theta}{2} + \mu_z\cos\gamma \\
-\dfrac{F_{2y_0}l^2}{2EI} + \dfrac{M_{2z_0}l}{EI} = \alpha_x\sin\gamma\cos\dfrac{\theta}{2} + \beta_y\sin\gamma\sin\dfrac{\theta}{2} + \mu_z\cos\gamma \\
-\dfrac{F_{3y_0}l^2}{2EI} + \dfrac{M_{3z_0}l}{EI} = \alpha_x\sin\gamma\cos\dfrac{\theta}{2} - \beta_y\sin\gamma\sin\dfrac{\theta}{2} + \mu_z\cos\gamma \\
-\dfrac{F_{4y_0}l^2}{2EI} + \dfrac{M_{4z_0}l}{EI} = \alpha_x\sin\gamma\cos\dfrac{\theta}{2} + \beta_y\sin\gamma\sin\dfrac{\theta}{2} + \mu_z\cos\gamma
\end{cases} \tag{7.70}$$

Suppose that the original lengths of cables are both $L$. While the cables are stretched and the connecting points of frame and cables move some distance back, the displacements of connecting points can be expressed with that of the fork frame; therefore, the extension of the cables, denoted by $\Delta L_1$ and $\Delta L_2$, can be expressed as

$$\begin{cases}
\Delta L_1 = -\left(-\alpha_x d_2 + \dfrac{\beta_y b_2}{2} + s_z\right) \\
\Delta L_2 = -\left(\alpha_x d_2 + \dfrac{\beta_y b_2}{2} + s_z\right)
\end{cases} \tag{7.71}$$

where $\Delta L_1$ and $\Delta L_2$ are always positive and the minus signs indicate the reverse direction relations of the extensions and the displacements of the connecting points.

**Table 7.4** Structure and dynamic parameters

| $h$(m) | $d_1$(m) | $d_2$(m) | $b_2$(m) | $EI$(Nm$^2$) | $GI_p$(Nm$^2$) | $EA$(N) | $l_0$(m) |
|---|---|---|---|---|---|---|---|
| 0.47 | 0.75 | 0.3 | 0.08 | $9.82 \times 10^5$ | $5.89 \times 10^5$ | $1.57 \times 10^9$ | 1.3 |
| $L$(m) | $\theta$ | $m_g$(kg) | $G_x$(N) | $G_y$(N) | $G_z$(N) | $m_0$(kg) | $G_0$(N) |
| 5.0 | $\frac{\pi}{6}$ | 1,000 | 0 | 0 | −9,800 | 200 | −1,960 |
| $x_0$(m) | $y_0$(m) | $J_z$(kgm$^2$) | $a_x$(ms$^{-2}$) | $a_y$(ms$^{-2}$) | $a_z$(ms$^{-2}$) | $\varepsilon_z$(s$^{-2}$) | |
| −0.2 | 0 | 150 | 0 | 0 | 0.5 | 0 | |

According to compatibility equation, there are

$$\begin{cases} f_1 = \dfrac{\Delta L_1}{L} EA = \dfrac{-\left(-\alpha_x d_2 + \frac{\beta_y b_2}{2} + s_z\right) EA}{L} \\[2ex] f_2 = \dfrac{\Delta L_2}{L} EA = \dfrac{-\left(\alpha_x d_2 + \frac{\beta_y b_2}{2} + s_z\right) EA}{L} \end{cases} \tag{7.72}$$

Associating Eqs. (7.58), (7.63), (7.64), (7.65), (7.70), and (7.72), we can establish a system of equations, which consists of 20 equations and 20 unknowns, including $F_{iy_0}(i = 1,2,3,4)$, $M_{ix_0}(i = 1,2,3,4)$, $M_{iz_0}(i = 1,2,3,4)$, $f_1$, $f_2$, $\alpha_x$, $\beta_y$, $\mu_z$, $s_x$, $s_y$, and $s_z$. The other quantities are either structure parameters or dynamic parameters; the system therefore has complicated analytic solutions. But in fact, some parameters are uncertain or changing within a specified interval during the implementation process. In order to investigate the dynamic characteristics of the lift mechanism in work, the principal parameters are proposed to change in an estimated interval. In this way, both the trends of variations of forces/moments and the extent of the elastic deformations can be estimated. As a result, the dynamic stiffness and strength of this mechanism can be analyzed.

### *7.5.2 Interval Analysis of Dynamic Stiffness*

Suppose that in the implementation process, cargoes are loaded with this lift mechanism. Because the ends of four connection rods generate elastic deformations under load, the fork frame has a micro-rotation around the $y$-direction or $x$-direction. The rotation possibly causes slide down for the cargoes and the rotation angle reflects the stiffness of this mechanism. A system of equations is established by associating Eqs. (7.58), (7.63), (7.64), (7.65), (7.70), and (7.72). Substituting the structure and dynamic parameters listed in Table 7.4, the system of equations can be solved.

For most of the working conditions and for the sake of simplicity, it is supposed that the load is exerted along the $z$-direction only, i.e., $G_x = 0$ and $G_x = 0$, and the cargoes are being lifted up straightly with an acceleration of $a_z = 0.5\,\text{ms}^{-2}$.

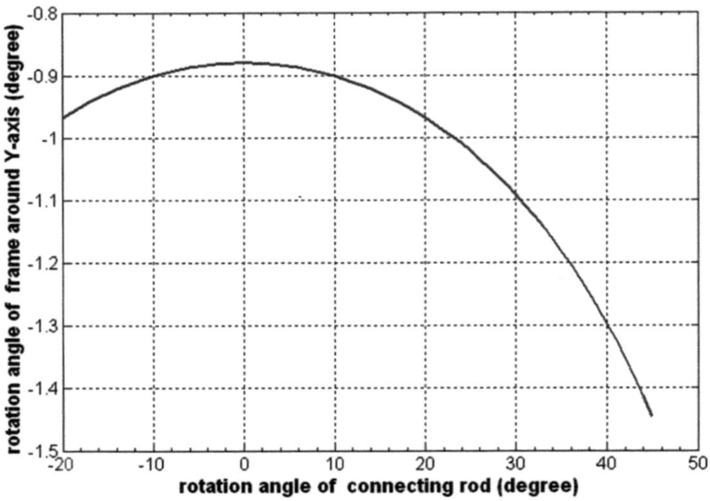


**Fig. 7.19** Rotation angle $\beta_y$ during up and down

It should be noticed that the position of cargoes' center of gravity is uncertain in the implementation process. That is, the parameters $x_G$, $y_G$, and $z_G$ are uncertain, but the ranges of them are determined. As is shown in Fig. 7.16, the forks are at the left side of the $yoz$-plane and the distance between them is designed as 0.2 m. Because the cargoes must be settled on the forks, it must be true that $x_G < 0.2$ m. Furthermore, according to the frame's size, the length of the forks is 1 m and the drivers always carried the cargoes as backward as possible, so that it is reasonable to suppose $x_G$ to be some value within the interval $[-0.7, -0.3]$ with the unit of m. Similarly, the distance between two forks is designed as 0.6 m and the cargoes are always carried in the middle, so $y_G$ is supposed to be a value in interval $[-0.3, 0.3]$ with the unit of m. When the frame is being lifted up and down, the included angle between the long connection rod and horizontal, i.e., $\gamma$, must be changing. As is shown in Fig. 7.16 and in accordance to the operation demand and vehicle layout, $\gamma$ is designed as $-20°$ at the lowest position and $45°$ at the highest position. So, $\gamma$ is a certain value within the interval $[-20°, 45°]$ during work.

The vertical stiffness is first investigated and it is reflected by the frame's micro-rotation angle around the $y$-direction, i.e., $\beta_y$. In order to investigate the variation trend of $\beta_y$ when the frame is being lifted up and down, it is supposed that the position of cargo's center of gravity is at a specific position that $x_G = -0.5$ and $y_G = 0$. Substituting the parameters in Table 7.4, solving the system of equations established in Sect. 7.5.1, and conducting the numerical simulation, we can find how $\beta_y$ changes as $\gamma$ is changing from $-20°$ to $45°$, which is shown in Fig. 7.19.

According to the characteristics of structure and the direction of load, it is obvious that at a specific position, $|\beta_y|$ will increase when $|G_z|$ increases, that is, $|\beta_y|$ has a positive correlation with $|G_z|$. In order to measure and evaluate the vertical

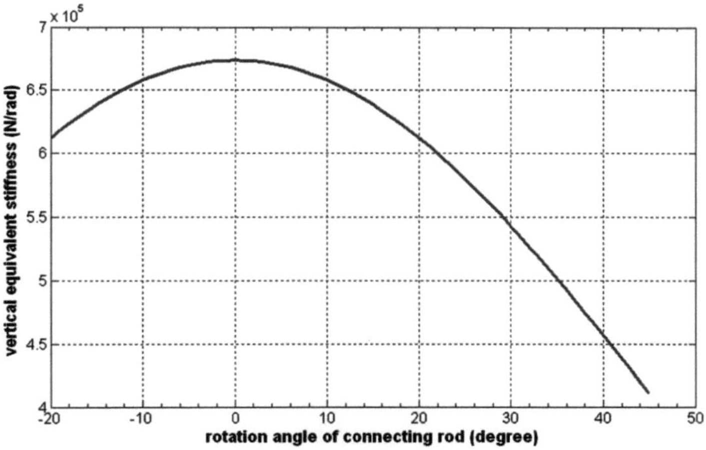


**Fig. 7.20** Equivalent vertical stiffness during lifting up and down

stiffness, a quantity that equals $\left|\frac{\Delta G_z}{\Delta \beta_y}\right|$ is defined as equivalent vertical stiffness and it is denoted by $W_v$, that is,

$$W_v = \left|\frac{\Delta G_z}{\Delta \beta_y}\right| \tag{7.73}$$

where $\Delta G_z$ indicates the micro-increment of $G_z$ and $\Delta \beta_y$ indicates the micro-increment of $\beta_y$.

If the position of center of gravity is $x_G = -0.5$ and $y_G = 0$ while $\gamma$ changes from $\gamma = -20°$ to $\gamma = 45°$, the equivalent vertical stiffness can be obtained by conducting a numerical simulation which is shown in Fig. 7.20. That is, $W_v = 6.74 \times 10^5 \text{N/rad}$ when $\gamma = 0°$ at the horizontal position and $W_v = 4.10 \times 10^5 \text{N/rad}$ when $\gamma = 45°$ at the highest position.

In fact, however, the parameter $x_G$, which is an important influencing factor of the stiffness, is uncertain within the range of $[-0.7, -0.3]$(m) as discussed above. So as to evaluate this mechanism, it is necessary to obtain all the possible results of equivalent vertical stiffness in all conditions. Therefore, the parameter $x_G$ and $\gamma$ are considered as intervals:

$$\begin{cases} [x_G] = [-0.7, -0.3] \\ [\gamma] = [-20°, 45°] \end{cases} \tag{7.74}$$

From the solutions discussed in Sect. 7.5.1 and the numerical simulations, the analytic expression of equivalent vertical stiffness can be expressed approximately as

$$W_v = \left|\frac{\Delta G_z}{\Delta \beta_y}\right| = \frac{8.7158 \times 10^7 \cos\gamma + 3.3474 \times 10^7 \cos^3\gamma}{-430.61 x_G - 36.207} \tag{7.75}$$

Because $x_G$ and $\gamma$ are intervals, Eq. (7.75) should be rewritten as

$$[W_v] = \frac{8.7158 \times 10^7 \cos[\gamma] + 3.3474 \times 10^7 \cos^3[\gamma]}{-430.61[x_G] - 36.207} \tag{7.76}$$

According to Eq. (7.76), it can be found that

$$[W_v] \subset \frac{8.7158 \times 10^7 [\cos\gamma] + 3.3474 \times 10^7 [\cos\gamma]^3}{-430.61[x_G] - 36.207} = \frac{f_1([\cos\gamma])}{f_2([x_G])} \tag{7.77}$$

where

$$\begin{cases} f_1([\cos\gamma]) = 8.7158 \times 10^7 [\cos\gamma] + 3.3474 \times 10^7 [\cos\gamma]^3 \\ f_2([x_G]) = -430.61[x_G] - 36.207 \end{cases}$$

Because $[\gamma] = [-20°, 45°]$, we obtain

$$\begin{cases} [\cos\gamma] = \left[\frac{\sqrt{2}}{2}, 1\right] = [\cos 45°, \cos 0°] \\ [\cos\gamma]^3 = \left[\frac{\sqrt{2}}{2}, 1\right]^3 = \left[\frac{\sqrt{2}}{4}, 1\right] = [\cos^3 45°, \cos^3 0°] \end{cases} \tag{7.78}$$

From Eq. (7.78), it is easy to find that the lower bound of the interval $f_1([\cos\gamma])$ occurs when $\gamma = 45°$ and its upper bound occurs when $\gamma = 0°$. On the other hand, $[x_G] = [-0.7, -0.3]$; we therefore gain

$$f_2([x_G]) = -430.61[x_G] - 36.207 = [92.976, 265.22] \tag{7.79}$$

Obviously neither of $f_1([\cos\gamma])$ and $f_2([x_G])$ contains zero. Therefore, in line with the elementary operations of interval analysis method and Eqs. (7.52), (7.53), and (7.77), we obtain

$$[W_v] \subset \frac{f_1([\cos\gamma])}{f_2([x_G])} = \left[\min\left(\frac{\underline{f_1}}{\underline{f_2}}, \frac{\underline{f_1}}{\bar{f_2}}, \frac{\bar{f_1}}{\underline{f_2}}, \frac{\bar{f_1}}{\bar{f_2}}\right), \quad \max\left(\frac{\underline{f_1}}{\underline{f_2}}, \frac{\underline{f_1}}{\bar{f_2}}, \frac{\bar{f_1}}{\underline{f_2}}, \frac{\bar{f_1}}{\bar{f_2}}\right)\right] \tag{7.80}$$

From Eq. (7.80), we can conduct a numerical simulation to calculate all the possible equivalent stiffness. It is found that $x_G$ plays a more important role in influencing the equivalent vertical stiffness compared with $\gamma$. Consequently, all the possible results of equivalent stiffness are plotted throughout the range of $x_G$ which is shown in Fig. 7.21. The gray area completely envelops the probabilistic result,

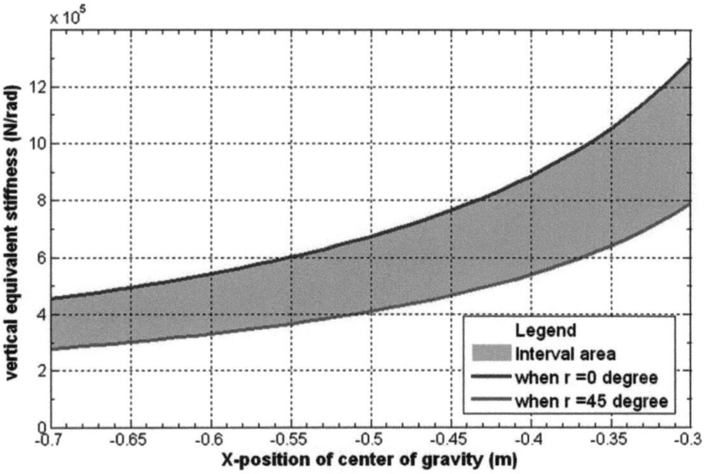


**Fig. 7.21** Interval area of equivalent vertical stiffness

i.e., the interval area of $W_y$. From Fig. 7.21 we can find that the equivalent vertical stiffness is about $5 - 9 \times 10^5$N/rad in most cases. While at the worst situation that $x_G = -0.7$ m and the parameter $\gamma$ changes from 0° to 45° during the frame being lifted, the equivalent vertical stiffness changes from $4.55 \times 10^5$N/rad to $2.77 \times 10^5$N/rad. The rotation angle around $y$-direction $\beta_y$ changes from $-1.28°$ to $-2.10°$. We can see that the stiffness is larger than $2.77 \times 10^5$N/rad at least and the rotation angle changes a little during up and down even at the worst situation. Consequently, this mechanism has very good longitudinal stability that the posture of frame will not change a lot during up and down.

When the center of gravity of cargoes does not lie in the longitudinal symmetrical plane of vehicle, i.e., $y_G \neq 0$, the frame will tilt and rotate with a micro-angle $\alpha_x$ around the $x$-direction. Therefore, the torsional rigidity needs to be investigated and it is reflected by the frame's micro-rotation angle around the $x$-direction, i.e., $\alpha_x$. Similarly to the above analysis, a quantity of $\left|\frac{\Delta G_z}{\Delta \alpha_x}\right|$ is defined as the equivalent torsional stiffness so as to evaluate the torsional rigidity, and it is denoted by $W_t$, that is,

$$W_t = \left|\frac{\Delta G_z}{\Delta \alpha_x}\right| \tag{7.81}$$

where $\Delta G_z$ indicates the micro-increment of $G_z$ and $\Delta \alpha_x$ indicates the micro-increment of $\alpha_x$.

The parameter $y_G$ is uncertain, but its changing range is $[-0.3, 0.3]$(m) as discussed above. Similarly, the parameters $y_G$ and $\gamma$ are considered as intervals, namely,

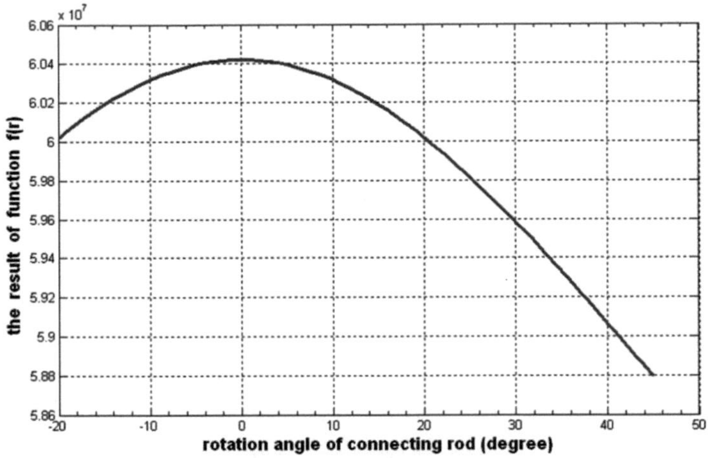


**Fig. 7.22** Result of function $f_3(\gamma)$

$$\begin{cases} [y_G] = [-0.3, 0.3] \\ [\gamma] = [-20°, 45°] \end{cases} \tag{7.82}$$

From Sect. 7.5.1, the analytic expression of equivalent torsional stiffness can be solved. For the sake of simplicity, it is expressed as

$$W_t = \left| \frac{\Delta G_z}{\Delta \alpha_x} \right| = \left| \frac{f_3(\gamma)}{y_G} \right| \tag{7.83}$$

where $f_3(\gamma)$ is a function that depends on the variable $\gamma$ and it is too complicated to be expanded.

Because $x_G$ and $\gamma$ are intervals, according to Eqs. (7.52) and (7.83), we gain that

$$[W_t] = \left| \frac{[f_3(\gamma)]}{[y_G]} \right| \tag{7.84}$$

From numerical simulations, the trend of $f_3(\gamma)$ is shown in Fig. 7.22. According to the calculation and Fig. 7.22, we can find that $f_3(\gamma)$ reaches its lower bound when $\gamma = 45°$ and reaches its upper bound when $\gamma = 0°$. $[f_3(\gamma)]$ does not contain zero.

On the other hand, $[y_G] = [-0.3, 0.3]$ contains zero. When $y_G = 0$, the torsional equivalent stiffness $W_t$ will be infinite. $[y_G]$ is hence considered as the union of two intervals and gets rid of zero, that is,

$$[y_G] = [y_{G1}] \cup [y_{G2}] \tag{7.85}$$

where

$$\begin{cases} [y_{G1}] = [-0.3,\, 0) \\ [y_{G2}] = (0,\, 0.3] \end{cases}$$

Therefore, we find that

$$[W_t] = \left|\frac{[f_3(\gamma)]}{[y_G]}\right| = \left|\frac{[f_3(\gamma)]}{[y_{G1}]}\right| \cup \left|\frac{[f_3(\gamma)]}{[y_{G2}]}\right| \tag{7.86}$$

According to the elementary operations of interval analysis method and Eqs. (7.52), (7.53), and (7.86), we have

$$\begin{aligned} [W_t] &= \left|\frac{[f_3(\gamma)]}{[y_{G1}]}\right| \cup \left|\frac{[f_3(\gamma)]}{[y_{G2}]}\right| \\ &= \left[\min\left(\frac{\underline{f}_3}{\underline{y}_{G1}}, \frac{\underline{f}_3}{\overline{y}_{G1}}, \frac{\overline{f}_3}{\underline{y}_{G1}}, \frac{\overline{f}_3}{\overline{y}_{G1}}\right),\quad \max\left(\frac{\underline{f}_3}{\underline{y}_{G1}}, \frac{\underline{f}_3}{\overline{y}_{G1}}, \frac{\overline{f}_3}{\underline{y}_{G1}}, \frac{\overline{f}_3}{\overline{y}_{G1}}\right)\right] \\ &\cup \left[\min\left(\frac{\underline{f}_3}{\underline{y}_{G2}}, \frac{\underline{f}_3}{\overline{y}_{G2}}, \frac{\overline{f}_3}{\underline{y}_{G2}}, \frac{\overline{f}_3}{\overline{y}_{G2}}\right),\quad \max\left(\frac{\underline{f}_3}{\underline{y}_{G2}}, \frac{\underline{f}_3}{\overline{y}_{G2}}, \frac{\overline{f}_3}{\underline{y}_{G2}}, \frac{\overline{f}_3}{\overline{y}_{G2}}\right)\right] \end{aligned} \tag{7.87}$$

From Eq. (7.87), we can conduct numerical simulation and calculate all the possible equivalent torsional stiffness. It is obvious that $y_G$ plays a key role in influencing the equivalent stiffness compared with $\gamma$. Consequently, all the possible results of $W_t$ are plotted throughout the range of $y_G$ and it is shown in Fig. 7.23. The gray area completely envelops the probabilistic result, i.e., the interval area of $W_t$. From Fig. 7.23 we can find that the equivalent torsional stiffness is larger than $2 \times 10^8$N/rad in most cases and tends to infinity when $y_G$ approaches to zero which means that there is no partial load on the folk. Moreover, the equivalent stiffness $W_t$ changes little during work as $\gamma$ changes within its range. Even at the worst situation that $y_G = -0.3$ m, the rotation angle around the $x$-direction $\alpha_x$ changes from 0.0028° to 0.0029°. We can see that the stiffness is very high and the rotation angle is tiny and changes little during up and down. Consequently, this mechanism has excellent transversal stability that the frame tilts little laterally.

The equivalent vertical and torsional stiffness has been simulated and calculated with determined structure parameters. In some cases, however, the structure parameters might be uncertain due to temperature change, material shrinkage, manufacturing errors, and so on. Self-internal force could be generated and the stiffness of the structure could be affected.

Assume that the parameter $l_0$, which stands for the distance between the revolute joint in support block to the revolute joint in frame at horizontal position, is uncertain but some value within the interval $[1.28,\, 1.32]$ with the unit of m due to manufacturing errors. Because the posture of the frame has to be adjusted to be

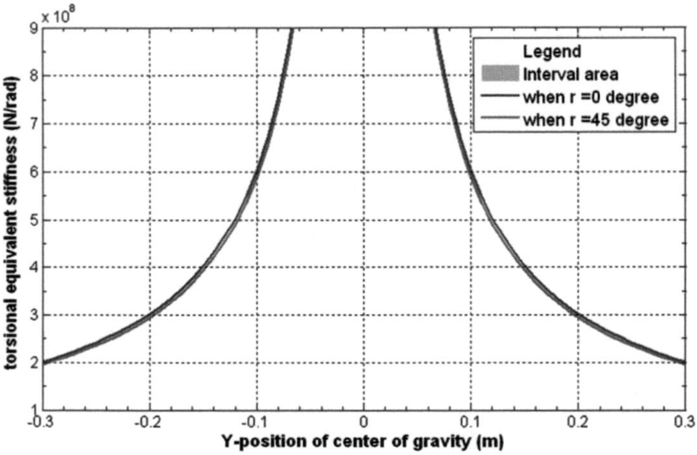


**Fig. 7.23** Interval area of equivalent torsional stiffness

vertical to the ground before used, it is also assumed that the parameters $l_0$ for all the four connection rods are the same. Therefore, the parameters $l_0$ and $l$ in Eqs. (7.66) and (7.70) will be no longer determined numerical values but intervals. They affect the solution of the system of equations, while the affections can also be interpreted as internal forces.

Similarly as above, the system of equations with interval parameters can be solved by interval analysis method. The equivalent vertical stiffness can be calculated by conducting a numerical simulation. All the possible results are plotted throughout the range of $x_G$ and it is shown in Fig. 7.24. The gray area completely envelops the probabilistic result, i.e., the interval area of equivalent vertical stiffness.

Comparing Fig. 7.24 with Fig. 7.21, we can find that the manufacturing errors enlarge the interval area of the result. The stiffness will be improved with the decreasing of parameter $l_0$, but the vertical heights and positions of center of gravity are still the most important factors in affecting the equivalent stiffness. We can also simulate how the stiffness is affected by other factors such as temperature change, material shrinkage, and so on. Some similar results can be obtained, but they are omitted for the sake of length.

From Fig. 7.19, it is reasonable to draw a conclusion that the rotation angles $\beta_y$ is far less than 5°. $\alpha_x$ and $\mu_z$ can be similarly gained and which is omitted for the sake of length. So the assumption that $\alpha_x$, $\beta_y$, and $\mu_z$ are micro-angles holds and the dynamic analysis model is reasonable. The numerical simulations have shown that the mechanism with overconstraints can provide very good vertical and torsional stiffness for lifting. It has high stability in both longitudinal and transversal directions because the redundant constraints of structure restrain the frame to move with only one degree of freedom. The indeterminate structure provides high

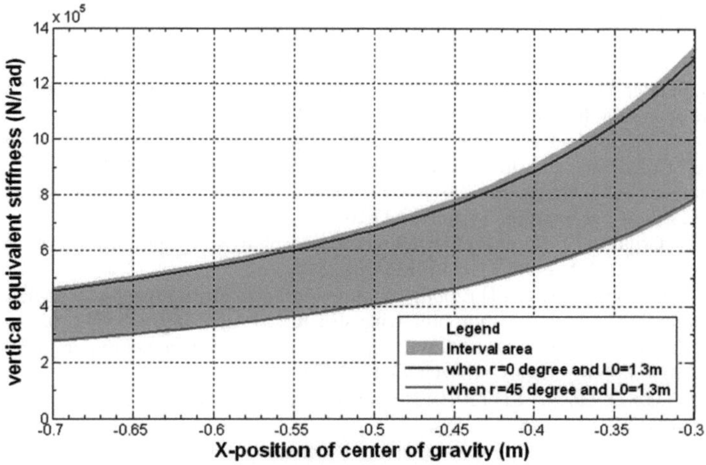


**Fig. 7.24** Interval area of the equivalent vertical stiffness with manufacturing errors

resistance to the rotation and lateral tilt of the frame. The lift mechanism proposed in this chapter also has a wide application in many other fields such as suspensions in a vehicle.

## 7.6 Conclusions

This chapter discusses the kinetostatics of double planar linkages by focusing on the foldable frame for a morphing wing and a lift mechanism with overconstraints through interval analysis method. The overconstraints of these mechanisms not only retain the advantages of high structural stiffness and strength as a structure but also allow the mechanisms to move smoothly. Because of the overconstraints, these mechanisms are indeterminate in structure; therefore, they are often considered as elastic bodies and the compatibility equations should be utilized in kinetostatic analysis. The elastic deformation of the links and displacement of a structural frame are then discussed. After that, a system of equations can be established. The dynamic responses of the mechanism could be described by solving the associated equations with determined structural and dynamic parameters. Equivalent transversal, vertical, and torsional stiffness could be discussed with intervals of the primary structure parameters. Numerical simulations were conducted to quantify the tendency and possible interval area of the equivalent stiffness. This method could also be used in the similar indeterminate problems such as robotics, deployable structures, or other mechanisms with overconstraints.

# References


1. Bae JS, Seigler TM, Inman DJ (2005) Aerodynamic and static aeroelastic characteristics of a variable-span morphing wing. J Aircr 42(2):528–534
2. Gamboa P, Vale J, Lau FJP, Suleman A (2009) Optimization of a morphing wing based on coupled aerodynamic and structural constraints. AIAA J 47(9):2087–2104
3. Wickenheiser AM, Garcia E (2007) Aerodynamic modeling of morphing wings using an extended lifting-line analysis. J Aircr 44(1):10–17
4. Popov AV, Grigorie LT, Botez R (2010) Real time morphing wing optimization validation using wind-tunnel tests. J Aircr 47(4):1346–1355
5. Inoyama D, Sanders BP, Joo JJ (2008) Topology optimization approach for the determination of the multiple-configuration morphing wing structure. J Aircr 45(6):1853–1863
6. Bae JS, Seigler TM, Inman DJ (2005) Computation of actuation power requirements for smart wings with morphing airfoils. AIAA J 43(12):2481–2487
7. Attar PJ, Tang D, Dowell EH (2010) Nonlinear aeroelastic study for folding wing structures. AIAA J 48(10):2187–2195
8. Pinero EP (1962) Expandable space framing. Progress Archit 43(6):154–155
9. Gantes C, Giakoumakis A, Vousvounis P (1997) Symbolic manipulation as a tool for design of deployable domes. Comput Struct 64(1–4):865–878
10. Jaksch S, Sedlak V (2011) A foldable umbrella structure-developments and experiences. Int J Space Struct 26(1):1–18
11. Chen Y, You Z (2009) Two-fold symmetrical 6R foldable frame and its bifurcations. Int J Solids Struct 46(25–26):4504–4514
12. Gutiérrez EM, Valcárcel JP (2004) Foldable systems based on bundle modules with quadrangular base. Int J Space Struct 19(3):155–165
13. Ragavan V, Amde AM (2001) An algorithm for nonlinear stability analysis of an expandable self-erecting structure. Comput Struct 79(27–30):2587–2893
14. De Temmerman IN, Mollaert M, Van Mele T, De Laet L (2007) Design and analysis of a foldable mobile shelter system. Int J Space Struct 22(3):161–168
15. Warnaar DB, Chew M (1995) Kinematic synthesis of deployable-foldable truss structures using graph theory, part 1: graph generation. ASME Trans J Mech Des 117(1):112–117
16. Tang D, Dowell EH (2008) Theoretical and experimental aeroelastic study for folding wing structures. J Aircr 45(4):1136–1147
17. Zhao JS, Chu FL, Feng ZJ (2009) The mechanism theory and application of deployable structures based on SLE. Mech Mach Theory 44(2):324–335
18. Wickenheiser AM, Garcia E (2011) Extended nonlinear lifting-line method for aerodynamic modeling of reconfigurable aircraft. J Aircr 48(5):1812–1817
19. Crossley WA, Skillen MD, Frommer JB, Roth BD (2011) Morphing aircraft sizing using design optimization. J Aircr 48(2):612–622
20. Ceccarelli M, Carbone G (2002) A stiffness analysis for CaPaMan (Cassino Parallel Manipulator). Mech Mach Theory 37(5):427–437
21. Szuminski P (2007) Determination of the stiffness of rolling kinematic pairs of manipulators. Mech Mach Theory 42(9):1082–1102
22. Pashkevich A, Klimchik A, Chablat D (2011) Enhanced stiffness modeling of manipulators with passive joints. Mech Mach Theory 46(5):662–677
23. Pashkevich A, Chablat D, Wenger P (2009) Stiffness analysis of over constrained parallel manipulators. Mech Mach Theory 44(5):966–982
24. Yue Y, Gao F, Zhao X, Ge QJ (2010) Relationship among input-force, payload, stiffness and displacement of a3-DOF perpendicular parallel micro-manipulator. Mech Mach Theory 45(5):756–771
25. Li Y, Xu Q (2008) Stiffness analysis for a 3-PUU parallel kinematic machine. Mech Mach Theory 43(2):186–200

26. Carbone G, Lim H, Takanishi A, Ceccarelli M (2006) Stiffness analysis of biped humanoid robot WABIAN-RIV. Mech Mach Theory 41(1):17–40
27. Jung HK, CraneIII CD, Roberts RG (2008) Stiffness mapping of compliant parallel mechanisms in a serial arrangements. Mech Mach Theory 43(3):271–284
28. Pigoski T, Groffis M, Duffy J (1998) Stiffness mappings employing different frames of reference. Mech Mach Theory 33(6):825–838
29. Hyun D, Yang HS, Park J, Shim Y (2010) Variable stiffness mechanism for human-friendly robots. Mech Mach Theory 45(6):880–897
30. Majou F, Gosselin C, Wenger P, Chablat D (2007) Parametric stiffness analysis of the Orthoglide. Mech Mach Theory 42(3):296–311
31. Moore RE (1996) Interval analysis. Prentice-Hall, New York
32. Moore RE (1979) Methods and applications of interval analysis. Siam, Philadelphia
33. Han C, Ruy K, Lee S (2007) Experimental study of a telescopic wing inside a channel. J Aircr 44(3):1027–1030
34. Samuel JB, Pines D (2007) Design and testing of a pneumatic telescopic wing for unmanned aerial vehicles. J Aircr 44(4):1088–1097
35. Wang FX, Han JR (1987) Theory of wing. Renmin Transportation Publishing House, Beijing, pp 101–102
36. Bedford A, Kenneth M (2000) Mechanics of materials. Prentice Hall, Upper Saddle River
37. Zhao JS, Chu FL, Feng ZJ (2008) Synthesis of rectilinear motion generating spatial mechanism with application to automotive suspension. ASME J Mech Des 130(6):065001
38. Wu J, Zhang Y, Chen L, Chen P, Qin G (2011) Uncertain analysis of vehicle handling using interval method. Int J Veh Des 56(1–4):81–105
39. Jaulin L (2001) Applied interval analysis: with examples in parameter and state estimation, robust control and robotics. Springer, New York
40. Sun L, Huang WM, Ding Z, Zhao Y, Wang CC, Purnawali H, Tang C (2012) Stimulus-responsive shape memory materials: a review. Mater Des 33(1):577–640
41. Horberry T, Larsson TJ, Johnston IA, Lambert J (2004) Forklift safety, traffic engineering and intelligent transport systems: a case study. Appl Ergon 35(6):575–581
42. Wang J-Y, Zhao J-S, Chu FL, Feng ZJ (2010) Innovative design of the lifting mechanisms for forklift trucks. Mech Mach Theory 45(12):1892–1896
43. Zhao J-S, Feng Z-J, Chu F (2009) Analytical theory of degrees of freedom for robot mechanisms. Science Press, Beijing

# Chapter 8
# Structural Dynamics of Planar Linkages

In this chapter, we propose a structural dynamics method for foldable linkages based on transfer matrix. The foldable stair and deployable wing are all typical planar linkages which are made up of a number of identical units. For each unit, every link is supposed to be an Euler-Bernoulli beam. Therefore, the dynamics of each segment beam between every two adjacent revolute joints can be precisely expressed by the transfer matrix of the segment with the variables of boundary conditions of the joints. In this way, the structural dynamics of the whole structure can be built using the least number of variables compared with the traditional methods. In addition, this algorithm avoids the problem of the traditional transfer-matrix method that the number of variables greatly increases when there are a huge number of cross joints within a structure.

## 8.1 Structural Dynamics Analysis of Deployable Structure

As stated in Chap. 3, the deployable structure is an important mechanism in engineering applications [1–4]. In fact, many deployable structural systems can often be found in some biological organisms in nature. The statics [5, 6], kinematics [7, 8], dynamics [9–11], and the force accommodation control operations [12] of the deployable structures are widely investigated.

To monitor the structural health, vibration-based methods are the most conventional approaches. Numerical evaluation of vibration-based methods was investigated for the damage assessment of cable-stayed bridges [13]. The performance of the method for real-life situations is highly dependent on the accuracy of the important modal features of the structure extracted from output-only vibration measurements [14]. Therefore, the structural dynamics of the foldable stair should be analyzed to reveal its vibration characteristics before real applications. Multi-input, single-output semi-active fuzzy controller for vibration control of seismically excited small-scale buildings is developed through integration of a set of model-based

J. Zhao et al., *Design of Special Planar Linkages*, Springer Tracts in Mechanical Engineering, DOI 10.1007/978-3-642-38448-6_8,

fuzzy controllers that are formulated in terms of linear matrix inequalities such that the global asymptotical stability is guaranteed and the performance on transient responses is also satisfied [15].

Homogenized beam model was proposed to reproduce the dynamic behavior of unconstrained layer damping beams with a thick viscoelastic layer [16]. A meta-heuristic algorithm [17] was proposed to solve the multi-objective combinatorial optimization problems. Four beam collapse modes were discussed for the progressive collapse analysis associated with debris loading generated when disengaged structural components fall into lower parts of the structure [18]. A family of explicit methods were investigated [19] to obtain unconditional stability and second-order accuracy. An integral type quadratic function [20] was proposed as the performance index to derive the optimal control policy for the earthquake-excited structures. Hybrid computational strategy of combining generic algorithm with a compatible local search operator [21] was proposed and the hybrid time-domain method was investigated [22] to obtain a higher accurate solution of structural dynamic problems. To effectively solve the structural dynamics, this chapter proposes a theoretically accurate method based on transfer matrix of dynamics. It takes advantage of the fact that the stair is made up of a number of identical units of SLE and establishes the structural dynamics through assembling the unit dynamics model based on Euler-Bernoulli beam assumption. And therefore, this method is accurate in theory.

A most common characteristic of the deployable structure is that it is made up of identical units, which is composed of pairs of bars connected at joint that allows compact and deployed configurations [23–25]. Kang and Miranda [26] proposed a numerical method to simulate the kinematic and dynamic activities of a crane. In some cases, the structural responses are in dependency of uncertain load processes and environmental influences [27]. Passive tuned mass damper for reducing the dynamic responses of structures under near-fault earthquake excitations was proposed to simulate the near-fault ground motion [28]. Polynomial chaos representation for identification of mechanical characteristics of instrumented structures was investigated to monitor the structures by providing more realistic models for mechanical behavior or input values for their parameters [29]. Among the deployable mechanisms, the large-span foldable structures based on scissorlike elements [3, 23] are widely used in engineering. A scissorlike element (SLE) is one planar II-Assur group [30] in mechanism theory. From a geometric viewpoint, the SLE is a pair of plane links that are connected at the intermediate point by a revolute joint which allows them to rotate freely about an axis perpendicular to their common plane but restrains all other degrees of freedom while their end points are hinged to the end points of other SLEs [31]. When one or more such elements are added to a mechanism, the number of actuators needed to control the mechanism will not change [30].

The optimal design of foldable structures based on structural steel analysis [32] ant colony optimization algorithm [33] is investigated and utilized in the foldable structures. Optimum design [34] in structures provides new and high-performance computing methods [35] for foldable structures. Hachem et al. [36]

surveyed biological systems in terms of their morphology, kinematics, and structural characteristics and discussed the similarities between biological systems and the existing deployable structures.

Applications of deployable structures in space structures and nonconventional structures have extensively propelled the research on the mechanics and design of large flexible structures [37]. The deployable structures were widely researched in space engineering with the increasing needs of complex space missions in the 1970s [38, 39]. Recently, deployable structures are often used in the situations requiring minimal damage to structural components during repeated deployment and fold [40]. In civil engineering, foldable stair seldom appears while the steel-reinforced concrete stairs are most popular in the current apartments. A structural concrete analysis using rigid-body-spring networks [41] and the numerical analysis of debonding mechanisms in FRP-strengthened RC beams [42] are widely probed for the applications in the concrete structures. With the emergency of high-stiffness, high-strength, and low-weight alloy material in recent years, the foldable staircases are becoming feasible in engineering. However, the structural dynamics of foldable stair should be different from those of the fixed structural or steel-reinforced concrete stairs. Consequently, this chapter focuses on the structural dynamics of a foldable stair and then discusses the dynamics algorithm in other complex foldable structures such as deployable wing frame which might be used in morphing wing.

## 8.2 Mechanism of the Foldable Staircase

As stated above, the primary unit of the staircase adopts the planar parallelogram four-bar mechanism, $C_0A_1C_1B_0$, shown in Fig. 8.1. So the beam, $A_1C_1B_1$, not only has the moving flexibility of a mechanism but also has the high stiffness and high strength of statically indeterminate structures when considering that every side of the stair has one such parallelogram four-bar mechanism connected by five axles

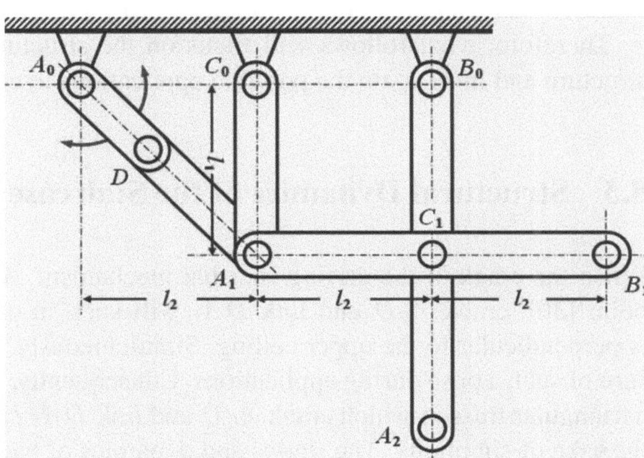


**Fig. 8.1** Driving mechanism and the first staircase of the foldable stair

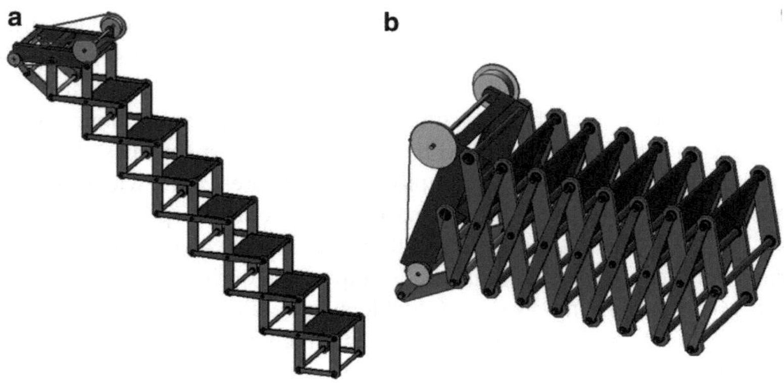

**Fig. 8.2** The deployed and folded stair (**a**) Deployed stair (**b**) Folded stair

which are represented by $A_1$, $C_1$, $B_1$, $B_0$, and $C_0$ in Fig. 8.1. Another four-bar mechanism, $A_0DA_1C_0$, is a driving crank-rocker mechanism which controls the deploying or folding of the whole stair.

The beams $A_1C_1B_1$ and $B_0C_1A_2$ are jointed at $C_1$ form a scissorlike element (SLE), $A_1B_1B_0A_2$. From the geometry point of view, SLE $A_1B_1B_0A_2$ shown in Fig. 8.1 is a pair of planar links that are connected by one pin at their common middle point, $C$, through a revolute joint which allows them to rotate freely about an axis perpendicular to their common plane but restrains all other degrees of freedom while their end points allow to be hinged to the end points of other SLEs. When one or more such elements are added to the mechanism shown in Fig. 8.1, the number of actuators needed to control the mechanism will not change. Figure 8.2 illustrates the deployable stair that consists of seven complete SLEs. This stair can be deployed for use and folded to save the occupied space after utilization. The stair forms a statically indeterminate structure after deploying because the ground and the roof as well will provide real supports. So both the statics and the dynamics of the foldable stair are different from those of the traditional stairs. The structural dynamics of the foldable stair cannot be directly analyzed with the general methods, either.

Therefore, what follows will focus on the structural dynamics of the foldable structure and investigate the possible applications in engineering.

## 8.3 Structural Dynamics of the Staircase

When the crank of the driving four-bar mechanism, $A_0DA_1C_0$, rotates to the dead point [30], crank $A_0D$ and link $DA_1$ will keep in one line and the rocker $C_0A_1$ is perpendicular to the upper ceiling. Simultaneously, a lock mechanism will make sure of such a pose during applications. Consequently, the driving mechanism forms a triangular truss in which crank $A_0D$ and link $DA_1$ can be assumed as one link for the sake of simplicity. The statics and dynamics of the staircase and the whole stair

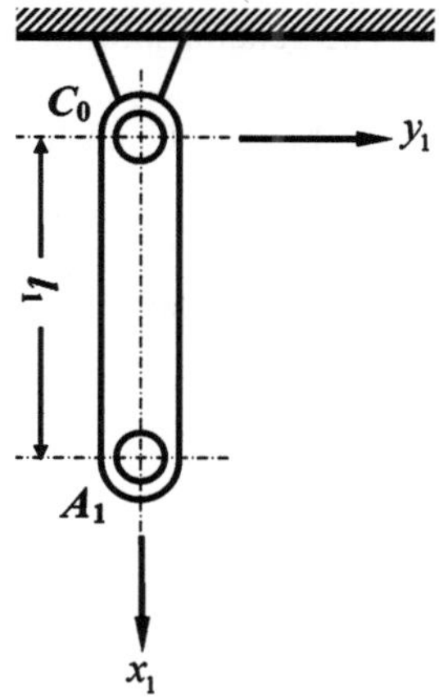


**Fig. 8.3** Primary coordinate system for the vertical rocker

should be comprehensively discussed. The general computing mechanics developed in the 1990s [43, 44] can be directly used to model the structural characteristics of the stair. What follows will first build the structural dynamics of this triangular truss.

### *8.3.1 Structural Dynamics of the Driving Mechanism*

Because the components of the foldable stair are all thin and long straight links, we can suppose that the links are all Euler-Bernoulli beams. Suppose that the $x_1$-axis is along the axis of the vertical rocker, $C_0A_1$, and $y_1$-axis is perpendicular rightward to the link. The primary coordinate frame for $C_0A_1$ is shown in Fig. 8.3.

The differential equation of bending vibration of an Euler-Bernoulli beam pivoted at both ends [45, 46] in Fig. 8.3 is

$$\frac{\partial^2}{\partial x^2}\left(EI_z\frac{\partial^2 y(x,t)}{\partial x^2}\right) + \rho A\frac{\partial^2 y(x,t)}{\partial t^2} = 0 \tag{8.1}$$

where $E$ denotes the Young's elastic modulus, $I_z$ denotes the inertial moment of the section about the $z$-direction, $\rho$ is the linear density of mass, and $A$ is the sectional area of the beam. Suppose

$$y(x,t) = Y(x)\sin(\omega t + \varphi) \tag{8.2}$$

where $Y(x)$ represents the shape function of bending vibration at $x$, $\omega$ represents the frequency of a sine vibration, and $\varphi$ represents the phase angle.

Assume that the mass is uniformly distributed within the whole link, Eq. (8.1) can be transformed into

$$\frac{d^4Y}{dx^4} - \frac{\rho A\omega^2}{EI_z}Y = 0 \tag{8.3}$$

The general solution of Eq. (8.3) is

$$Y(x) = C_1 \cosh(\lambda x) + C_2 \sinh(\lambda x) + C_3 \cos(\lambda x) + C_4 \sin(\lambda x) \tag{8.4}$$

where $\lambda = \sqrt[4]{\frac{\rho A \omega^2}{EI_z}}$, $C_i (i = 1, 2, 3, 4)$ can be determined by the initial boundary conditions of the beam. Because the Euler-Bernoulli beam satisfies

$$\begin{cases} \Theta_z = \dfrac{\mathrm{d}Y}{\mathrm{d}x} \\ M_z = EI \dfrac{\mathrm{d}\Theta_z}{\mathrm{d}x} \\ Q_y = \dfrac{\mathrm{d}M_z}{\mathrm{d}x} \end{cases} \tag{8.5}$$

Substituting Eq. (8.4) into (8.5) and associating Eq. (8.4) gains

$$\boldsymbol{\xi}_{4\times1}(x) = \mathbf{B}_{4\times4}(x) \begin{bmatrix} C_1 \\ C_2 \\ C_3 \\ C_4 \end{bmatrix} \tag{8.6}$$

where

$$\boldsymbol{\xi}_{4\times1}(x) = \begin{bmatrix} Y(x) & \Theta_z(x) & Q_y(x) & M_z(x) \end{bmatrix}^{\mathrm{T}} \tag{8.7}$$

$$\mathbf{B}_{4\times4}(x) = \begin{bmatrix} \cosh(\lambda x) & \sinh(\lambda x) & \cos(\lambda x) & \sin(\lambda x) \\ \lambda \sinh(\lambda x) & \lambda \cosh(\lambda x) & -\lambda \sin(\lambda x) & \lambda \cos(\lambda x) \\ EI\lambda^3 \sinh(\lambda x) & EI\lambda^3 \cosh(\lambda x) & EI\lambda^3 \sin(\lambda x) & -EI\lambda^3 \cos(\lambda x) \\ EI\lambda^2 \cosh(\lambda x) & EI\lambda^2 \sinh(\lambda x) & -EI\lambda^2 \cos(\lambda x) & -EI\lambda^2 \sin(\lambda x) \end{bmatrix} \tag{8.8}$$

When $x = 0$, Eq. (8.8) can be simplified as

$$\mathbf{B}_{4\times4}(0) = \begin{bmatrix} 1 & 0 & 1 & 0 \\ 0 & \lambda & 0 & \lambda \\ 0 & EI\lambda^3 & 0 & -EI\lambda^3 \\ EI\lambda^2 & 0 & -EI\lambda^2 & 0 \end{bmatrix} \tag{8.9}$$

Substituting Eq. (8.9) into (8.6), we gain

$$\begin{bmatrix} C_1 \\ C_2 \\ C_3 \\ C_4 \end{bmatrix} = \mathbf{B}_{4\times4}^{-1}(0)\boldsymbol{\xi}_{4\times1}(0) \tag{8.10}$$

where

$$\boldsymbol{\xi}_{(4\times1)}(0) = \left[\, Y(0)\ \Theta_z(0)\ Q_y(0)\ M_z(0) \,\right]^{\mathrm{T}} \tag{8.11}$$

$$\mathbf{B}_{4\times4}^{-1}(0) = \begin{bmatrix} \frac{1}{2} & 0 & 0 & \frac{1}{2EI\lambda^2} \\ 0 & \frac{1}{2\lambda} & \frac{1}{2EI\lambda^3} & 0 \\ \frac{1}{2} & 0 & 0 & -\frac{1}{2EI\lambda^2} \\ 0 & \frac{1}{2\lambda} & -\frac{1}{2EI\lambda^3} & 0 \end{bmatrix} \tag{8.12}$$

Substituting Eq. (8.10) into (8.6) generates

$$\boldsymbol{\xi}_{4\times1}(x) = \mathbf{H}_{4\times4}(x)\, \boldsymbol{\xi}_{4\times1}(0) \tag{8.13}$$

where

$$\mathbf{H}_{4\times4}(x) = \mathbf{B}_{4\times4}(x)\mathbf{B}_{4\times4}^{-1}(0)$$

$$= \begin{bmatrix} \frac{\cosh(\lambda x) + \cos(\lambda x)}{2} & \frac{\sinh(\lambda x) + \sin(\lambda x)}{2\lambda} & \frac{\sinh(\lambda x) - \sin(\lambda x)}{2EI\lambda^3} & \frac{\cosh(\lambda x) - \cos(\lambda x)}{2EI\lambda^2} \\ \frac{\lambda[\sinh(\lambda x) - \sin(\lambda x)]}{2} & \frac{\cosh(\lambda x) + \cos(\lambda x)}{2} & \frac{\cosh(\lambda x) - \cos(\lambda x)}{2EI\lambda^2} & \frac{\sinh(\lambda x) + \sin(\lambda x)}{2EI\lambda} \\ \frac{EI\lambda^3[\sinh(\lambda x) + \sin(\lambda x)]}{2} & \frac{EI\lambda^2[\cosh(\lambda x) - \cos(\lambda x)]}{2} & \frac{\cosh(\lambda x) + \cos(\lambda x)}{2} & \frac{\lambda[\sinh(\lambda x) - \sin(\lambda x)]}{2} \\ \frac{EI\lambda^2[\cosh(\lambda x) - \cos(\lambda x)]}{2} & \frac{EI\lambda[\sinh(\lambda x) - \sin(\lambda x)]}{2} & \frac{\sinh(\lambda x) + \sin(\lambda x)}{2\lambda} & \frac{\cosh(\lambda x) + \cos(\lambda x)}{2} \end{bmatrix} \tag{8.14}$$

Suppose

$$\begin{cases} R(x) = \dfrac{\cosh x + \cos x}{2} \\ S(x) = \dfrac{\sinh x + \sin x}{2} \\ U(x) = \dfrac{\cosh x - \cos x}{2} \\ V(x) = \dfrac{\sinh x - \sin x}{2} \end{cases} \tag{8.15}$$

Substituting Eq. (8.15) into (8.14) yields

$$\mathbf{H}_{4\times4}(x) = \begin{bmatrix} R(\lambda x) & \dfrac{S(\lambda x)}{\lambda} & \dfrac{V(\lambda x)}{EI\lambda^3} & \dfrac{U(\lambda x)}{EI\lambda^2} \\ \lambda V(\lambda x) & R(\lambda x) & \dfrac{U(\lambda x)}{EI\lambda^2} & \dfrac{S(\lambda x)}{EI\lambda} \\ EI\lambda^3 S(\lambda x) & EI\lambda^2 U(\lambda x) & R(\lambda x) & \lambda V(\lambda x) \\ EI\lambda^2 U(\lambda x) & EI\lambda V(\lambda x) & \dfrac{S(\lambda x)}{\lambda} & R(\lambda x) \end{bmatrix} \tag{8.16}$$

The differential equation of longitudinal vibration of the link shown in Fig. 8.3 is

$$EA\frac{\partial^2 Xc}{\partial x^2} - \rho A\frac{\partial^2 X(x,t)}{\partial t^2} = 0 \tag{8.17}$$

Suppose

$$X(x,t) = X(x)\sin(\omega t + \varphi) \tag{8.18}$$

where $X(x)$ denotes the longitudinal vibration shape function of $x$.

Substituting Eq. (8.18) into (8.17), we obtain the general solution of the shape function of the longitudinal vibration mode of the beam

$$X = D_1 \cos(kx) + D_2 \sin(kx) \tag{8.19}$$

where $k = \omega\sqrt{\frac{\rho}{E}}$ and $D_i\,(i = 1, 2)$ can be determined from the initial conditions.

The axial inner force of the link is

$$Q_x = -EA\frac{\partial X}{\partial x} = D_1 EAk\sin(kx) - D_2 EAk\cos(kx) \tag{8.20}$$

Associating Eqs. (8.19) and (8.20), we obtain

$$\boldsymbol{\xi}_{2\times1}(x) = \mathbf{B}_{2\times2}(x)\begin{bmatrix} D_1 \\ D_2 \end{bmatrix} \tag{8.21}$$

where $\xi_{2\times1}(x) = \begin{bmatrix} X \\ Q_x \end{bmatrix}$, $\mathbf{B}_{2\times2}(x) = \begin{bmatrix} \cos(kx) & \sin(kx) \\ EAk\sin(kx) & -EAk\cos(kx) \end{bmatrix}$. Consequently, we gain that

$$\begin{bmatrix} D_1 \\ D_2 \end{bmatrix} = \mathbf{B}_{2\times2}^{-1}(0)\boldsymbol{\xi}_{2\times1}(0) \tag{8.22}$$

where $\mathbf{B}_{2\times2}(0) = \begin{bmatrix} 1 & 0 \\ 0 & -EAk \end{bmatrix}$ and $\boldsymbol{\xi}_{2\times1}(0) = \begin{bmatrix} X(0) \\ Q_x(0) \end{bmatrix}$.

Substituting Eq. (8.22) into (8.21) results in

$$\boldsymbol{\xi}_{2\times1}(x) = \mathbf{H}_{2\times2}(x)\boldsymbol{\xi}_{2\times1}(0) \tag{8.23}$$

where

$$\mathbf{H}_{2\times2}(x) = \mathbf{B}_{2\times2}(x)\mathbf{B}_{2\times2}^{-1}(0) = \begin{bmatrix} \cos(kx) & -\dfrac{\sin(kx)}{EAk} \\ EAk\sin(kx) & \cos(kx) \end{bmatrix} \tag{8.24}$$

When $x = l_1$, $\mathbf{H}_{2\times2}(x)$ denotes the transfer matrix from the state vector of $x = 0$ to the state vector of $x = l_1$. Within the $xoy$-plane, the link has both bending and longitudinal vibrations. Suppose that the state vector of the $i$th section is denoted by $(\boldsymbol{\xi}_{6\times1})_i = \begin{bmatrix} X & Y & \Theta_z & Q_x & Q_y & M_z \end{bmatrix}_i^{\mathrm{T}}$, from Eqs. (8.16) and (8.24), the $(i+1)$th state vector of the link can be expressed as

$$(\boldsymbol{\xi}_{6\times1})_{i+1} = \mathbf{H}_{6\times6}(x)(\boldsymbol{\xi}_{6\times1})_i \tag{8.25}$$

where

$$\mathbf{H}_{6\times6}(x) = \begin{bmatrix} \cos(kx) & 0 & 0 & -\dfrac{\sin(kx)}{EAk} & 0 & 0 \\ 0 & R(\lambda x) & \dfrac{S(\lambda x)}{\lambda} & 0 & \dfrac{V(\lambda x)}{EI\lambda^3} & \dfrac{U(\lambda x)}{EI\lambda^2} \\ 0 & \lambda V(\lambda x) & R(\lambda x) & 0 & \dfrac{U(\lambda x)}{EI\lambda^2} & \dfrac{S(\lambda x)}{EI\lambda} \\ EAk\sin(kx) & 0 & 0 & \cos(kx) & 0 & 0 \\ 0 & EI\lambda^3 S(\lambda x) & EI\lambda^2 U(\lambda x) & 0 & R(\lambda x) & \lambda V(\lambda x) \\ 0 & EI\lambda^2 U(\lambda x) & EI\lambda V(\lambda x) & 0 & \dfrac{S(\lambda x)}{\lambda} & R(\lambda x) \end{bmatrix} \tag{8.26}$$

With Eq. (8.25), the state vector of the $i$th joint is transferred to the $(i+1)$th joint. For example, the state vector of joint $C_0$ shown in Fig. 8.4 is now transferred to joint $A_1$ in the primary coordinate frame, $x_3A_1y_3$.

$$(\boldsymbol{\xi}_{6\times1})_{A_{11}} = \mathbf{R}_{6\times6}(\alpha_{13})\mathbf{H}_{6\times6}(l_1)(\boldsymbol{\xi}_{6\times1})_{C_0} \tag{8.27}$$

where

$$\mathbf{R}_{6\times6}(\alpha_{13}) = \begin{bmatrix} \cos\alpha_{13} & \sin\alpha_{13} & 0 & 0 & 0 & 0 \\ -\sin\alpha_{13} & \cos\alpha_{13} & 0 & 0 & 0 & 0 \\ 0 & 0 & 1 & 0 & 0 & 0 \\ 0 & 0 & 0 & \cos\alpha_{13} & \sin\alpha_{13} & 0 \\ 0 & 0 & 0 & -\sin\alpha_{13} & \cos\alpha_{13} & 0 \\ 0 & 0 & 0 & 0 & 0 & 1 \end{bmatrix} \tag{8.28}$$

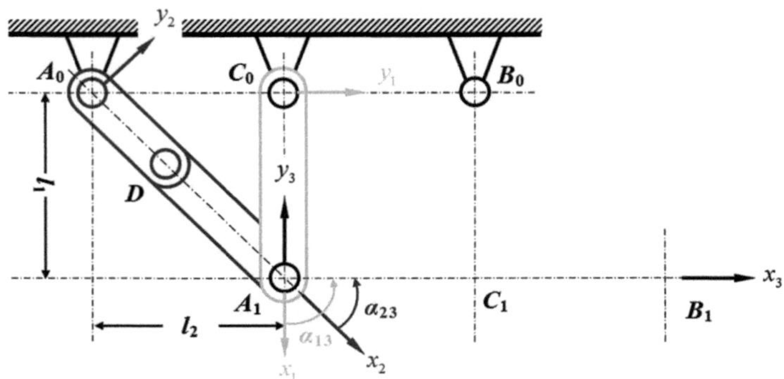


**Fig. 8.4** Primary coordinate system for link $A_0DA_1$

and $\mathbf{R}_{6\times6}(\alpha_{13})$ denotes the transformation matrix from $x_1A_1y_1$ to $x_3A_1y_3$, $\alpha_{13} = \frac{\pi}{2}$ denotes the subtended angle from $x_1$-axis to $x_3$-axis. $(\boldsymbol{\xi}_{6\times1})_{A_{11}} = \left[X_{A_1}\ Y_{A_1}\ \Theta_{zA_{11}}\ Q_{xA_{11}}\ Q_{yA_{11}}\ M_{zA_1}\right]_3^{\mathrm{T}}$, $[\cdot]_i (i = 1, 2, \ldots)$ represents that the state vectors are expressed in the $i$th coordinate frame, and $(\boldsymbol{\xi}_{6\times1})_{C_0} = \left[X_{C_0}\ Y_{C_0}\ \Theta_{zC_0}\ Q_{xC_0}\ Q_{yC_0}\ M_{zC_0}\right]_3^{\mathrm{T}}$.

Substituting Eqs. (8.26) and (8.28) into (8.27) yields

$$(\boldsymbol{\xi}_{6\times1})_{A_{11}} = \begin{bmatrix} 0 & R(\lambda l_1) & \dfrac{S(\lambda l_1)}{\lambda} & 0 & \dfrac{V(\lambda l_1)}{EI\lambda^3} & \dfrac{U(\lambda l_1)}{EI\lambda^2} \\ -\cos(kl_1) & 0 & 0 & \dfrac{\sin(kl_1)}{EAk} & 0 & 0 \\ 0 & \lambda V(\lambda l_1) & R(\lambda l_1) & 0 & \dfrac{U(\lambda l_1)}{EI\lambda^2} & \dfrac{S(\lambda l_1)}{EI\lambda} \\ 0 & EI\lambda^3 S(\lambda l_1) & EI\lambda^2 U(\lambda l_1) & 0 & R(\lambda l_1) & \lambda V(\lambda l_1) \\ -EAk\sin(kl_1) & 0 & 0 & -\cos kl_1 & 0 & 0 \\ 0 & EI\lambda^2 U(\lambda l_1) & EI\lambda V(\lambda l_1) & 0 & \dfrac{S(\lambda l_1)}{\lambda} & R(\lambda l_1) \end{bmatrix} (\boldsymbol{\xi}_{6\times1})_{C_0} \tag{8.29}$$

However, because $C_0$ is pivot jointed, the known initial conditions for joint $C_0$ are

$$\begin{cases} X_{C_0} = 0 \\ Y_{C_0} = 0 \\ M_{zC_0} = 0 \end{cases} \tag{8.30}$$

Equation (8.29) can be rewritten with the known criteria expressed by Eq. (8.30) as

$$(\boldsymbol{\xi}_{6\times1})_{A_{11}} = \begin{bmatrix} X_{A_{11}} \\ Y_{A_{11}} \\ \Theta_{zA_{11}} \\ Q_{xA_{11}} \\ Q_{yA_{11}} \\ M_{zA_{11}} \end{bmatrix}_3 = \begin{bmatrix} \frac{S(\lambda l_1)}{\lambda} & 0 & \frac{V(\lambda l_1)}{EI\lambda^3} \\ 0 & \frac{\sin(kl_1)}{EAk} & 0 \\ R(\lambda l_1) & 0 & \frac{U(\lambda l_1)}{EI\lambda^2} \\ EI\lambda^2 U(\lambda l_1) & 0 & R(\lambda l_1) \\ 0 & -\cos(kl_1) & 0 \\ EI\lambda V(\lambda l_1) & 0 & \frac{S(\lambda l_1)}{\lambda} \end{bmatrix} \begin{bmatrix} \Theta_{zC_0} \\ Q_{xC_0} \\ Q_{yC_0} \end{bmatrix} \tag{8.31}$$

Suppose that the primary coordinate frame of the link, $A_0DA_1$, is similarly set, namely, the $x$-axis is along the axial direction of the link and the $y$-axis passes through the fixed joint $A_0$ and is perpendicular upward to $A_0DA_1$. We can therefore write out the transfer expressions from the state vector of joint $A_0$ to joint $A_1$ in the primary coordinate frame, $x_3A_1y_3$, shown in Fig. 8.4:

$$(\boldsymbol{\xi}_{6\times1})_{A_{12}} = \mathbf{R}_{6\times6}(\alpha_{23})\mathbf{H}_{6\times6}(L)(\boldsymbol{\xi}_{6\times1})_{A_0} \tag{8.32}$$

where $(\boldsymbol{\xi}_{6\times1})_{A_0} = \left[X_{A_0}\ Y_{A_0}\ \Theta_{zA_0}\ Q_{xA_0}\ Q_{yA_0}\ M_{zA_0}\right]_3^{\mathrm{T}}$, $L = \frac{l_1}{\sin\alpha_{23}}$ and here $\alpha_{23} = \frac{\pi}{4}$ represents the subtended angle between the $x_2$-axis and the$x_3$-axis.

Expanding Eq. (8.32), we gain that

$$(\boldsymbol{\xi}_{6\times1})_{A_{12}} = \frac{\sqrt{2}}{2} \begin{bmatrix} \cos(kL) & R(\lambda L) & \frac{S(\lambda L)}{\lambda} & -\frac{\sin(kL)}{EAk} & \frac{V(\lambda L)}{EI\lambda^3} & \frac{U(\lambda L)}{EI\lambda^2} \\ -\cos(kL) & R(\lambda L) & \frac{S(\lambda L)}{\lambda} & \frac{\sin(kL)}{EAk} & \frac{V(\lambda L)}{EI\lambda^3} & \frac{U(\lambda L)}{EI\lambda^2} \\ 0 & \sqrt{2}\lambda V(\lambda L) & \sqrt{2}R(\lambda L) & 0 & \frac{\sqrt{2}U(\lambda L)}{EI\lambda^2} & \frac{\sqrt{2}S(\lambda L)}{EI\lambda} \\ EAk\sin(kL) & EI\lambda^3 S(\lambda L) & EI\lambda^2 U(\lambda L) & \cos(kL) & R(\lambda L) & \lambda V(\lambda L) \\ -EAk\sin(kL) & EI\lambda^3 S(\lambda L) & EI\lambda^2 U(\lambda L) & -\cos(kL) & R(\lambda L) & \lambda V(\lambda L) \\ 0 & \sqrt{2}EI\lambda^2 U(\lambda L) & \sqrt{2}EI\lambda V(\lambda L) & 0 & \frac{\sqrt{2}S(\lambda L)}{\lambda} & \sqrt{2}R(\lambda L) \end{bmatrix} (\boldsymbol{\xi}_{6\times1})_{A_0} \tag{8.33}$$

The known initial conditions for joint $A_0$ are

$$\begin{cases} X_{A_0} = 0 \\ Y_{A_0} = 0 \\ M_{zA_0} = 0 \end{cases} \tag{8.34}$$

With the known criteria of Eq. (8.34), Eq. (8.33) can be simplified as

$$(\boldsymbol{\xi}_{6\times1})_{A_{12}} = \begin{bmatrix} X_{A_{12}} \\ Y_{A_{12}} \\ \Theta_{zA_{12}} \\ Q_{xA_{12}} \\ Q_{yA_{12}} \\ M_{zA_{12}} \end{bmatrix}_3 = \frac{\sqrt{2}}{2} \begin{bmatrix} \frac{S(\lambda L)}{\lambda} & -\frac{\sin(kL)}{EAk} & \frac{V(\lambda L)}{EI\lambda^3} \\ \frac{S(\lambda L)}{\lambda} & \frac{\sin(kL)}{EAk} & \frac{V(\lambda L)}{EI\lambda^3} \\ \sqrt{2}R(\lambda L) & 0 & \frac{\sqrt{2}U(\lambda L)}{EI\lambda^2} \\ EI\lambda^2 U(\lambda L) & \cos(kL) & R(\lambda L) \\ EI\lambda^2 U(\lambda L) & -\cos(kL) & R(\lambda L) \\ \sqrt{2}EI\lambda V(\lambda L) & 0 & \frac{\sqrt{2}S(\lambda L)}{\lambda} \end{bmatrix} \begin{bmatrix} \Theta_{zA_0} \\ Q_{xA_0} \\ Q_{yA_0} \end{bmatrix} \tag{8.35}$$

Because $A_1$ is pivot jointed, the following relation must hold:

$$M_{zA_{11}} = M_{zA_{12}} = 0 \tag{8.36}$$

Substituting Eq. (8.31) into (8.36) yields

$$EI\lambda V(\lambda l_1)\Theta_{zC_0} + \frac{S(\lambda l_1)}{\lambda} Q_{yC_0} = 0 \tag{8.37}$$

From Eq. (8.37), we obtain that

$$\Theta_{zC_0} = -\frac{S(\lambda l_1)}{EI\lambda^2 V(\lambda l_1)} Q_{yC_0} \tag{8.38}$$

So $\Theta_{zC_0}$ is the function of $Q_{yC_0}$. Substituting Eq. (8.38) into (8.31) yields

$$(\boldsymbol{\xi}_{6\times1})_{A_{11}} = \begin{bmatrix} X_{A_{11}} \\ Y_{A_{11}} \\ \Theta_{zA_{11}} \\ Q_{xA_{11}} \\ Q_{yA_{11}} \\ M_{zA_{11}} \end{bmatrix}_3 = \mathbf{P}_{C_0A_1} \begin{bmatrix} Q_{xC_0} \\ Q_{yC_0} \end{bmatrix} \tag{8.39}$$

Where

$$\mathbf{P}_{C_0A_{11}} = \begin{bmatrix} 0 & \dfrac{V^2(\lambda l_1) - S^2(\lambda l_1)}{EI\lambda^3 V(\lambda l_1)} \\ \dfrac{\sin(kl_1)}{EAk} & 0 \\ 0 & \dfrac{U(\lambda l_1)V(\lambda l_1) - R(\lambda l_1)S(\lambda l_1)}{EI\lambda^2 V(\lambda l_1)} \\ 0 & \dfrac{R(\lambda l_1)V(\lambda l_1) - S(\lambda l_1)U(\lambda l_1)}{V(\lambda l_1)} \\ -\cos(kl_1) & 0 \\ 0 & 0 \end{bmatrix}.$$

Substituting Eq. (8.35) into (8.36) yields

$$EI\lambda V(\lambda L)\Theta_{zA_0} + \frac{S(\lambda L)}{\lambda} Q_{yA_0} = 0 \tag{8.40}$$

From Eq. (8.40), we obtain

$$\Theta_{zA_0} = -\frac{S(\lambda L)}{EI\lambda^2 V(\lambda L)} Q_{yA_0} \tag{8.41}$$

So $\Theta_{zA_0}$ is the function of $Q_{yA_0}$. Substituting Eq. (8.41) into (8.35) yields

$$(\boldsymbol{\xi}_{6\times1})_{A_{12}} = \begin{bmatrix} X_{A_{12}} \\ Y_{A_{12}} \\ \Theta_{zA_{12}} \\ Q_{xA_{12}} \\ Q_{yA_{12}} \\ M_{zA_{12}} \end{bmatrix}_3 = \mathbf{P}_{A_0A_{12}} \begin{bmatrix} Q_{xA_0} \\ Q_{yA_0} \end{bmatrix} \tag{8.42}$$

where

$$\mathbf{P}_{A_0A_{12}} = \frac{\sqrt{2}}{2} \begin{bmatrix} -\dfrac{\sin(kL)}{EAk} & \dfrac{V^2(\lambda L) - S^2(\lambda L)}{EI\lambda^3 V(\lambda L)} \\ \dfrac{\sin(kL)}{EAk} & \dfrac{V^2(\lambda L) - S^2(\lambda L)}{EI\lambda^3 V(\lambda L)} \\ 0 & \dfrac{\sqrt{2}\,[U(\lambda L)V(\lambda L) - R(\lambda L)S(\lambda L)]}{EI\lambda^2 V(\lambda L)} \\ \cos(kL) & \dfrac{R(\lambda L)V(\lambda L) - S(\lambda L)U(\lambda L)}{V(\lambda L)} \\ -\cos(kL) & \dfrac{R(\lambda L)V(\lambda L) - S(\lambda L)U(\lambda L)}{V(\lambda L)} \\ 0 & 0 \end{bmatrix}.$$

Because links $C_0DA_1$ and $B_0A_1$ are pivot jointed at $A_1$, $(\xi_{6\times1})_{A_{11}}$ and $(\xi_{6\times1})_{A_{12}}$ must have identical position coordinates when they are transformed into the coordinate frame of $x_3A_1y_3$.

$$\begin{bmatrix} X_{A_1} \\ Y_{A_1} \end{bmatrix}_3 = \begin{bmatrix} X_{A_{11}} \\ Y_{A_{11}} \end{bmatrix}_3 = \begin{bmatrix} X_{A_{12}} \\ Y_{A_{12}} \end{bmatrix}_3 \tag{8.43}$$

Therefore, associating Eqs. (8.39) and (8.42), we obtain

$$\begin{bmatrix} 0 & \dfrac{V^2(\lambda l_1) - S^2(\lambda l_1)}{EI\lambda^3 V(\lambda l_1)} \\ \dfrac{\sin(kl_1)}{EAk} & 0 \end{bmatrix} \begin{bmatrix} Q_{xC_0} \\ Q_{yC_0} \end{bmatrix} = \frac{\sqrt{2}}{2} \begin{bmatrix} -\dfrac{\sin(kL)}{EAk} & \dfrac{V^2(\lambda L) - S^2(\lambda L)}{EI\lambda^3 V(\lambda L)} \\ \dfrac{\sin(kL)}{EAk} & \dfrac{V^2(\lambda L) - S^2(\lambda L)}{EI\lambda^3 V(\lambda L)} \end{bmatrix} \begin{bmatrix} Q_{xA_0} \\ Q_{yA_0} \end{bmatrix} \tag{8.44}$$

As a result, $\begin{bmatrix} Q_{xC_0} \\ Q_{yC_0} \end{bmatrix}$ can be expressed as the function of $\begin{bmatrix} Q_{xA_0} \\ Q_{yA_0} \end{bmatrix}$:

$$\begin{bmatrix} Q_{xC_0} \\ Q_{yC_0} \end{bmatrix} = \mathbf{P}_{C_0A_0} \begin{bmatrix} Q_{xA_0} \\ Q_{yA_0} \end{bmatrix} \tag{8.45}$$

Where

$$\mathbf{P}_{C_0A_0} = \frac{\sqrt{2}}{2} \begin{bmatrix} \dfrac{\sin(kL)}{\sin(kl_1)} & \dfrac{Ak\left[V^2(\lambda L) - S^2(\lambda L)\right]}{I\lambda^3 V(\lambda L)\sin(kl_1)} \\ -\dfrac{I\lambda^3 V(\lambda l_1)\sin(kL)}{Ak\left[V^2(\lambda l_1) - S^2(\lambda l_1)\right]} & \dfrac{V(\lambda l_1)\left[V^2(\lambda L) - S^2(\lambda L)\right]}{V(\lambda L)\left[V^2(\lambda l_1) - S^2(\lambda l_1)\right]} \end{bmatrix}.$$

Substituting Eq. (8.45) into (8.39) generates

$$(\boldsymbol{\xi}_{6\times1})_{A_{11}} = \begin{bmatrix} X_{A_{11}} \\ Y_{A_{11}} \\ \Theta_{zA_{11}} \\ Q_{xA_{11}} \\ Q_{yA_{11}} \\ M_{zA_{11}} \end{bmatrix}_3 = \mathbf{P}_{C_0A_1}\mathbf{P}_{C_0A_0} \begin{bmatrix} Q_{xA_0} \\ Q_{yA_0} \end{bmatrix} \tag{8.46}$$

Because $A_1$ is pivot jointed by three links, the bending moments and rotational angles cannot be transferred between each other while keeping the same values of $x$- and $y$-displacements. As a result, associating Eqs. (8.42) and (8.46), we get

$$(\boldsymbol{\xi}_{6\times1})_{A_1}^{\text{output}} = \begin{bmatrix} X_{A_1} \\ Y_{A_1} \\ \Theta_{zA_1} \\ Q_{xA_1} \\ Q_{yA_1} \\ M_{zA_1} \end{bmatrix}_3 = \mathbf{P}_D \begin{bmatrix} Q_{xA_0} \\ Q_{yA_0} \end{bmatrix} \tag{8.47}$$

where $\mathbf{P}_D = \mathbf{P}_{A_0A_{12}} + \mathbf{E}_1\mathbf{P}_{C_0A_1}\mathbf{P}_{C_0A_0}$ and $\mathbf{E}_1 = \begin{bmatrix} 0&0&0&0&0&0 \\ 0&0&0&0&0&0 \\ 0&0&0&0&0&0 \\ 0&0&0&1&0&0 \\ 0&0&0&0&1&0 \\ 0&0&0&0&0&0 \end{bmatrix}$.

Equation (8.43) expresses the state vector of joint $A_1$ under the associated inputs of links $C_0A_1$ and $A_0DA_1$ with the unknown variables of joint $A_0$, namely, $\begin{bmatrix} Q_{xA_0} & Q_{yA_0} \end{bmatrix}^{\mathrm{T}}$.

The dynamic properties of the driving mechanism under the working position shown in Fig. 8.4 can be obtained when other links are omitted. Under this condition, $\begin{bmatrix} Q_{xA_1} \\ Q_{yA_1} \end{bmatrix} = 0$, namely,

$$\mathbf{T}_Q\mathbf{P}_D \begin{bmatrix} Q_{xA_0} \\ Q_{yA_0} \end{bmatrix} = 0 \tag{8.48}$$

where $\mathbf{T}_Q = \begin{bmatrix} 0&0&0&1&0&0 \\ 0&0&0&0&1&0 \end{bmatrix}$ represents force intercepting operation on a vector.

Because $\varsigma_{F_{A_0}} = \begin{bmatrix} Q_{xA_0} \\ Q_{yA_0} \end{bmatrix}$ is a nonzero 2-dimension vector, we immediately obtain the frequency determinant:

$$\left|\mathbf{T}_Q\mathbf{P}_D\right| = 0 \tag{8.49}$$

where $|\cdot|$ denotes the determinant of "·".

Solving Eq. (8.49), we obtain the master frequency of the driving mechanism in the working position.

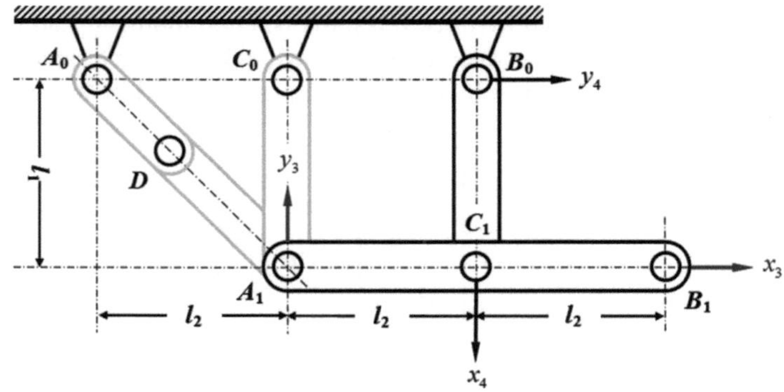


**Fig. 8.5** Coordinate system for the vertical beam $B_0C_1$

## 8.3.2 Structural Dynamics of the Driving Mechanism and One Staircase

In a similar way, we can investigate the transfer matrix of the state vector of link $A_1C_1B_1$.

For the horizontal beam, $A_1C_1B_1$, two ends of which are all free of bending moments, the slope of beam $A_1C_1B_1$, denoted by $\theta_{H_1}$, is an independent variable. As a result, the state vector for the horizontal beam $A_1C_1B_1$ can be rewritten as

$$(\boldsymbol{\xi}_{6\times1})_{A_1} = \begin{bmatrix} X_{A_{11}} \\ Y_{A_{11}} \\ \theta_{H_1} \\ \\ Q_{xA_{11}} \\ Q_{yA_{11}} \\ M_{zA_{11}} \end{bmatrix}_3 = \mathbf{P}_D \begin{bmatrix} Q_{xA_0} \\ Q_{yA_0} \end{bmatrix} + \theta_{H_1}\boldsymbol{e}_6(3) \tag{8.50}$$

where $\boldsymbol{e}_6(3) = [0\ 0\ 1\ 0\ 0\ 0]^{\mathrm{T}}$.

With Eq. (8.25), we can transfer the associated state vector of $A_1$, represented by Eq. (8.50), along the $x_3$-direction, shown in Fig. 8.5, to joint $C_1$:

$$(\boldsymbol{\xi}_{6\times1})_{C_{11}} = \begin{bmatrix} X_{C_{11}} \\ Y_{C_{11}} \\ \Theta_{zC_{11}} \\ \\ Q_{xC_{11}} \\ Q_{yC_{11}} \\ M_{zC_{11}} \end{bmatrix}_3 = \mathbf{P}_{A_1C_1} \begin{bmatrix} Q_{xA_0} \\ Q_{yA_0} \end{bmatrix} + \theta_{H_1}\mathbf{H}_{6\times6}(l_2)\boldsymbol{e}_6(3) \tag{8.51}$$

where $\mathbf{P}_{A_1C_1} = \mathbf{H}_{6\times6}(l_2)\mathbf{P}_D$.

Obviously, the vertical link $B_0C_1$ shown in Fig. 8.5 will provide two forces along the $x_3$- and $y_3$-axes, respectively, to joint $C_1$. For the sake of simplicity, we assume that the joint forces exerted to $A_1C_1B_1$ by link $B_0C_1$ are

$$\boldsymbol{F}_{C_{11}} = \begin{bmatrix} F_{xC_{11}} \\ F_{yC_{11}} \end{bmatrix} \tag{8.52}$$

where $F_{xC_{11}}$ represents the force in $x_3$-direction and $F_{yC_{11}}$ represents the force in $y_3$-direction. So the output state vector of joint $C_1$ along link $A_1C_1B_1$ should be

$$(\boldsymbol{\xi}_{6\times1})_{C_{11}}^{\text{output}} = \begin{bmatrix} X_{C_{11}} \\ Y_{C_{11}} \\ \Theta_{zC_{11}} \\ Q_{xC_{11}} \\ Q_{yC_{11}} \\ M_{zC_{11}} \end{bmatrix}_3^{\text{output}} = \mathbf{P}_{A_1C_1}\begin{bmatrix} Q_{xA_0} \\ Q_{yA_0} \end{bmatrix} + \mathbf{T}_Q^{\mathrm{T}}\boldsymbol{F}_{C_{11}} + \theta_{H_1}\mathbf{H}_{6\times6}(l_2)\boldsymbol{e}_6(3) \tag{8.53}$$

With Eq. (8.25), we can transfer the state vector, $(\boldsymbol{\xi}_{6\times1})_{C_{11}}^{\text{output}}$, represented by Eq. (8.53), along the $x_3$-direction, shown in Fig. 8.5, to joint $B_1$:

$$\begin{aligned}(\boldsymbol{\xi}_{6\times1})_{B_1}^{\text{output}} &= \begin{bmatrix} X_{B_{11}} \\ Y_{B_{11}} \\ \Theta_{zB_{11}} \\ Q_{xB_{11}} \\ Q_{yB_{11}} \\ M_{zB_{11}} \end{bmatrix}_3 \\ &= \mathbf{H}_{6\times6}(l_2)\mathbf{P}_{A_1C_1}\begin{bmatrix} Q_{xA_0} \\ Q_{yA_0} \end{bmatrix} + \mathbf{H}_{6\times6}(l_2)\mathbf{T}_Q^{\mathrm{T}}\boldsymbol{F}_{C_{11}} + \theta_{H_1}\mathbf{H}_{6\times6}^2(l_2)\boldsymbol{e}_6(3)\end{aligned} \tag{8.54}$$

Suppose that the state vector of joint $B_0$ is denoted by

$$(\boldsymbol{\xi}_{6\times1})_{B_0} = \begin{bmatrix} X_{B_0} & Y_{B_0} & \Theta_{zB_0} & Q_{xB_0} & Q_{yB_0} & M_{zB_0} \end{bmatrix}_4^{\mathrm{T}} \tag{8.55}$$

The known initial conditions for joint $B_0$ are

$$\begin{cases} X_{B_0} = 0 \\ Y_{B_0} = 0 \\ M_{B_0} = 0 \end{cases} \tag{8.56}$$

The vertical input state vector for joint $C_1$ from joint $B_0$ can be similarly denoted by

$$(\boldsymbol{\xi}_{6\times1})_{C_{12}} = \begin{bmatrix} X_{C_{12}} \\ Y_{C_{12}} \\ \Theta_{zC_{12}} \\ \\ Q_{xC_{12}} \\ Q_{yC_{12}} \\ M_{zC_{12}} \end{bmatrix}_4 = \mathbf{P}_{B_0} \begin{bmatrix} \Theta_{zB_0} \\ Q_{xB_0} \\ Q_{yB_0} \end{bmatrix} \tag{8.57}$$

Where $\mathbf{P}_{B_0} = \begin{bmatrix} 0 & -\frac{\sin(kl_1)}{EAk} & 0 \\ \frac{S(\lambda l_1)}{\lambda} & 0 & \frac{V(\lambda l_1)}{EI\lambda^3} \\ R(\lambda l_1) & 0 & \frac{U(\lambda l_1)}{EI\lambda^2} \\ 0 & \cos(kl_1) & 0 \\ EI\lambda^2 U(\lambda l_1) & 0 & R(\lambda l_1) \\ EI\lambda V(\lambda l_1) & 0 & \frac{S(\lambda l_1)}{\lambda} \end{bmatrix}$.

Similarly, the vertical link $B_0C_1$ shown in Fig. 8.5 will bear two reaction forces along the $x_4$- and $y_4$-axes from the horizontal link $A_1C_1B_1$ at joint $C_1$

$$\boldsymbol{F}_{C_{12}} = -\mathbf{A}\left(-\frac{\pi}{2}\right)\boldsymbol{F}_{C_{11}} = \mathbf{A}\left(\frac{\pi}{2}\right)\begin{bmatrix} F_{xC_{11}} \\ F_{yC_{11}} \end{bmatrix} \tag{8.58}$$

where $\mathbf{A}(\beta) = \begin{bmatrix} \cos\beta & \sin\beta \\ -\sin\beta & \cos\beta \end{bmatrix}_{\beta=\frac{\pi}{2}} = \begin{bmatrix} 0 & 1 \\ -1 & 0 \end{bmatrix}$ denotes a rotational transformation about $z$-axis for $\beta$ angle.

So the output state vector of joint $C_1$ along the $x_4$-direction should be

$$(\boldsymbol{\xi}_{6\times1})^{\text{output}}_{C_{12}} = \begin{bmatrix} X_{C_{12}} \\ Y_{C_{12}} \\ \Theta_{zC_{12}} \\ \\ Q_{xC_{12}} \\ Q_{yC_{12}} \\ M_{zC_{12}} \end{bmatrix}^{\text{output}}_4 = \mathbf{P}_{B_0} \begin{bmatrix} \Theta_{zB_0} \\ Q_{xB_0} \\ Q_{yB_0} \end{bmatrix} + \mathbf{T}_Q^{\mathrm{T}}\mathbf{A}\left(\frac{\pi}{2}\right)\begin{bmatrix} F_{xC_{11}} \\ F_{yC_{11}} \end{bmatrix} \tag{8.59}$$

Because $C_1$ is pivot jointed, the following relation must hold from Fig. 8.5:

$$\begin{bmatrix} X_{C_{11}} \\ Y_{C_{11}} \end{bmatrix}^{\text{output}}_3 = \begin{bmatrix} X_{C_{12}} \\ Y_{C_{12}} \end{bmatrix}^{\text{output}}_3 = \mathbf{A}\left(\frac{\pi}{2}\right)\begin{bmatrix} X_{C_{12}} \\ Y_{C_{12}} \end{bmatrix}^{\text{output}}_4 \tag{8.60}$$

In addition, both joint $B_1$ and joint $C_1$ shown in Fig. 8.5 are all free of external forces. Therefore, we obtain

$$\begin{cases} \begin{bmatrix} Q_{xB_{11}} \\ Q_{yB_{11}} \\ M_{zB_{11}} \end{bmatrix}_3 = 0 \\ \begin{bmatrix} Q_{xC_{12}} \\ Q_{yC_{12}} \\ M_{zC_{12}} \end{bmatrix}_4 = 0 \end{cases} \tag{8.61}$$

Associating Eqs. (8.60) and (8.61), and substituting Eqs. (8.53), (8.54) and (8.59) into them, yields

$$\begin{bmatrix} X_{C_{11}} - Y_{C_{12}} \\ Y_{C_{11}} + X_{C_{12}} \\ \hline Q_{xB_{11}} \\ Q_{yB_{11}} \\ M_{zB_{11}} \\ \hline Q_{xC_{12}} \\ Q_{yC_{12}} \\ M_{zC_{12}} \end{bmatrix} = \mathbf{T}_{8\times 8} \begin{bmatrix} Q_{xA_0} \\ Q_{yA_0} \\ \hline F_{xC_{11}} \\ F_{yC_{11}} \\ \hline \theta_{H_1} \\ \hline \Theta_{zB_0} \\ Q_{xB_0} \\ Q_{yB_0} \end{bmatrix} = 0 \tag{8.62}$$

where

$$\mathbf{T}_{8\times 8} = \begin{bmatrix} [\mathbf{T}_C \mathbf{P}_{A_1C_1}]_{2\times 2} & \left[\mathbf{T}_C \mathbf{T}_Q^{\mathrm{T}} - \mathbf{A}\left(\frac{\pi}{2}\right)\mathbf{T}_C \mathbf{T}_Q^{\mathrm{T}} \mathbf{A}\left(\frac{\pi}{2}\right)\right]_{2\times 2} \\ [\mathbf{E}_{3\times 6}\mathbf{H}_{6\times 6}(l_2)\mathbf{P}_{A_1C_1}]_{3\times 2} & \left[\mathbf{E}_{3\times 6}\mathbf{H}_{6\times 6}(l_2)\mathbf{T}_Q^{\mathrm{T}}\right]_{3\times 2} \\ [0]_{3\times 2} & \left[\mathbf{E}_{3\times 6}\mathbf{T}_Q^{\mathrm{T}}\mathbf{A}\left(\frac{\pi}{2}\right)\right]_{3\times 2} \end{bmatrix}$$

$$\times \begin{bmatrix} [\mathbf{T}_C\mathbf{H}_{6\times 6}(l_2)\boldsymbol{e}_6(3)]_{2\times 1} & -\left[A\left(\frac{\pi}{2}\right) T_C P_{B_0}\right]_{2\times 3} \\ \left[\mathbf{E}_{3\times 6}\mathbf{H}_{6\times 6}^2(l_2)\boldsymbol{e}_6(3)\right]_{3\times 1} & [0]_{3\times 3} \\ [0]_{3\times 1} & [\mathbf{E}_{3\times 6}\mathbf{P}_{B_0}]_{3\times 3} \end{bmatrix},$$

$\mathbf{T}_C = \begin{bmatrix} 1\,0\,0\,0\,0\,0 \\ 0\,1\,0\,0\,0\,0 \end{bmatrix}$ represents an intercepting operation on a vector,

$\mathbf{T}_Q = \begin{bmatrix} 0\,0\,0\,1\,0\,0 \\ 0\,0\,0\,0\,1\,0 \end{bmatrix}$ represents the force intercepting operation on a vector,

$\mathbf{E}_{3\times 6} = \begin{bmatrix} 0\,0\,0\,1\,0\,0 \\ 0\,0\,0\,0\,1\,0 \\ 0\,0\,0\,0\,0\,1 \end{bmatrix}$ represents an intercepting operation on a matrix, and

$[0]_{m\times n}(m, n = 2, 3, \ldots)$ denotes the $m \times n$-dimension zero matrix.

**Table 8.1** Parameters of the links of the foldable stair

| $\rho(\text{kg/m}^3)$ | $E(\text{Pa})$ | $A(\text{m}^2)$ | $I(\text{m}^4)$ | $l_1(\text{m})$ | $l_2(\text{m})$ |
|---|---|---|---|---|---|
| 7, 850 | $2.06 \times 10^{11}$ | $1.2 \times 10^{-3}$ | $3.6 \times 10^{-7}$ | 0.3 | 0.3 |

**Table 8.2** Comparisons for the first staircase

| Order | Simulations in ANSYS (9 Hz) | Numerical results (9 Hz) | Relative errors (9 %) |
|---|---|---|---|
| 1 | 348.1891 | 349.2178 | 0.5843 |
| 2 | 668.5764 | 670.6153 | 0.3050 |
| 3 | 848.6745 | 853.3251 | 0.6666 |
| 4 | 1,498.9546 | 1,521.7122 | 1.5182 |
| 5 | 1,638.8323 | 1,665.0472 | 1.5996 |
| 6 | 1,675.2331 | 1,700.7297 | 1.5220 |

As a result, we obtain the frequency determinant of the staircase shown in Fig. 8.5:

$$|\mathbf{T}| = 0 \tag{8.63}$$

where $\mathbf{T} = \mathbf{T}_{8\times8}$ for the case shown in Fig. 8.5.

With a set of parameters shown in Table 8.1, we can obtain the numerical results from Eq. (8.63).

Simulations through ANSYS are also executed. Comparisons between theoretical results and the simulations are list in Table 8.2.

Comparisons indicate that the numerical results based on the theoretical analysis are very close to the simulation results through ANSYS. For the sake of clarity, the corresponding normal vibration modes obtained from ANSYS are illustrated in Fig. 8.6, where the bold curves indicate the normal vibration modes, the hidden lines indicate the undeformed shapes, the larger triangles indicate the constrained degrees of freedom (9-DOFs), and the smaller triangles represent the coupled DOFs. This conversion will be adopted in the following Fig. 8.6 obtained from ANSYS in this chapter.

## 8.4 Structural Dynamics of the Foldable Stairs

Beamlet transforms provide an effective, relatively easy, and fast method for the extraction of curvilinear features such as cracks in pavement images [47]. Similarly, the foldable stair can be thought of as the serial connection of the driving four-bar mechanism, $A_0DA_1C_0$, and a number of identical scissorlike elements, $A_{i+1}C_{i+1}B_{i+1} - B_iC_{i+1}A_{i+2}(i = 0, 1, 2, \ldots)$, shown in Fig. 8.7. Therefore, we can investigate the transfer of the state vectors through these SLEs.

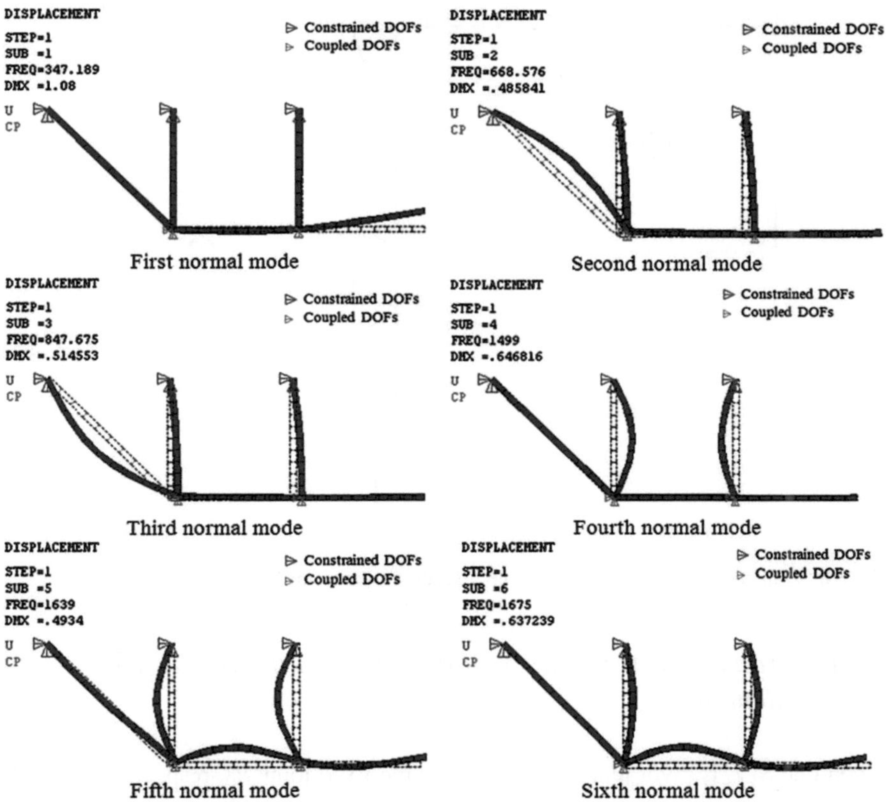


**Fig. 8.6** The first six normal modes of the driving mechanism and the first staircase

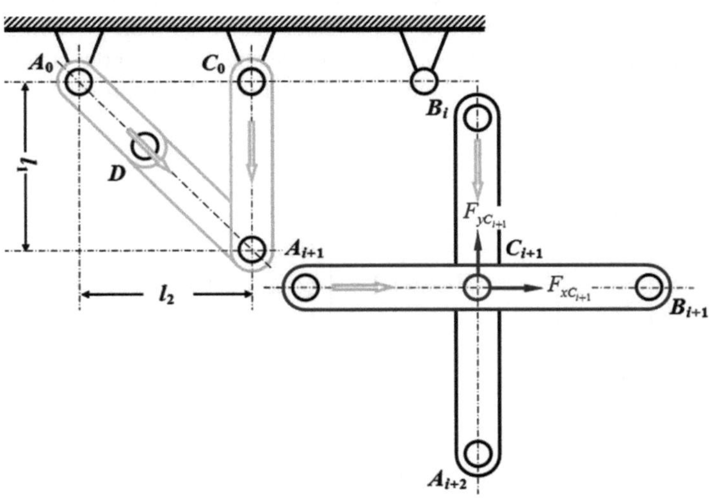


**Fig. 8.7** Transfer of the state vectors for a scissorlike element

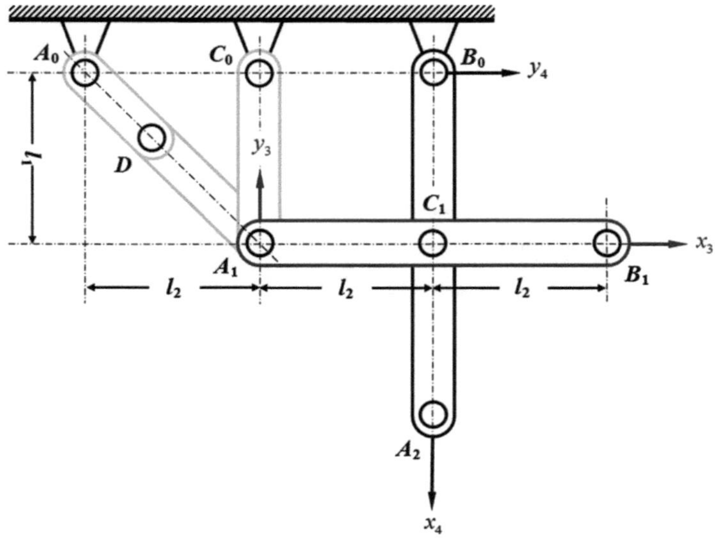


**Fig. 8.8** Transfer of the state vectors for the first scissorlike element

### *8.4.1 Transfer of the State Vectors Through the Scissorlike Elements*

When one SLE is pivot jointed with $A_1$ and the fixed joint $B_0$, respectively, which is shown in Fig. 8.8, the transfer algorithm for link $B_0C_1$ is also obtained similarly to Eqs. (8.53) and (8.54).

With a process discussed in Sect. 8.3.2, we can obtain the state vector of $B_1$, represented by Eq. (8.54) and the output state vector of joint $C_1$ along the $x_4$-direction represented by Eq. (8.59). Next, we investigate the output state vector of joint $A_2$ along the $x_4$-direction which is shown in Fig. 8.8. According to Eq. (8.25), we immediately gain

$$(\boldsymbol{\xi}_{6\times1})_{A_2}^{\text{output}} = \begin{bmatrix} X_{A_{21}} \\ Y_{A_{21}} \\ \Theta_{zA_{21}} \\ Q_{xA_{21}} \\ Q_{yA_{21}} \\ M_{zA_{21}} \end{bmatrix}_4^{\text{output}} = \mathbf{H}_{6\times6}(l_1)\mathbf{P}_{B_0}\begin{bmatrix} \Theta_{zB_0} \\ Q_{xB_0} \\ Q_{yB_0} \end{bmatrix} + \mathbf{H}_{6\times6}(l_1)\mathbf{T}_Q^{\mathrm{T}}\mathbf{A}\left(\frac{\pi}{2}\right)\begin{bmatrix} F_{xC_{11}} \\ F_{yC_{11}} \end{bmatrix} \tag{8.64}$$

where $\mathbf{H}_{6\times6}(l_1)$ represents the transfer matrix from joint $C_1$ to joint $A_2$ in the coordinate frame $x_4B_0y_4$.

Because $C_1$ is pivot jointed, relation (8.60) must also hold. Besides, both joint $B_1$ and joint $A_2$ shown in Fig. 8.8 are all free of external forces. Therefore, we have

$$\begin{cases} \begin{bmatrix} Q_{xB_{11}} \\ Q_{yB_{11}} \\ M_{zB_{11}} \end{bmatrix}_3 = 0 \\ \begin{bmatrix} Q_{xA_{21}} \\ Q_{yA_{21}} \\ M_{zA_{21}} \end{bmatrix}_4 = 0 \end{cases} \tag{8.65}$$

Associating Eqs. (8.60) and (8.65), and substituting Eqs. (8.53), (8.54), (8.59), and (8.64) into them, yields

$$\begin{bmatrix} X_{C_{11}} - Y_{C_{12}} \\ Y_{C_{11}} + X_{C_{12}} \\ \hline Q_{xB_{11}} \\ Q_{yB_{11}} \\ M_{zB_{11}} \\ \hline Q_{xA_{21}} \\ Q_{yA_{21}} \\ M_{zA_{21}} \end{bmatrix} = \mathbf{T}_{8\times8} \begin{bmatrix} Q_{xA_0} \\ Q_{yA_0} \\ \hline F_{xC_{11}} \\ F_{yC_{11}} \\ \hline \theta_{H_1} \\ \hline \Theta_{zB_0} \\ Q_{xB_0} \\ Q_{yB_0} \end{bmatrix} = 0 \tag{8.66}$$

where

$$\mathbf{T}_{8\times8} = \begin{bmatrix} [\mathbf{T}_C \mathbf{P}_{A_1C_1}]_{2\times2} & \left[\mathbf{T}_C \mathbf{T}_Q^{\mathrm{T}} - \mathbf{A}\left(\frac{\pi}{2}\right) \mathbf{T}_C \mathbf{T}_Q^{\mathrm{T}} \mathbf{A}\left(\frac{\pi}{2}\right)\right]_{2\times2} \\ [\mathbf{E}_{3\times6}\mathbf{H}_{6\times6}(l_2)\mathbf{P}_{A_1C_1}]_{3\times2} & \left[\mathbf{E}_{3\times6}\mathbf{H}_{6\times6}(l_2)\mathbf{T}_Q^{\mathrm{T}}\right]_{3\times2} \\ [0]_{3\times2} & \left[\mathbf{E}_{3\times6}\mathbf{H}_{6\times6}(l_1)\mathbf{T}_Q^{\mathrm{T}}\mathbf{A}\left(\frac{\pi}{2}\right)\right]_{3\times2} \end{bmatrix}$$
$$\times \begin{bmatrix} [\mathbf{T}_C \mathbf{H}_{6\times6}(l_2)\boldsymbol{e}_6(3)]_{2\times1} & -\left[\mathbf{A}\left(\frac{\pi}{2}\right)\mathbf{T}_C\mathbf{P}_{B_0}\right]_{2\times3} \\ \left[\mathbf{E}_{3\times6}\mathbf{H}_{6\times6}^2(l_2)\boldsymbol{e}_6(3)\right]_{3\times1} & [0]_{3\times3} \\ [0]_{3\times1} & [\mathbf{E}_{3\times6}\mathbf{H}_{6\times6}(l_1)\mathbf{P}_{B_0}]_{3\times3} \end{bmatrix} \tag{8.67}$$

As a result, we obtain the frequency equation of the staircase shown in Fig. 8.8 by substituting Eq. (8.67) into (8.63).

With a set of parameters shown in Table 8.1, we can obtain the numerical results from Eq. (8.63). In addition, simulations through ANSYS are also executed. Comparisons between theoretical results and the simulations are listed in Table 8.3.

**Table 8.3** Comparisons for the first SLE

| Order | Simulations in ANSYS (Hz) | Numerical results (Hz) | Relative errors (%) |
|---|---|---|---|
| 1 | 324.3277 | 325.8857 | 0.4804 |
| 2 | 348.4973 | 349.5361 | 0.5867 |
| 3 | 673.7953 | 676.7905 | 0.4445 |
| 4 | 836.1196 | 842.4708 | 0.7596 |
| 5 | 1,529.8605 | 1,553.6069 | 1.5522 |
| 6 | 1,568.0679 | 1,578.9125 | 0.6916 |

Comparisons indicate that the numerical results based on the theoretical model are very close to the simulation results through ANSYS. For the sake of clarity, the corresponding normal modes obtained from ANSYS are illustrated in Fig. 8.9.

## 8.4.2 Structural Dynamics of the Foldable Stair

Suppose that the $i$th$(i = 1, 2, \ldots)$ SLE is pivot jointed with the driving mechanism correspondingly, which is shown in Fig. 8.10. Two ends of the vertical link, $B_iC_{i+1}A_{i+2}$, and the horizontal link, $A_{i+1}C_{i+1}B_{i+1}$, are all free of bending moments. As a result, the slopes of links $B_iC_{i+1}A_{i+2}$ and $A_{i+1}C_{i+1}B_{i+1}$, denoted by $\theta_{V_i}$ and $\theta_{H_{i+1}}$, respectively, are two independent variables.

For the vertical link, $A_1C_1B_1$, two ends of which are all free of bending moments, the slope of beam $B_iC_{i+1}A_{i+2}$, denoted by $\theta_{V_i}$, is a new independent variable. As a result, the input state vector for the vertical link $B_iC_{i+1}A_{i+2}$ can be rewritten as

$$(\boldsymbol{\xi}_{6\times 1})_{B_i}^{\text{input}} = \mathbf{T}_{6\times 6}\left(-\frac{\pi}{2}\right)\begin{bmatrix} X_{B_{i1}} \\ Y_{B_{i1}} \\ 0 \\ Q_{xB_{i1}} \\ Q_{yB_{i1}} \\ 0 \end{bmatrix}_{2i+1}^{\text{output}} + \theta_{V_i}\boldsymbol{e}_6(3) \tag{8.68}$$

Similarly, for the horizontal beam, $A_{i+1}B_{i+1}C_{i+1}$, we can induce a new slope, denoted by $\theta_{H_{i+1}}$. The input state vector for joint $A_{i+1}$ in the $2(i+1)$th coordinate frame will be

$$(\boldsymbol{\xi}_{6\times 1})_{A_{i+1}}^{\text{input}} = \mathbf{T}_{6\times 6}\left(\frac{\pi}{2}\right)\begin{bmatrix} X_{A_{i1}} \\ Y_{A_{i1}} \\ 0 \\ Q_{xA_{i1}} \\ Q_{yA_{i1}} \\ 0 \end{bmatrix}_{2(i+1)}^{\text{output}} + \theta_{H_{i+1}}\boldsymbol{e}_6(3) \tag{8.69}$$

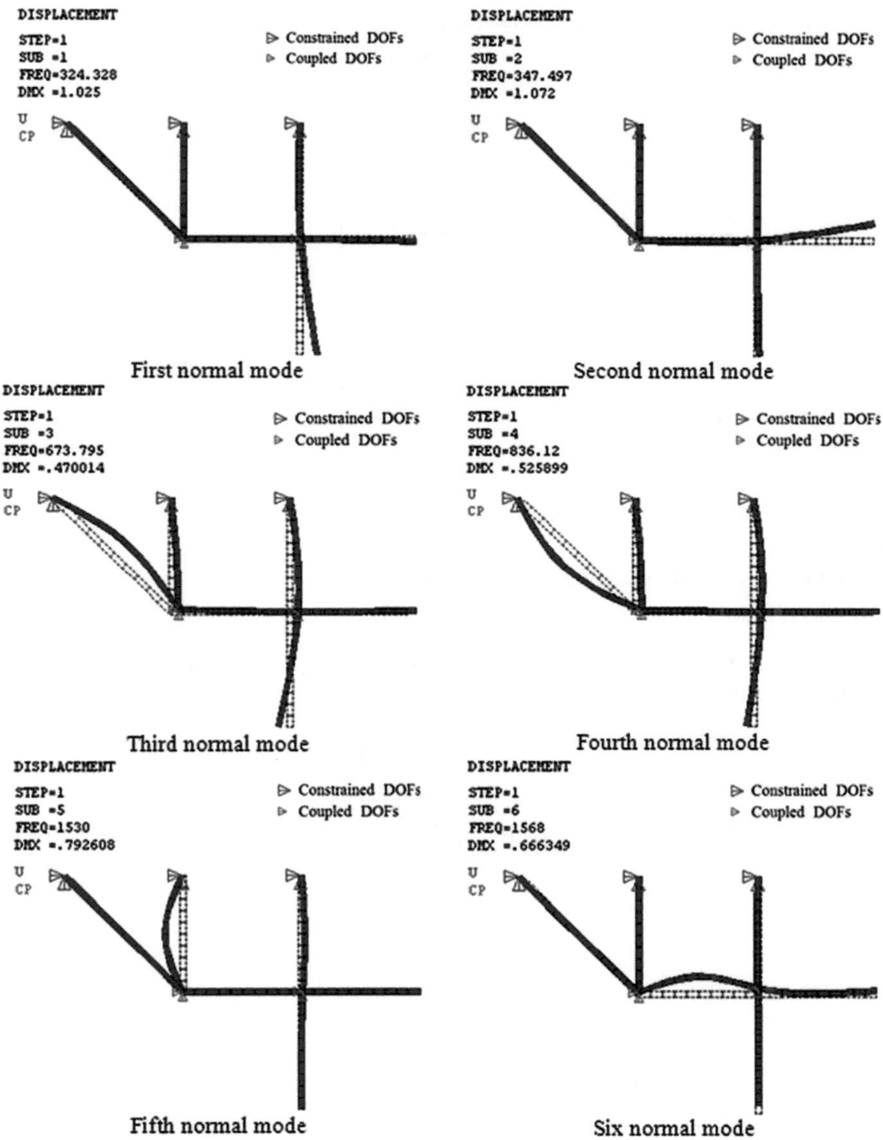


**Fig. 8.9** The first six normal modes of the driving mechanism and the first SLE

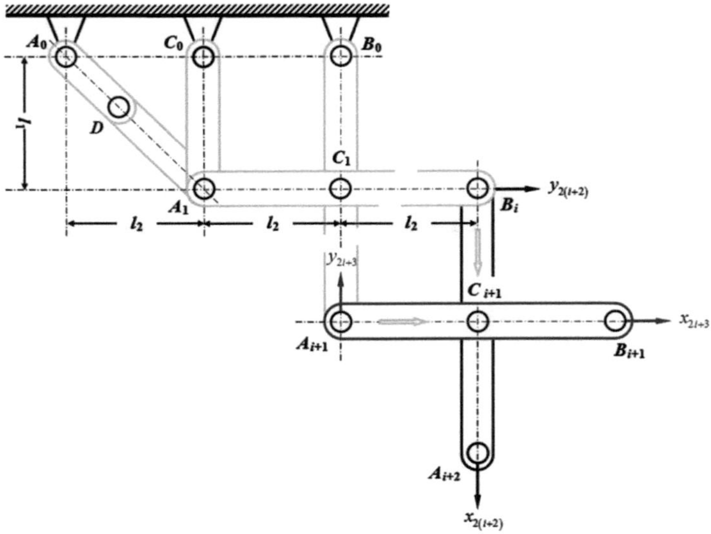


**Fig. 8.10** Transfer of the state vector for the $i$th SLE

With Eq. (8.25), the state vectors expressed by Eqs. (8.68) and (8.69) will be transferred into the cross joint $C_{i+1}$:

$$\left[(\boldsymbol{\xi}_{6\times1})^{\text{input}}_{C_{i+1}}\right]_{2(i+2)} = -\mathbf{H}_{6\times6}(l_1)\mathbf{T}_{6\times6}\left(\frac{\pi}{2}\right)\begin{bmatrix} X_{B_{i1}} \\ Y_{B_{i1}} \\ 0 \\ Q_{xB_{i1}} \\ Q_{yB_{i1}} \\ 0 \end{bmatrix}^{\text{output}}_{2i+1} + \theta_{V_i}\mathbf{H}_{6\times6}(l_1)\boldsymbol{e}_6(3) \tag{8.70}$$

$$\left[(\boldsymbol{\xi}_{6\times1})^{\text{input}}_{C_{i+1}}\right]_{2i+3} = \mathbf{H}_{6\times6}(l_2)\mathbf{T}_{6\times6}\left(\frac{\pi}{2}\right)\begin{bmatrix} X_{A_{i1}} \\ Y_{A_{i1}} \\ 0 \\ Q_{xA_{i1}} \\ Q_{yA_{i1}} \\ 0 \end{bmatrix}^{\text{output}}_{2(i+1)} + \theta_{H_{i+1}}\mathbf{H}_{6\times6}(l_2)\boldsymbol{e}_6(3) \tag{8.71}$$

The vertical link $B_iC_{i+1}A_{i+2}$ shown in Fig. 8.10 will exert two forces along the $x_{2i+3}$- and $y_{2i+3}$-axes, respectively, on joint $C_{i+1}$ of the horizontal beam $A_{i+1}C_{i+1}B_{i+1}$. For the sake of simplicity, we assume that the joint forces provided to $A_{i+1}C_{i+1}B_{i+1}$ by link $B_iC_{i+1}A_{i+2}$ are

$$\boldsymbol{F}_{C_{(i+1)1}} = \begin{bmatrix} F_{xC_{(i+1)1}} \\ F_{yC_{(i+1)1}} \end{bmatrix} \tag{8.72}$$

where $F_{xC_{(i+1)1}}$ represents the force in the $x_{2i+3}$-direction and $F_{yC_{(i+1)1}}$ represents the force in the $y_{2i+3}$-direction. In accordance with Eq. (8.71), the output state vector of joint $C_{i+1}$ along link $A_{i+1}C_{i+1}B_{i+1}$ should be

$$\left[(\boldsymbol{\xi}_{6\times1})^{\text{output}}_{C_{(i+1)1}}\right]_{2i+3} = \begin{bmatrix} X_{C_{i1}} \\ Y_{C_{i1}} \\ \Theta_{zC_{i1}} \\ Q_{xC_{i1}} \\ Q_{yC_{i1}} \\ M_{zC_{i1}} \end{bmatrix}^{\text{output}}_{2i+3} = \mathbf{H}_{6\times6}(l_2)\mathbf{T}_{6\times6}\left(\frac{\pi}{2}\right)\mathbf{E}_3 \begin{bmatrix} X_{A_{i1}} \\ Y_{A_{i1}} \\ \Theta_{zA_{i1}} \\ Q_{xA_{i1}} \\ Q_{yA_{i1}} \\ M_{zA_{i1}} \end{bmatrix}^{\text{output}}_{2(i+1)}$$
$$+ \theta_{H_{i+1}}\mathbf{H}_{6\times6}(l_2)\boldsymbol{e}_6(3) + \mathbf{T}_Q^{\mathrm{T}} \begin{bmatrix} F_{xC_{(i+1)1}} \\ F_{yC_{(i+1)1}} \end{bmatrix} \tag{8.73}$$

where $\mathbf{E}_3 = \begin{bmatrix} 1&0&0&0&0&0 \\ 0&1&0&0&0&0 \\ 0&0&0&0&0&0 \\ 0&0&0&1&0&0 \\ 0&0&0&0&1&0 \\ 0&0&0&0&0&0 \end{bmatrix}$ represents a 6-dimension intercepting operation on a vector.

Similarly, the vertical link $B_iC_{i+1}$ shown in Fig. 8.10 will bear two reaction forces along the $x_{i+5}$- and $y_{i+5}$-axes from the horizontal link $A_{i+1}C_{i+1}B_{i+1}$ at joint $C_{i+1}$

$$\boldsymbol{F}_{C_{2(i+2)2}} = -\mathbf{A}\left(-\frac{\pi}{2}\right)\boldsymbol{F}_{C_{(2i+3)1}} = \mathbf{A}\left(\frac{\pi}{2}\right)\boldsymbol{F}_{C_{(2i+3)1}} \tag{8.74}$$

So according to Eq. (8.70), the output state vector of joint $C_{i+1}$ along the $x_{2(i+2)}$-direction should be

$$\left[(\boldsymbol{\xi}_{6\times1})^{\text{output}}_{C_{(i+1)2}}\right]_{2(i+2)} = \begin{bmatrix} X_{C_{(i+1)2}} \\ Y_{C_{(i+1)2}} \\ \Theta_{zC_{(i+1)2}} \\ Q_{xC_{(i+1)2}} \\ Q_{yC_{(i+1)2}} \\ M_{zC_{(i+1)2}} \end{bmatrix}^{\text{output}}_{2(i+2)} = -\mathbf{H}_{6\times6}(l_1)\mathbf{T}_{6\times6}\left(\frac{\pi}{2}\right)\mathbf{E}_3 \begin{bmatrix} X_{B_{i1}} \\ Y_{B_{i1}} \\ \Theta_{zB_{i1}} \\ Q_{xB_{i1}} \\ Q_{yB_{i1}} \\ M_{zB_{i1}} \end{bmatrix}^{\text{output}}_{2i+1} + \theta_{V_i}\mathbf{H}_{6\times6}(l_1)\boldsymbol{e}_6(3) + \mathbf{T}_Q^{\text{T}}\mathbf{A}\left(\frac{\pi}{2}\right)\begin{bmatrix} F_{xC_{(i+1)1}} \\ F_{yC_{(i+1)1}} \end{bmatrix} \tag{8.75}$$

With Eq. (8.25), the state vector, $\left[(\chi_{6\times1})^{\text{output}}_{C_{(i+1)1}}\right]_{2i+3}$ and $\left[(\chi_{6\times1})^{\text{output}}_{C_{(i+1)2}}\right]_{2(i+2)}$, can be transferred to joints $B_{i+1}$ and $A_{i+2}$:

$$\left[(\boldsymbol{\xi}_{6\times1})^{\text{output}}_{B_{(i+1)1}}\right]_{2i+3} = \begin{bmatrix} X_{B_{(i+1)1}} \\ Y_{B_{(i+1)1}} \\ \Theta_{zB_{(i+1)1}} \\ Q_{xB_{(i+1)1}} \\ Q_{yB_{(i+1)1}} \\ M_{zB_{(i+1)1}} \end{bmatrix}^{\text{output}}_{2i+3} = \mathbf{H}^2_{6\times6}(l_2)\mathbf{T}_{6\times6}\left(\frac{\pi}{2}\right)\mathbf{E}_3 \begin{bmatrix} X_{A_{i1}} \\ Y_{A_{i1}} \\ \Theta_{zA_{i1}} \\ Q_{xA_{i1}} \\ Q_{yA_{i1}} \\ M_{zA_{i1}} \end{bmatrix}^{\text{output}}_{2(i+1)} + \theta_{H_{i+1}}\mathbf{H}^2_{6\times6}(l_2)\boldsymbol{e}_6(6) \tag{8.76}$$

$$\left[(\boldsymbol{\xi}_{6\times1})^{\text{output}}_{A_{(i+2)1}}\right]_{2(i+2)} = \begin{bmatrix} X_{A_{(i+2)1}} \\ Y_{A_{(i+2)1}} \\ \Theta_{zA_{(i+2)1}} \\ Q_{xA_{(i+2)1}} \\ Q_{yA_{(i+2)1}} \\ M_{zA_{(i+2)1}} \end{bmatrix}^{\text{output}}_{2(i+2)} = -\mathbf{H}^2_{6\times6}(l_1)\mathbf{T}_{6\times6}\left(\frac{\pi}{2}\right)\mathbf{E}_3 \begin{bmatrix} X_{B_{i1}} \\ Y_{B_{i1}} \\ \Theta_{zB_{i1}} \\ Q_{xB_{i1}} \\ Q_{yB_{i1}} \\ M_{zB_{i1}} \end{bmatrix}^{\text{output}}_{2i+1} + \theta_{V_i}\mathbf{H}^2_{6\times6}(l_1)\boldsymbol{e}_6(6) \tag{8.77}$$

The known initial conditions for joints $B_{i+1}$ and $A_{i+2}$ are

$$\begin{cases} X_{B_{(i+1)1}} = 0 \\ Y_{B_{(i+1)1}} = 0 \\ M_{B_{(i+1)1}} = 0 \\ X_{A_{(i+2)1}} = 0 \\ Y_{B_{(i+2)1}} = 0 \\ M_{B_{(i+2)1}} = 0 \end{cases} \tag{8.78}$$

Because $C_{i+1}$ is pivot jointed, the following relation must hold from Fig. 8.10:

$$\begin{bmatrix} X_{C_{(i+1)1}} \\ Y_{C_{(i+1)1}} \end{bmatrix}_{2i+3}^{\text{output}} - \mathbf{A}\left(\frac{\pi}{2}\right)\begin{bmatrix} X_{C_{(i+1)2}} \\ Y_{C_{(i+1)2}} \end{bmatrix}_{2(i+2)}^{\text{output}} = 0 \tag{8.79}$$

Associating Eqs. (8.78) and (8.79), and substituting Eqs. (8.73), (8.75), (8.76), and (8.77) into Eqs. (8.78) and (8.79), we can obtain the constraint for frequency of the extended stair. When $i = 1$, we obtain

$$\begin{bmatrix} X_{C_{11}} - Y_{C_{12}} \\ Y_{C_{11}} + X_{C_{12}} \\ \hline M_{zB_{11}} \\ \hline M_{zA_{21}} \\ \hline X_{C_{21}} - Y_{C_{22}} \\ Y_{C_{21}} + X_{C_{22}} \\ \hline Q_{xB_{21}} \\ Q_{yB_{21}} \\ M_{zB_{21}} \\ \hline Q_{xA_{31}} \\ Q_{yA_{31}} \\ M_{zA_{31}} \end{bmatrix} = \mathbf{T}_{12\times 12} \begin{bmatrix} Q_{xA_0} \\ Q_{yA_0} \\ \hline F_{xC_{11}} \\ F_{yC_{11}} \\ \hline \theta_{H_1} \\ \hline \Theta_{zB_0} \\ Q_{xB_0} \\ Q_{yB_0} \\ \hline F_{xC_{21}} \\ F_{yC_{21}} \\ \hline \theta_{V_1} \\ \hline \theta_{H_2} \end{bmatrix} = 0 \tag{8.80}$$

where $\mathbf{T}_{12\times 12} = [\mathbf{T}_{12\times 5}\ \mathbf{T}_{12\times 7}]$ and

$$
\mathbf{T}_{12\times5} = \begin{bmatrix}
[\mathbf{T}_C\mathbf{P}_{A_1C_1}]_{2\times2} \\
\left[\boldsymbol{e}_3^{\mathrm{T}}(3)\mathbf{E}_{3\times6}\mathbf{H}_{6\times6}(l_2)\mathbf{P}_{A_1C_1}\right]_{1\times2} \\
[0]_{1\times2} \\
\left[\mathbf{A}\left(\frac{\pi}{2}\right)\mathbf{T}_C\mathbf{H}_{6\times6}(l_1)\mathbf{T}_{6\times6}\left(\frac{\pi}{2}\right)\mathbf{E}_3\mathbf{H}_{6\times6}(l_2)\mathbf{P}_{A_1C_1}\right]_{2\times2} \\
[0]_{3\times2} \\
\left[-\mathbf{E}_{3\times6}\mathbf{H}_{6\times6}^2(l_1)\mathbf{T}_{6\times6}\left(\frac{\pi}{2}\right)\mathbf{E}_3\mathbf{H}_{6\times6}(l_2)\mathbf{P}_{A_1C_1}\right]_{3\times2}
\end{bmatrix.
$$

$$
\begin{matrix}
\left[\mathbf{T}_C\mathbf{T}_Q^{\mathrm{T}} - \mathbf{A}\left(\frac{\pi}{2}\right)\mathbf{T}_C\mathbf{T}_Q^{\mathrm{T}}\mathbf{A}\left(\frac{\pi}{2}\right)\right]_{2\times2} \\
\left[\boldsymbol{e}_3^{\mathrm{T}}(3)\mathbf{E}_{3\times6}\mathbf{H}_{6\times6}(l_2)\mathbf{T}_Q^{\mathrm{T}}\right]_{1\times2} \\
\left[\boldsymbol{e}_3^{\mathrm{T}}(3)\mathbf{E}_{3\times6}\mathbf{H}_{6\times6}(l_1)\mathbf{T}_Q^{\mathrm{T}}\mathbf{A}\left(\frac{\pi}{2}\right)\right]_{1\times2} \\
\begin{bmatrix}
\mathbf{T}_C\mathbf{H}_{6\times6}(l_2)\mathbf{T}_{6\times6}\left(\dfrac{\pi}{2}\right)\mathbf{E}_3\mathbf{H}_{6\times6}(l_1)\mathbf{T}_Q^{\mathrm{T}}\mathbf{A}\left(\dfrac{\pi}{2}\right) \\
+\mathbf{A}\left(\dfrac{\pi}{2}\right)\mathbf{T}_C\mathbf{H}_{6\times6}(l_1)\mathbf{T}_{6\times6}\left(\dfrac{\pi}{2}\right)\mathbf{E}_3\mathbf{H}_{6\times6}(l_2)\mathbf{T}_Q^{\mathrm{T}}
\end{bmatrix}_{2\times2} \\
\left[\mathbf{E}_{3\times6}\mathbf{H}_{6\times6}^2(l_2)\mathbf{T}_{6\times6}\left(\frac{\pi}{2}\right)\mathbf{E}_3\mathbf{H}_{6\times6}(l_1)\mathbf{T}_Q^{\mathrm{T}}\mathbf{A}\left(\frac{\pi}{2}\right)\right]_{3\times2} \\
\left[-\mathbf{E}_{3\times6}\mathbf{H}_{6\times6}^2(l_1)\mathbf{T}_{6\times6}\left(\frac{\pi}{2}\right)\mathbf{E}_3\mathbf{H}_{6\times6}(l_2)\mathbf{T}_Q^{\mathrm{T}}\right]_{3\times2}
\end{matrix}
$$

$$
\begin{matrix}
[\mathbf{T}_C\mathbf{H}_{6\times6}(l_2)\boldsymbol{e}_6(3)]_{2\times1} \\
\left[\boldsymbol{e}_3^{\mathrm{T}}(3)\mathbf{E}_{3\times6}\mathbf{H}_{6\times6}^2(l_2)\boldsymbol{e}_6(3)\right]_{1\times1} \\
[0]_{1\times1} \\
\left[\mathbf{A}\left(\frac{\pi}{2}\right)\mathbf{T}_C\mathbf{H}_{6\times6}(l_1)\mathbf{T}_{6\times6}\left(\frac{\pi}{2}\right)\mathbf{E}_3\mathbf{H}_{6\times6}^2(l_2)\boldsymbol{e}_6(3)\right]_{2\times1} \\
[0]_{3\times1} \\
\left[-\mathbf{E}_{3\times6}\mathbf{H}_{6\times6}^2(l_1)\mathbf{T}_{6\times6}\left(\frac{\pi}{2}\right)\mathbf{E}_3\mathbf{H}_{6\times6}^2(l_2)\boldsymbol{e}_6(3)\right]_{3\times1}
\end{matrix}\Bigg]
$$

$$
\mathbf{T}_{12\times7}
$$

$$
= \begin{bmatrix}
-\left[\mathbf{A}\left(\frac{\pi}{2}\right)\mathbf{T}_C\mathbf{P}_{B_0}\right]_{2\times3} & [0]_{2\times2} \\
[0]_{1\times3} & [0]_{1\times2} \\
\left[\boldsymbol{e}_3^{T}(3)\mathbf{E}_{3\times6}\mathbf{H}_{6\times6}(l_1)\mathbf{P}_{B_0}\right]_{1\times3} & [0]_{1\times2} \\
\left[\mathbf{T}_C\mathbf{H}_{6\times6}(l_2)\mathbf{T}_{6\times6}\left(\frac{\pi}{2}\right)\mathbf{E}_3\mathbf{H}_{6\times6}(l_1)\mathbf{P}_{B_0}\right]_{2\times3} & \left[\mathbf{T}_C\mathbf{T}_Q^{\mathrm{T}}-\mathbf{A}\left(\frac{\pi}{2}\right)\mathbf{T}_C\mathbf{T}_Q^{\mathrm{T}}\mathbf{A}\left(\frac{\pi}{2}\right)\right]_{2\times2} \\
\left[\mathbf{E}_{3\times6}\mathbf{H}_{6\times6}^2(l_2)\mathbf{T}_{6\times6}\left(\frac{\pi}{2}\right)\mathbf{E}_3\mathbf{H}_{6\times6}(l_1)\mathbf{P}_{B_0}\right]_{3\times3} & \left[E_{3\times6}\mathbf{H}_{6\times6}(l_2)\mathbf{T}_Q^{\mathrm{T}}\right]_{3\times2} \\
[0]_{3\times3} & \left[\mathbf{E}_{3\times6}\mathbf{H}_{6\times6}(l_1)\mathbf{T}_Q^{\mathrm{T}}\mathbf{A}\left(\frac{\pi}{2}\right)\right]_{3\times2}
\end{bmatrix.
$$

$$
\begin{matrix}
[0]_{2\times1} & [0]_{2\times1} \\
[0]_{1\times1} & [0]_{1\times1} \\
[0]_{1\times1} & [0]_{1\times1} \\
\left[-\mathbf{A}\left(\frac{\pi}{2}\right)\mathbf{T}_C\mathbf{H}_{6\times6}(l_1)\boldsymbol{e}_6(3)\right]_{2\times1} & [\mathbf{T}_C\mathbf{H}_{6\times6}(l_2)\boldsymbol{e}_6(3)]_{2\times1} \\
[0]_{3\times1} & \left[\mathbf{E}_{3\times6}\mathbf{H}_{6\times6}^2(l_2)\boldsymbol{e}_6(3)\right]_{3\times1} \\
\left[\mathbf{E}_{3\times6}\mathbf{H}_{6\times6}^2(l_1)\boldsymbol{e}_6(3)\right]_{3\times1} & [0]_{3\times1}
\end{matrix}\Bigg] \tag{8.81}
$$

**Table 8.4** Comparisons for the 8-level stair with no terminal constraint

| Order | Simulations in ANSYS (Hz) | Numerical results (Hz) | Relative errors (%) |
|---|---|---|---|
| 1 | 5.9418 | 5.9444 | 0.0444 |
| 2 | 5.9750 | 5.9763 | 0.0212 |
| 3 | 36.4033 | 36.4226 | 0.0530 |
| 4 | 36.7037 | 36.7250 | 0.0580 |
| 5 | 98.4504 | 98.5620 | 0.1145 |
| 6 | 98.4513 | 98.5647 | 0.1151 |

With a set of parameters shown in Table 8.1, we can obtain the numerical results by substituting Eq. (8.81) into (8.63). Suppose that there is an eight-level stair.

When the last stair is free of any constraint, the whole deployed stair is a structural cantilever and the last level's constraints are given by the following equation set:

$$\begin{cases} Q_{xC_{81}} = 0 \\ Q_{yC_{81}} = 0 \\ M_{zC_{81}} = 0 \\ Q_{xC_{82}} = 0 \\ Q_{yC_{82}} = 0 \\ M_{zC_{81}} = 0 \\ X_{C_{81}} - Y_{C_{82}} = 0 \\ Y_{C_{81}} + X_{C_{82}} = 0 \end{cases} \tag{8.82}$$

Comparisons between theoretical results and the simulations through ANSYS are listed in Table 8.4.

Comparisons reveal that the numerical results based on the theoretical equations are very close to the simulation results in ANSYS. For the sake of clarity, the corresponding normal modes obtained from ANSYS are presented in Fig. 8.11.

When the last stair is fixed, the last level's constraints are given by the following equation set:

$$\begin{cases} Q_{x\,C_{81}} = 0 \\ Q_{y\,C_{81}} = 0 \\ M_{zC_{81}} = 0 \\ X_{C_{82}} = 0 \\ Y_{yC_{82}} = 0 \\ M_{zC_{82}} = 0 \\ X_{C_{81}} - Y_{C_{82}} = 0 \\ Y_{C_{81}} + X_{C_{82}} = 0 \end{cases} \tag{8.83}$$

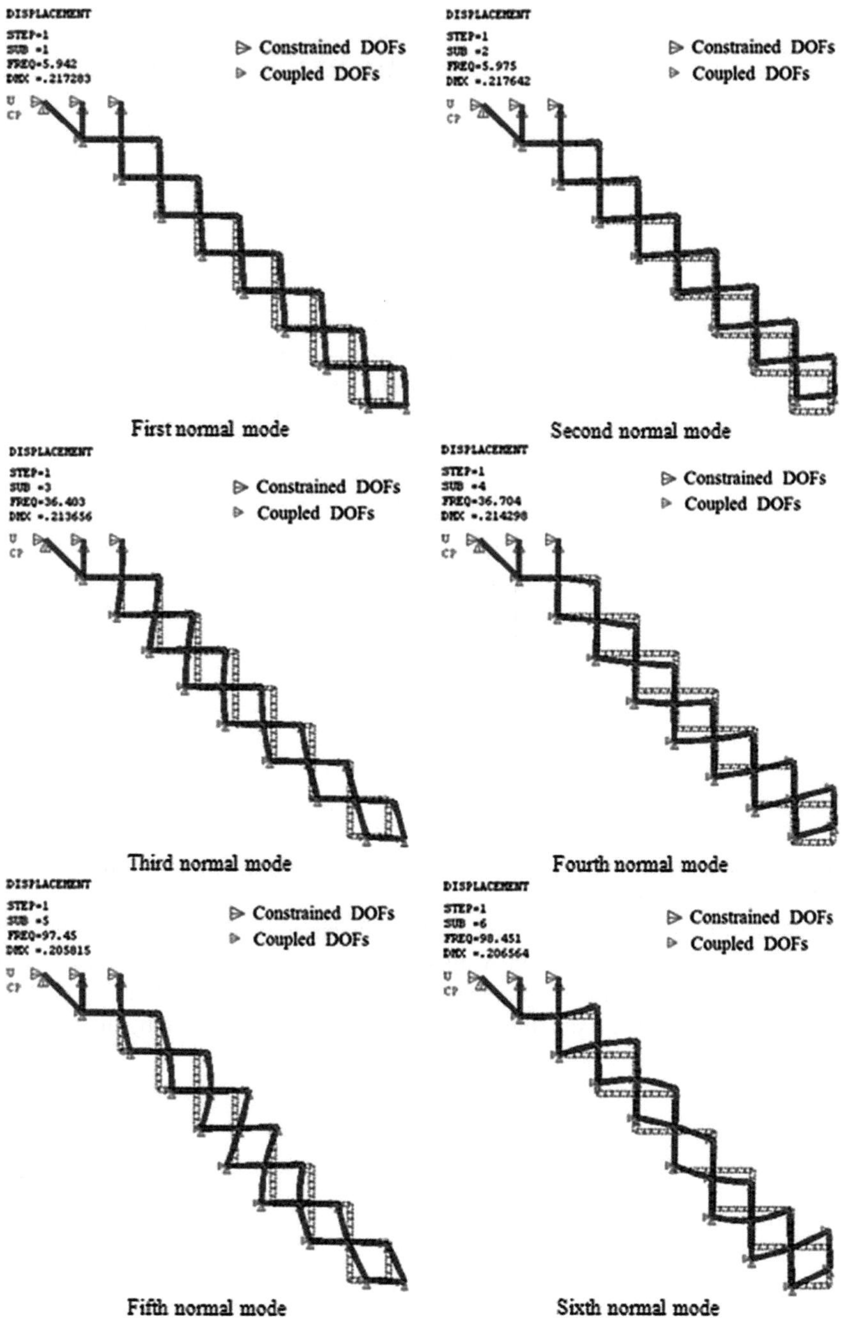


**Fig. 8.11** The first six normal modes of an 8-level stair with no terminal constraint

**Table 8.5** Comparisons for the 8-level foldable stair with fixed terminal constraints

| Order | Simulations in ANSYS (Hz) | Numerical results (Hz) | Relative errors (%) |
|---|---|---|---|
| 1 | 25.8640 | 25.8706 | 0.0257 |
| 2 | 43.9559 | 43.9666 | 0.0242 |
| 3 | 80.9052 | 80.9624 | 0.0704 |
| 4 | 118.1243 | 118.2335 | 0.0937 |
| 5 | 159.2604 | 159.4892 | 0.1436 |
| 6 | 215.3054 | 215.7345 | 0.1993 |

**Table 8.6** Comparisons for the 15-level stair with no terminal constraints

| Order | Simulations in ANSYS (Hz) | Numerical results (Hz) | Relative errors (%) |
|---|---|---|---|
| 1 | 1.5942 | 1.5947 | 0.0334 |
| 2 | 1.5984 | 1.6011 | 0.0169 |
| 3 | 9.9304 | 9.9345 | 0.0408 |
| 4 | 9.9633 | 9.9663 | 0.0299 |
| 5 | 28.4610 | 28.4701 | 0.0333 |
| 6 | 28.5695 | 28.5784 | 0.0322 |

Comparisons between the theoretical results and the simulations through ANSYS are listed in Table 8.5.

Comparisons reveal that the numerical results based on the theoretical equations are very close to the simulation results in ANSYS. For the sake of clarity, the corresponding normal modes obtained from ANSYS are presented in Fig. 8.12.

With a set of parameters shown in Table 8.1 and suppose that there is a 15-level stair, we can obtain the numerical results by substituting Eq. (8.81) into (8.63). When the last stair is free of any constraint, the whole deployed stair is a structural cantilever and the last level's constraints are given by the following equation set:

$$\begin{cases} Q_{xC_{151}} = 0 \\ Q_{yC_{151}} = 0 \\ M_{zC_{151}} = 0 \\ Q_{xC_{152}} = 0 \\ Q_{yC_{152}} = 0 \\ M_{zC_{151}} = 0 \\ X_{C_{151}} - Y_{C_{152}} = 0 \\ Y_{C_{151}} + X_{C_{152}} = 0 \end{cases} \tag{8.84}$$

Comparisons between theoretical results and the simulations through ANSYS are listed in Table 8.6.

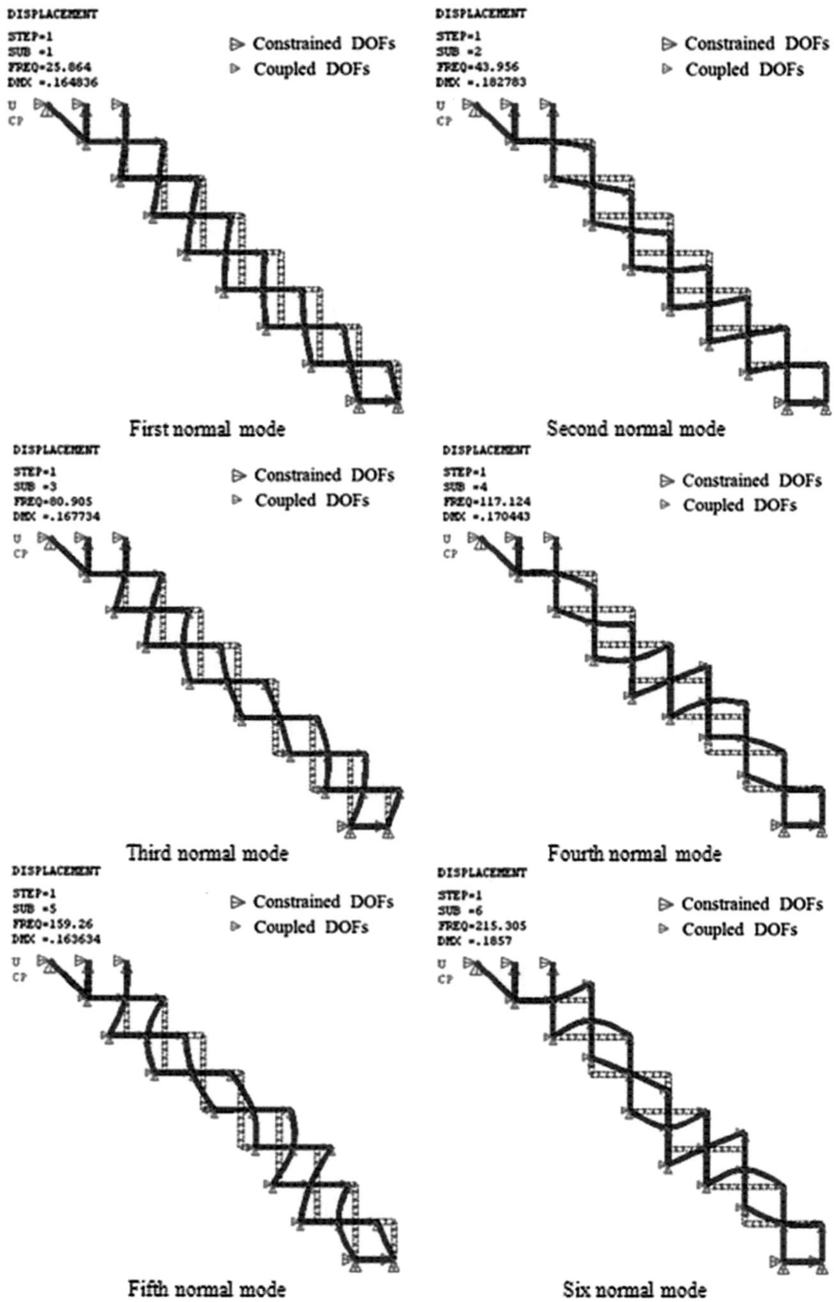


**Fig. 8.12** The first six normal modes of an 8-level foldable stair with fixed terminal constraints

**Table 8.7** Comparisons for the 15-level foldable stair with fixed terminal constraints

| Order | Simulations in ANSYS (Hz) | Numerical results (Hz) | Relative errors (%) |
|---|---|---|---|
| 1 | 6.9779 | 6.9805 | 0.0378 |
| 2 | 10.9263 | 10.9276 | 0.0117 |
| 3 | 22.3960 | 22.4026 | 0.0297 |
| 4 | 29.8760 | 29.8829 | 0.0232 |
| 5 | 45.9681 | 45.9862 | 0.0394 |
| 6 | 58.6706 | 58.6968 | 0.0455 |

When the last stair is fixed, the last level's constraints are given by the following equations:

$$\begin{cases} Q_{xC_{151}} = 0 \\ Q_{yC_{151}} = 0 \\ M_{zC_{151}} = 0 \\ X_{C_{152}} = 0 \\ Y_{yC_{152}} = 0 \\ M_{zC_{152}} = 0 \\ X_{C_{151}} - Y_{C_{152}} = 0 \\ Y_{C_{151}} + X_{C_{152}} = 0 \end{cases} \tag{8.85}$$

Comparisons between theoretical results and the simulations through ANSYS are listed in Table 8.7.

Comparisons reveal that the numerical results based on the theoretical equations are very close to the simulation results in ANSYS. Table 8.7 also indicates that the 15-level stair can be used to connect two adjacent floors whose distance is as far as 4.5 m.

## 8.5 Structural Dynamics of a Unit for the Foldable Wing Frame

Suppose that the actuator of the morphing wing is mounted on the revolute joint $R_1$. The deployable frame will be fixed when the relative angle of the two links of joint $R_1$ is specified. The structural dynamics of the frame can be established by analyzing the dynamics of every link.

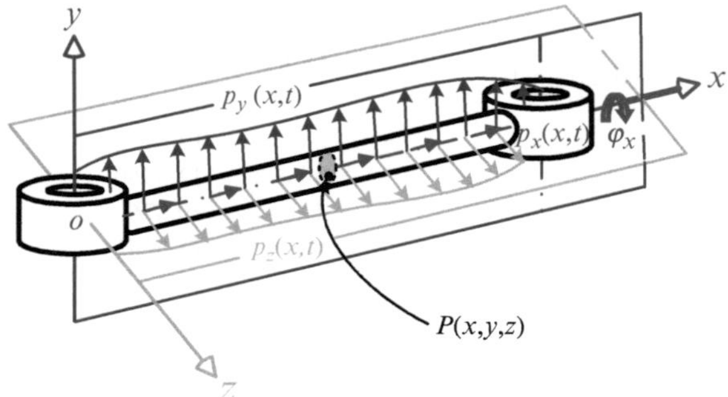


**Fig. 8.13** Multi-direction vibrations of a link

## *8.5.1 Structural Dynamics of a Link*

Because the components of the foldable frame are all thin and long straight links, we can suppose that the links are all Euler-Bernoulli beams. So for the sake of simplicity, the $x_i$-axis is set to superimpose on the axis of the $i$th $(i = 1, 2, \ldots)$ link, the $y_i$-axis is set to superimpose on the axis of one end revolute joint of the $i$th $(i = 1, 2, \ldots)$ link, and the $z_i$-axis is determined in accordance with the right-hand rule.

The link shown in Fig. 8.13 is a slender uniform rod. The displacements of a point on the central axis of the link are expressed as

$$\begin{cases} u = u(x,t) \\ v = v(x,t) \\ w = w(x,t) \end{cases} \tag{8.86}$$

where $u(x,t)$ represents the displacement component of the point in the $x$-direction, $v(x,t)$ represents the displacement component of the point in the $y$-direction, and $w(x,t)$ represents the displacement component of the point in the $z$-direction.

So for any point $P$ on the cross section of the round beam shown in Fig. 8.13, the displacements can be represented by

$$\begin{cases} u_A = u(x,t) - y\dfrac{\partial v}{\partial x} - z\dfrac{\partial w}{\partial x} \\ v_A = v(x,t) - r\varphi_x \cos\alpha \\ w_A = w(x,t) + r\varphi_x \sin\alpha \end{cases} \tag{8.87}$$

where $r$ denotes the distance from point $P$ to the center, denoted by $o_1$, of the cross section; $\varphi_x$ stands for the torsional angle around the $x$-axis; and $\alpha$ is the angle from the $z$-axis to the vector $\overline{o_1 P}$.

The axial strain and stress in the $x$-direction at point $P$ are

$$\begin{cases} \varepsilon_x = \dfrac{\partial u_A}{\partial x} = \dfrac{\partial u(x,t)}{\partial x} - y\dfrac{\partial^2 v}{\partial x^2} - z\dfrac{\partial^2 w}{\partial x^2} \\ \sigma_x = E\varepsilon_x \end{cases} \tag{8.88}$$

where $\varepsilon_x$ denotes the $x$-axial strain at point $P$, $\sigma_x$ indicates the stress in the $x$-direction at point $P$, and $E$ is the Young's elastic modulus of the material of the link.

The torsional strain about the $x$-axis and the shearing stress at point $P$ are

$$\gamma_x = r\frac{\partial \varphi_x}{\partial x}, \quad \tau_A = G\gamma_x = Gr\frac{\partial \varphi_x}{\partial x} \tag{8.89}$$

where $\gamma_x$ denotes the torsional strain about the $x$-axis, $\tau_x$ represents the shearing stress at point $P$, and $G$ is the torsional elastic modulus of the material of the link.

Hereto, the kinetic and the potential energy of the round beam shown in Fig. 8.13 can be obtained

$$\begin{cases} K = \dfrac{1}{2}\displaystyle\iiint_V \rho\left[\left(\frac{\partial u_A}{\partial t}\right)^2 + \left(\frac{\partial v_A}{\partial t}\right)^2 + \left(\frac{\partial w_A}{\partial t}\right)^2\right]\mathrm{d}V \\ \quad = \dfrac{1}{2}\displaystyle\int_0^l \rho\left\{\iint\limits_A\left[\left(\frac{\partial u}{\partial t} - y\frac{\partial^2 v}{\partial t\partial x} - z\frac{\partial^2 w}{\partial t\partial x}\right)^2 + \left(\frac{\partial v}{\partial t} - \frac{\partial \varphi_x}{\partial t}r\cos\alpha\right)^2\right.\right. \\ \quad \left.\left. + \left(\frac{\partial w}{\partial t} + \frac{\partial \varphi_x}{\partial t}r\sin\alpha\right)^2\right]\mathrm{d}y\mathrm{d}z\right\}\mathrm{d}x \\ U = \dfrac{1}{2}\displaystyle\iiint_V (\sigma_x\varepsilon_x + \tau_x\gamma_x)\mathrm{d}V \\ \quad = \dfrac{1}{2}\displaystyle\int_0^l \left\{\begin{array}{l} E\displaystyle\iint\limits_A\left[\frac{\partial u(x,t)}{\partial x}\right]^2\mathrm{d}y\mathrm{d}z - 2E\iint\limits_A \frac{\partial u(x,t)}{\partial x}\left(y\frac{\partial^2 v}{\partial x^2} + z\frac{\partial^2 w}{\partial x^2}\right)\mathrm{d}y\mathrm{d}z \\ + EI_z\left(\dfrac{\partial^2 v}{\partial x^2}\right)^2 + 2EI_{yz}\dfrac{\partial^2 v}{\partial x^2}\dfrac{\partial^2 w}{\partial x^2} + EI_y\left(\dfrac{\partial^2 w}{\partial x^2}\right)^2 + GI_x\left(\dfrac{\partial \varphi_x}{\partial x}\right)^2 \end{array}\right\}\mathrm{d}x \end{cases} \tag{8.90}$$

where $K$ represents the kinetic energy, $U$ represents the potential energy, $\rho$ is the linear density of mass of the link, $A$ is the sectional area of the beam which is assumed to be constant for the link shown in Fig. 8.13, and $I_y = \iint_A z^2\mathrm{d}A$ and $I_z = \iint_A y^2\mathrm{d}A$ which represent the moments of inertia about the $y$- and $z$-directions, respectively, and $I_{yz} = \iint_A yz\mathrm{d}A$ which is the product inertia of the section, $I_x = \iint_A \left(y^2 + z^2\right)\mathrm{d}A$ which represents the polar moment of inertia.

The virtual work done by the distributed loads $p_x(x,t)$, $p_y(x,t)$, and $p_z(x,t)$ and the bending moments $M_y(0,t)$, $M_z(0,t)$, $M_y(l,t)$, and $M_z(l,t)$ and the shearing forces $Q_y(0,t)$, $Q_z(0,t)$, $Q_y(l,t)$, and $Q_z(l,t)$ and the tensile forces $Q_x(0,t)$ and $Q_x(l,t)$ and the torques $T_x(0,t)$ and $T_x(l,t)$ should be

$$\begin{aligned}\delta W = &\int_0^l (p_x\delta u + p_y\delta v + p_z\delta w + T_x\delta\varphi)\mathrm{d}x\\ &+ Q_y(0,t)\delta v(0) + Q_z(0,t)\delta w(0) - Q_y(l,t)\delta v(l) - Q_z(l,t)\delta w(l)\\ &- Q_x(0,t)\delta u(0) + Q_x(l,t)\delta u(l)\\ &- M_y(0,t)\delta\left(\frac{\partial w(0)}{\partial x}\right) + M_z(0,t)\delta\left(\frac{\partial v(0)}{\partial x}\right) + M_y(l,t)\delta\left(\frac{\partial w(l)}{\partial x}\right)\\ &- M_z(l,t)\delta\left(\frac{\partial v(l)}{\partial x}\right)\\ &- T_x(0,t)\delta\left(\frac{\partial \varphi(0)}{\partial x}\right) + T_x(l,t)\delta\left(\frac{\partial \varphi(l)}{\partial x}\right)\end{aligned} \tag{8.91}$$

According to Hamilton's principle, we immediately gain

$$\delta\int_{t_1}^{t_2}(K-U)\mathrm{d}t + \int_{t_1}^{t_2}\delta W\mathrm{d}t = 0 \tag{8.92}$$

Substituting Eqs. (8.90) and (8.91) into (8.92) yields

$$\begin{aligned}&\delta\int_{t_1}^{t_2}\frac{1}{2}\int_0^l \rho\left\{\iint\limits_A\left[\left(\frac{\partial u}{\partial t} - y\frac{\partial^2 v}{\partial t\partial x} - z\frac{\partial^2 w}{\partial t\partial x}\right)^2 + \left(\frac{\partial v}{\partial t} - \frac{\partial\varphi_x}{\partial t}r\cos\alpha\right)^2\right.\right.\\ &\qquad\left.\left.+\left(\frac{\partial w}{\partial t} + \frac{\partial\varphi_x}{\partial t}r\sin\alpha\right)^2\right]\mathrm{d}y\mathrm{d}z\right\}\mathrm{d}x\mathrm{d}t\\ &-\delta\int_{t_1}^{t_2}\frac{1}{2}\int_0^l\left\{\begin{aligned}&E\iint\limits_A\left[\frac{\partial u(x,t)}{\partial x}\right]^2\mathrm{d}y\mathrm{d}z - 2E\iint\limits_A\frac{\partial u(x,t)}{\partial x}\left(y\frac{\partial^2 v}{\partial x^2} + z\frac{\partial^2 w}{\partial x^2}\right)\mathrm{d}y\mathrm{d}z\\ &+EI_y\left(\frac{\partial^2 w}{\partial x^2}\right)^2 + 2EI_{yz}\frac{\partial^2 v}{\partial x^2}\frac{\partial^2 w}{\partial x^2} + EI_z\left(\frac{\partial^2 v}{\partial x^2}\right)^2 + GI_x\left(\frac{\partial\varphi_x}{\partial x}\right)^2\end{aligned}\right\}\mathrm{d}x\mathrm{d}t\end{aligned}$$

$$
+\int_{t_1}^{t_2}\left[\begin{array}{l}
\int_0^l (p_x\delta u + p_y\delta v + p_z\delta w)\mathrm{d}x \\
+ Q_y(0,t)\delta v(0) + Q_z(0,t)\delta w(0) - Q_y(l,t)\delta v(l) - Q_z(l,t)\delta w(l) \\
- Q_x(0,t)\delta u(0) + Q_x(l,t)\delta u(l) \\
-M_y(0,t)\delta\left(\dfrac{\partial w(0)}{\partial x}\right) + M_z(0,t)\delta\left(\dfrac{\partial v(0)}{\partial x}\right) + M_y(l,t)\delta\left(\dfrac{\partial w(l)}{\partial x}\right) \\
- M_z(l,t)\delta\left(\dfrac{\partial v(l)}{\partial x}\right) \\
- T_x(0,t)\delta\left(\dfrac{\partial \varphi(0)}{\partial x}\right) + T_x(l,t)\delta\left(\dfrac{\partial \varphi(l)}{\partial x}\right)
\end{array}\right]\mathrm{d}t = 0 \tag{8.93}
$$

Considering the boundary conditions, Eq. (8.93) presents the dynamics equation sets

$$
\left[\begin{array}{l}
EA\dfrac{\partial u^2(x,t)}{\partial x^2} - \rho A\dfrac{\partial^2 u}{\partial t^2} + \rho S_z\dfrac{\partial^3 v}{\partial t^2\partial x} + \rho S_y\dfrac{\partial^3 w}{\partial t^2\partial x} - ES_z\dfrac{\partial^3 v}{\partial x^3} - ES_y\dfrac{\partial^3 w}{\partial x^3} + p_x \\
\rho I_{yz}\dfrac{\partial^4 w}{\partial x^2\partial t^2} + \rho I_z\dfrac{\partial^4 v}{\partial x^2\partial t^2} - \rho S_z\dfrac{\partial^3 u}{\partial x\partial t^2} - \rho A\dfrac{\partial^2 v}{\partial t^2} + \rho S_y\dfrac{\partial^2 \varphi_x}{\partial t^2} - EI_{yz}\dfrac{\partial^4 w}{\partial x^4} \\
\quad - EI_z\dfrac{\partial^4 v}{\partial x^4} + ES_z\dfrac{\partial u^3(x,t)}{\partial x^3} + p_y \\
\rho I_y\dfrac{\partial^4 w}{\partial x^2\partial t^2} + \rho I_{yz}\dfrac{\partial^4 v}{\partial x^2\partial t^2} - \rho S_y\dfrac{\partial^3 u}{\partial x\partial t^2} - \rho A\dfrac{\partial^2 w}{\partial t^2} - \rho S_z\dfrac{\partial^2 \varphi_x}{\partial t^2} - EI_y\dfrac{\partial^4 w}{\partial x^4} \\
\quad - EI_{yz}\dfrac{\partial^4 v}{\partial x^4} + ES_y\dfrac{\partial u^3(x,t)}{\partial x^3} + p_z \\
GI_x\dfrac{\partial^2 \varphi_x}{\partial x^2} + \rho S_y\dfrac{\partial^2 v}{\partial t^2} - \rho I_x\dfrac{\partial^2 \varphi_x}{\partial t^2} - \rho S_z\dfrac{\partial^2 w}{\partial t^2} + T_x
\end{array}\right] = 0 \tag{8.94}
$$

where $S_y = \iint\limits_A z\mathrm{d}y\mathrm{d}z$, which denotes the first moment of a cross section about the $y$-axis; $S_z = \iint\limits_A y\mathrm{d}y\mathrm{d}z$, which denotes the first moment of a cross section about the $z$-axis; $I_y = \iint\limits_A z^2\mathrm{d}y\mathrm{d}z$, which denotes the inertia moment of a cross section about the $y$-axis; $I_z = \iint\limits_A y^2\mathrm{d}y\mathrm{d}z$, which denotes the inertia moment of a cross section about the $z$-axis; $I_{yz} = \iint\limits_A yz\mathrm{d}y\mathrm{d}z$, which denotes the product of inertia of a cross section perpendicular to the $x$-axis; and $I_x = I_y + I_z$, which denotes the polar moment of inertial about the $x$-axis.

Suppose that the link shown in Fig. 8.13 is a uniform round beam and its external loads satisfy that $p_x = p_y = p_z = T_x = 0$. Immediately, we gain

$$\begin{cases} I_{yz} = 0 \\ S_y = 0 \\ S_z = 0 \end{cases} \tag{8.95}$$

As a result, Eq. (8.94) can be simplified as

$$\begin{cases} E\dfrac{\partial u^2(x,t)}{\partial x^2} - \rho\dfrac{\partial^2 u}{\partial t^2} = 0 \\ EI_z\dfrac{\partial^4 v}{\partial x^4} - \rho I_z\dfrac{\partial^4 v}{\partial x^2 \partial t^2} + \rho A\dfrac{\partial^2 v}{\partial t^2} = 0 \\ EI_y\dfrac{\partial^4 w}{\partial x^4} - \rho I_y\dfrac{\partial^4 w}{\partial x^2 \partial t^2} + \rho A\dfrac{\partial^2 w}{\partial t^2} = 0 \\ G\dfrac{\partial^2 \varphi_x}{\partial x^2} - \rho\dfrac{\partial^2 \varphi_x}{\partial t^2} = 0 \end{cases} \tag{8.96}$$

Therefore, the differential equations in equation (8.96) are decoupled. The first one is the differential equation of longitudinal vibration in the $x$-direction for the link shown in Fig. 8.13, the second and third are the transverse vibration equations in the $xoy$- and $xoz$-planes, individually, and the last one is the torsional vibration equation about the $x$-direction.

For the differential equation of longitudinal vibration in the $x$-direction, the solution can be supposed as

$$X(x,t) = X(x)\sin(\omega t + \theta) \tag{8.97}$$

where $X(x)$ denotes the mode function of $x$ for the longitudinal vibration, $\omega$ denotes the angular frequency, and $\theta$ stands for the phase of the vibration.

Substituting Eq. (8.97) into the first one of Eq. (8.96), we immediately obtain the general solution of the shape function of the longitudinal vibration mode of the beam

$$X(x) = C_1\cos(kx) + C_2\sin(kx) \tag{8.98}$$

where $k = \omega\sqrt{\frac{\rho}{E}}$ which represents the frequency of mode function of longitudinal vibration of the beam, and $C_1$ and $C_2$ can be determined from the initial conditions.

The axial inner force of the link is therefore obtained from Eq. (8.97):

$$Q_x = -EA\frac{\partial X}{\partial x} = C_1 EAk\sin(kx) - C_2 EAk\cos(kx) \tag{8.99}$$

For the second equation of Eq. (8.96), we can also suppose

$$y(x,t) = Y(x)\sin(\omega t + \theta) \tag{8.100}$$

where $y(x,t)$ represents the vibration function of $x$ for the transversal vibration in the $xoy$-plane, $Y(x)$ represents the shape function of bending vibration in the $xoy$-plane at $x$, $\omega$ represents the frequency of a sine vibration, and $\theta$ represents the phase angle. Also assume that the mass is uniformly distributed within the whole link, the second equation of Eq. (8.96) can be transformed into

$$\frac{d^4Y(x)}{dx^4} + \frac{\rho}{E}\omega^2\frac{d^2Y(x)}{dx^2} - \frac{\rho A\omega^2}{EI_z}Y(x) = 0 \tag{8.101}$$

The general solution of Eq. (8.101) is

$$Y(x) = C_3\cosh(\lambda_1 x) + C_4\sinh(\lambda_1 x) + C_5\cos(\lambda_2 x) + C_6\sin(\lambda_2 x) \tag{8.102}$$

where $\lambda_1 = \sqrt{\frac{\sqrt{\left(\frac{\rho}{E}\omega^2\right)^2 + \frac{4\rho A\omega^2}{EI_z}} - \frac{\rho}{E}\omega^2}{2}}$, and $\lambda_2 = \sqrt{\frac{\sqrt{\left(\frac{\rho}{E}\omega^2\right)^2 + \frac{4\rho A\omega^2}{EI_z}} + \frac{\rho}{E}\omega^2}{2}}$; $C_i (i = 3, 4, 5, 6)$ can be determined by the initial boundary conditions of the beam. The Euler-Bernoulli beam satisfies

$$\begin{cases} \Theta_z = \dfrac{\mathrm{d}Y}{\mathrm{d}x} \\ M_z = -EI_z\dfrac{\mathrm{d}\Theta_z}{\mathrm{d}x} \\ Q_y = -\dfrac{\mathrm{d}M_z}{\mathrm{d}x} \end{cases} \tag{8.103}$$

where $\Theta_z$ denotes the deflection of the beam in the $z$-direction, $M_z$ represents the moment about the $z$-direction, and $Q_y$ stands for the shearing force in the $y$-direction.

Similarly, in the $xoz$-plane, the general solution of the bending vibration of the Euler-Bernoulli beam represented by the third differential equation of equation set (8.96) pivoted at both ends [45] is

$$Z(x) = C_7\cosh(\xi_1 x) + C_8\sinh(\xi_1 x) + C_9\cos(\xi_2 x) + C_{10}\sin(\xi_2 x) \tag{8.104}$$

where $Z(x)$ represents the shape function of bending vibration in the $xoz$-plane at $x$, $\xi_1 = \sqrt{\frac{\sqrt{\left(\frac{\rho}{E}\omega^2\right)^2 + \frac{4\rho A\omega^2}{EI_y}} - \frac{\rho}{E}\omega^2}{2}}$, and $\xi_2 = \sqrt{\frac{\sqrt{\left(\frac{\rho}{E}\omega^2\right)^2 + \frac{4\rho A\omega^2}{EI_y}} + \frac{\rho}{E}\omega^2}{2}}$ and

$$\begin{cases} \Theta_y = -\dfrac{\mathrm{d}Z}{\mathrm{d}x} \\ M_y = -EI_y\dfrac{\mathrm{d}\Theta_y}{\mathrm{d}x} \\ Q_z = \dfrac{\mathrm{d}M_y}{\mathrm{d}x} \end{cases} \tag{8.105}$$

For the fourth differential equation of torsional vibration around the $x$-direction, the solution can be supposed as

$$\phi(x,t) = \Theta(x)\sin(\omega t + \varphi) \tag{8.106}$$

where $\phi(x,t)$ denotes the torsional angle about the $x$-axis at $(x,t)$ and $\Theta(x)$ denotes the shape function of longitudinal vibration at $x$.

Substituting Eq. (8.106) into the last one of Eq. (8.96), we immediately obtain the general solution of the shape function of the longitudinal vibration mode of the beam

$$\Theta_x = C_{11}\cos(\varsigma x) + C_{12}\sin(\varsigma x) \tag{8.107}$$

where $\Theta_x$ denotes the angle mode function of torsional vibration around the $x$-axis, $\varsigma = \omega\sqrt{\frac{\rho}{G}}$ which denotes the frequency of mode function of torsional vibration of the beam, and $C_{11}$ and $C_{12}$ can be determined from the initial conditions.

The axial inner torque of the link is therefore obtained from Eq. (8.106):

$$M_x = -GI_x\frac{\partial\phi}{\partial x} = C_{11}GI_x\varsigma\sin(\varsigma x) - C_{12}GI_x\varsigma\cos(\varsigma x) \tag{8.108}$$

Suppose that the state vector is denoted by

$$\boldsymbol{b}_{12}(x) = \begin{bmatrix} X & Y & Z & \Theta_x & \Theta_y & \Theta_z & M_x & M_y & M_z & Q_x & Q_y & Q_z \end{bmatrix}^{\mathrm{T}} \tag{8.109}$$

and the constant coefficient vector is denoted by

$$\boldsymbol{c}_{12} = \begin{bmatrix} C_1 & C_2 & C_3 & C_4 & C_5 & C_6 & C_7 & C_8 & C_9 & C_{10} & C_{11} & C_{12} \end{bmatrix}^{\mathrm{T}} \tag{8.110}$$

So the state vector can be represented as

$$\boldsymbol{b}_{12}(x) = \mathbf{B}_{12\times 12}(x)\boldsymbol{c}_{12} \tag{8.111}$$

where $\mathbf{B}_{12\times 12}(x) = \begin{bmatrix} \mathbf{B}_{11}(x) & \mathbf{B}_{12}(x) \\ \mathbf{B}_{21}(x) & \mathbf{B}_{22}(x) \end{bmatrix}$ which stands for the state vector matrix, and

$$\mathbf{B}_{11}(x)=\begin{bmatrix}\cos(kx) & \sin(kx) & 0 & 0 & 0 & 0\\ 0 & 0 & \cosh(\lambda_1 x) & \sinh(\lambda_1 x) & \cos(\lambda_2 x) & \sin(\lambda_2 x)\\ 0 & 0 & 0 & 0 & 0 & 0\\ 0 & 0 & 0 & 0 & 0 & 0\\ 0 & 0 & 0 & 0 & 0 & 0\\ 0 & 0 & \lambda_1\sinh(\lambda_1 x) & \lambda_1\cosh(\lambda_1 x) & -\lambda_2\sin(\lambda_2 x) & \lambda_2\cos(\lambda_2 x)\end{bmatrix},$$

$$\mathbf{B}_{12}(x)=\begin{bmatrix}0 & 0 & 0 & 0 & 0 & 0\\ 0 & 0 & 0 & 0 & 0 & 0\\ \cosh(\xi_1 x) & \sinh(\xi_1 x) & \cos(\xi_2 x) & \sin(\xi_2 x) & 0 & 0\\ 0 & 0 & 0 & 0 & \cos(\varsigma x) & \sin(\varsigma x)\\ -\xi_1\sinh(\xi_1 x) & -\xi_1\cosh(\xi_1 x) & +\xi_2\sin(\xi_2 x) & -\xi_2\cos(\xi_2 x) & 0 & 0\\ 0 & 0 & 0 & 0 & 0 & 0\end{bmatrix},$$

$$\mathbf{B}_{21}(x)=\begin{bmatrix}0 & 0 & 0 & 0\\ 0 & 0 & 0 & 0\\ 0 & 0 & -EI_z\lambda_1^2\cosh(\lambda_1 x) & -EI_z\lambda_1^2\sinh(\lambda_1 x)\\ EAk\sin(kx) & -EAk\cos(kx) & 0 & 0\\ 0 & 0 & EI_z\lambda_1^3\sinh(\lambda_1 x) & EI_z\lambda_1^3\cosh(\lambda_1 x)\\ 0 & 0 & 0 & 0\end{bmatrix.}$$

$$\left.\begin{matrix}0 & 0\\ 0 & 0\\ +EI_z\lambda_2^2\cos(\lambda_2 x) & +EI_z\lambda_2^2\sin(\lambda_2 x)\\ 0 & 0\\ EI_z\lambda_2^3\sin(\lambda_2 x) & -EI_z\lambda_2^3\cos(\lambda_2 x)\\ 0 & 0\end{matrix}\right],$$

$$\mathbf{B}_{22}(x)=\begin{bmatrix}0 & 0 & 0 & 0\\ EI_y\xi_1^2\cosh(\xi_1 x) & EI_y\xi_1^2\sinh(\xi_1 x) & -EI_y\xi_2^2\cos(\xi_2 x) & -EI_y\xi_2^2\sin(\xi_2 x)\\ 0 & 0 & 0 & 0\\ 0 & 0 & 0 & 0\\ 0 & 0 & 0 & 0\\ EI_y\xi_1^3\sinh(\xi_1 x) & EI_y\xi_1^3\cosh(\xi_1 x) & EI_y\xi_2^3\sin(\xi_2 x) & -EI_y\xi_2^3\cos(\xi_2 x)\end{matrix.}$$

$$\left.\begin{matrix}GI_x\varsigma\sin(\varsigma x) & -GI_x\varsigma\cos(\varsigma x)\\ 0 & 0\\ 0 & 0\\ 0 & 0\\ 0 & 0\\ 0 & 0\end{matrix}\right].$$

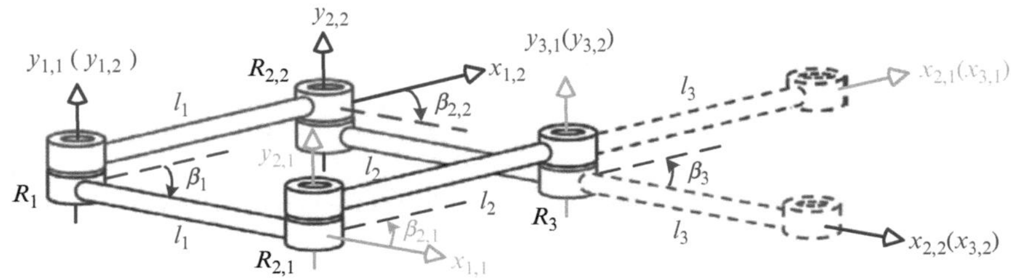


**Fig. 8.14** One quadrilateral element

The initial state vector of the left jointed point is known, so substituting $x = 0$ into Eq. (8.111) presents

$$\boldsymbol{b}_{12}(0) = \mathbf{B}_{12\times12}(0)\boldsymbol{c}_{12}$$

Immediately, we get

$$\boldsymbol{c}_{12} = \mathbf{B}^{-1}_{12\times12}(0)\boldsymbol{b}_{12}(0) \tag{8.112}$$

where $\mathbf{B}^{-1}_{12\times12}(0)$ is the inverse of matrix $\mathbf{B}_{12\times12}(0)$.

Substituting Eq. (8.112) in (8.111) yields

$$\boldsymbol{b}_{12}(x) = \mathbf{H}_{12\times12}(x)\boldsymbol{b}_{12}(0) \tag{8.113}$$

where $\mathbf{H}_{12\times12}(x) = \mathbf{B}_{12\times12}(x)\mathbf{B}^{-1}_{12\times12}(0)$, which represents the transfer matrix of the link from the state vector of the known point at $x = 0$ to that of the point at any $x$.

With Eq. (8.113), the state vector of the right end of a rod whose length is $l$ can be expressed as

$$\boldsymbol{b}_{12}(l) = \mathbf{H}_{12\times12}(l)\boldsymbol{b}_{12}(0) \tag{8.114}$$

Assume that the state vector of the $i$th section on the link is denoted by $[\boldsymbol{b}_{12}]_i = \left[X\ Y\ Z\ \Theta_x\ \Theta_y\ \Theta_z\ M_x\ M_y\ M_z\ Q_x\ Q_y\ Q_z\right]_i^{\mathrm{T}}$, the $(i+1)$th state vector of the link can be expressed as

$$[\boldsymbol{b}_{12}]_{i+1} = \mathbf{H}_{12\times12}(l_i)[\boldsymbol{b}_{12}]_i \tag{8.115}$$

where $l_i$ represents the distance between the $i$th section and the $(i + 1)$th section.

With Eq. (8.115), the state vector of the $i$th joint is transferred to the $(i + 1)$th joint. For example, the state vector of joint $R_1$ shown in Fig. 8.14 is now transferred to joint $R_{2,2}$ in the primary coordinate frame, $R_1$-$x_1y_1z_1$.

$$[\boldsymbol{b}_{12}]^{1,1}_{R_{2,1}} = \mathbf{H}_{12\times12}(l_1)[\boldsymbol{b}_{12}]^{1,1}_{R_1} \tag{8.116}$$

where the superscript "1,1" indicates that the coordinates are represented in the $x_{1,1}y_{1,1}z_{1,1}$-frame which will be simply denoted by "1,1-coordinate frame" in what follows.

### 8.5.2 Structural Dynamics of a Unit

In a similar way, we can investigate the transfer matrix of the state vector of the whole deployable frame of the morphing wing. Based on the above analysis, the whole deployable frame and the structure can be similarly modeled.

In 2,1-coordinate, the state vector represented with Eq. (8.116) will be

$$[\boldsymbol{b}_{12}]_{R_{2,1}}^{2,1} = R_{12\times12}(\beta_{2,1})[\boldsymbol{b}_{12}]_{R_{2,1}}^{1,1} \tag{8.117}$$

where

$$\mathbf{R}_{12\times12}(\beta_{2,1}) = \operatorname{diag}\left(\mathbf{R}_y(\beta_{2,1})\ \mathbf{R}_y(\beta_{2,1})\ \mathbf{R}_y(\beta_{2,1})\ \mathbf{R}_y(\beta_{2,1})\right) \tag{8.118}$$

and diag $(\cdot)$ represents a matrix whose diagonal elements are composed of "$\cdot$", and

$$\mathbf{R}_y(\beta_{2,1}) = \begin{bmatrix} \cos\beta_{2,1} & 0 & -\sin\beta_{2,1} \\ 0 & 1 & 0 \\ \sin\beta_{2,1} & 0 & \cos\beta_{2,1} \end{bmatrix} \tag{8.119}$$

which denotes the transformation matrix from the 1,1-coordinate frame to the 2,2-coordinate frame by rotating $\beta_{2,1}$ around $y_{2,1}$-axis passing through joint $R_{2,1}$.

For link $R_{2,1}R_3$, the state vector of joint $R_3$ can be obtained via Eq. (8.115):

$$[\boldsymbol{b}_{12}]_{R_3}^{2,1} = H_{12\times12}(l_2)[\boldsymbol{b}_{12}]_{R_{2,1}}^{2,1} = H_{R_1R_{2,1}R_3}\,[\boldsymbol{b}_{12}]_{R_1}^{1,1} \tag{8.120}$$

where $\mathbf{H}_{R_1R_{2,1}R_3} = \mathbf{H}_{12\times12}(l_2)\mathbf{R}_{12\times12}(-\beta_{2,1})\mathbf{H}_{12\times12}(l_1)$.

Equation (8.120) indicates that the state vector of joint $R_1$, $[\boldsymbol{b}_{12}]_{R_1}^{1,1}$ represented in the 1,1-coordinate frame, is now transformed to that of joint $R_3$, $[\boldsymbol{b}_{12}]_{R_3}^{2,1}$ represented in the 2,1-coordinate frame along the chain $R_1R_{2,1}R_3$. In addition, the state vector of joint $R_3$ can also be obtained via chain $R_1R_{2,2}R_3$:

$$[\boldsymbol{b}_{12}]_{R_3}^{2,2} = \mathbf{H}_{R_1R_{2,2}R_3}[\boldsymbol{b}_{12}]_{R_1}^{1,2} \tag{8.121}$$

where $\mathbf{H}_{R_1R_{2,2}R_3} = \mathbf{H}_{12\times12}(l_2)\mathbf{R}_{12\times12}\ (-\beta_{2,2})\mathbf{H}_{12\times12}(l_1)$ and $\beta_{2,2}$ indicates the rotational angle from the 1,2-coordinate frame to the 2,2-coordinate frame around the $y_{2,2}$-axis passing through joint $R_{2,2}$ which are shown in Fig. 8.14.

Because joint $R_3$ is jointed with two chains, namely, chain $R_1R_{2,1}R_3$ and chain $R_1R_{2,2}R_3$, the output state vector should be the combination of $[\boldsymbol{b}_{12}]_{R_3}^{2,1}$ and $[\boldsymbol{b}_{12}]_{R_3}^{2,2}$.

But before combining them, the two state vectors should be transformed into the same coordinate frame. For the sake of simplicity, we can transform $[\boldsymbol{b}_{12}]_{R_3}^{2,2}$ into that of the 2,1-coordinate frame:

$$(\boldsymbol{b}_{12})_{R_3}^{2,2\to 2,1} = \mathbf{R}_{12\times 12}(\beta_3)[\boldsymbol{b}_{12}]_{R_3}^{2,2} \tag{8.122}$$

where $[\boldsymbol{b}_{12}]_{R_3}^{2,2\to 2,1}$ denotes the expression of the state vector of joint $R_3$ at the 2,1-coordinate frame transferred by chain $R_1R_{2,2}R_3$ and $\beta_3$ represents the rotational angle from the 2,2-coordinate frame to the 2,1-coordinate frame around the $y_{31}$-axis.

The boundary conditions for joint $R_1$ in the 1,1-coordinate frame and 1,2-frame are

$$\begin{cases} [\boldsymbol{b}_{12}]_{R_1}^{1,1} = & \left[0\,0\,0\,0\,\Theta_y\;0\;M_x\;0\;M_z\;Q_x\;Q_y\;Q_z\right]^{\mathrm{T}} \\ [\boldsymbol{b}_{12}]_{R_1}^{1,2} = & \left[0\,0\,0\,0\,\Theta_y\;0\;M_x\;0\;M_z\;Q_x\;Q_y\;Q_z\right]^{\mathrm{T}} \end{cases} \tag{8.123}$$

Equation (8.123) indicates that the columns of $\mathbf{H}_{R_1R_{2,1}R_3}$ (or $\mathbf{H}_{R_1R_{2,2}R_3}$) corresponding to the zero rows of $[\boldsymbol{b}_{12}]_{R_1}^{1,1}$ (or $[\boldsymbol{b}_{12}]_{R_1}^{1,2}$) can be removed in the following calculations. Therefore, a trimming operation matrix is particularly defined:

$$\mathbf{M}_{6\times 12}(5,7,9,10,11,12) = \begin{bmatrix} 0&0&0&0&1&0&0&0&0&0&0&0 \\ 0&0&0&0&0&0&1&0&0&0&0&0 \\ 0&0&0&0&0&0&0&0&1&0&0&0 \\ 0&0&0&0&0&0&0&0&0&1&0&0 \\ 0&0&0&0&0&0&0&0&0&0&1&0 \\ 0&0&0&0&0&0&0&0&0&0&0&1 \end{bmatrix} \tag{8.124}$$

where $\mathbf{M}_{6\times 12}$ $(\cdot)$ is a matrix of $6\times 12$ the "1" entry of each row of which is defined by "·" and the other elements are all zeros. Therefore, Eqs. (8.121) and (8.122) can be equivalently written as

$$\begin{cases} [\boldsymbol{b}_{12}]_{R_3}^{2,1} = \left[\mathbf{H}_{R_1R_{2,1}R_3}\right]_{12\times 6}[\boldsymbol{b}_6]_{R_1}^{1,1} \\ [\boldsymbol{b}_{12}]_{R_3}^{2,2\to 2,1} = \left[\mathbf{H}_{R_1R_{2,2}R_3}\right]_{12\times 6}[\boldsymbol{b}_6]_{R_1}^{1,2} \end{cases} \tag{8.125}$$

where

$$\begin{cases} \left[\mathbf{H}_{R_1R_{2,1}R_3}\right]_{12\times 6} = \left[\mathbf{H}_{R_1R_{2,1}R_3}\right]_{12\times 12}[\mathbf{M}_{6\times 12}(5,7,9,10,11,12)]^{\mathrm{T}} \\ [\boldsymbol{b}_6]_{R_1}^{1,1} = [\mathbf{M}_{6\times 12}(5,7,9,10,11,12)]\,[\boldsymbol{b}_{12}]_{R_1}^{1,1} \\ [\mathbf{H}_{R_1R_{22}R_3}]_{12\times 6} = [\mathbf{H}_{R_1R_{22}R_3}]_{12\times 12}[\mathbf{M}_{6\times 12}(5,7,9,10,11,12)]^{\mathrm{T}} \\ [\boldsymbol{b}_6]_{R_1}^{1,2} = [\mathbf{M}_{6\times 12}(5,7,9,10,11,12)]\,[\boldsymbol{b}_{12}]_{R_1}^{1,2} \end{cases} \tag{8.126}$$

The two state vectors expressed with Eq. (8.125) depict the state of joint $R_3$, and therefore, the deflection/displacement and slope components of the two equations should be identical correspondingly:

$$[\mathbf{M}_{6\times12}(1,2,3,4,5,6)]\,[\boldsymbol{b}_{12}]_{R_3}^{2,1} = [\mathbf{M}_{6\times12}(1,2,3,4,5,6)]\,[\boldsymbol{b}_{12}]_{R_3}^{2,2\to2,1} \tag{8.127}$$

Substituting Eq. (8.125) into (8.127) yields

$$[\boldsymbol{b}_6]_{R_1}^{1,2} = \mathbf{E}^{1,2\to1,1}[\boldsymbol{b}_6]_{R_1}^{1,1} \tag{8.128}$$

where

$$\mathbf{E}^{1,2\to1,1} = \left\{[\mathbf{M}_{6\times12}(1,2,3,4,5,6)]\left[\mathbf{H}_{R_1R_{2,2}R_3}\right]_{12\times6}\right\}^{-1}[\mathbf{M}_{6\times12}(1,2,3,4,5,6)] \left[\mathbf{H}_{R_1R_{2,1}R_3}\right]_{12\times6} \tag{8.129}$$

Substituting Eq. (8.128) into the second equation of Eq. (8.125) presents

$$[\boldsymbol{b}_{12}]_{R_3}^{2,2\to2,1} = \left[\mathbf{H}_{R_1R_{2,2}R_3}\right]_{12\times6}\mathbf{E}^{1,2\to1,1}[\boldsymbol{b}_{12}]_{R_1}^{1,1} \tag{8.130}$$

Equations (8.120) and (8.130) indicate that both the state vector $[\boldsymbol{b}_{12}]_{R_3}^{2,1}$ and the state vector $[\boldsymbol{b}_{12}]_{R_3}^{2,2\to2,1}$ are expressed uniquely by $[\boldsymbol{b}_6]_{R_1}^{1,1}$. So, the output state vector of joint $R_3$ should be that the deflection and slope components are those of any one of $[\boldsymbol{b}_{12}]_{R_3}^{2,1}$ or $[\boldsymbol{b}_{12}]_{R_3}^{2,2\to2,1}$ but the sheering force and moment components are the summations of the corresponding items. Consequently, we immediately gain

$$\left[[\boldsymbol{b}_{12}]_{R_3}^{2,1}\right]_o = \left\{\left[\mathbf{H}_{R_1R_{2,2}R_3}\right]_{12\times6}\mathbf{E}^{1,2\to1,1} + \left[\frac{[0]_{6\times12}}{\mathbf{M}_{6\times12}(7,8,9,10,11,12)}\right]_{12\times12} \left[\mathbf{H}_{R_1R_{2,1}R_3}\right]_{12\times6}\right\}[\boldsymbol{b}_6]_{R_1}^{1,1} \tag{8.131}$$

where $[0]_{6\times12}$ denotes zero matrix of $6\times12$.

The connecting rectangle frame units could be made rigid enough so that it is here supposed to be a rigid body. As shown in Fig. 8.15, the state vector of the rigid rectangle frame is represented by that of its geometry center:

$$[\boldsymbol{b}_{12}]_{O_1}^{F_1} = \left[X\ Y\ Z\ \Theta_x\ \Theta_y\ \Theta_z\ M_x\ M_y\ M_z\ Q_x\ Q_y\ Q_z\right]^{\mathrm{T}} \tag{8.132}$$

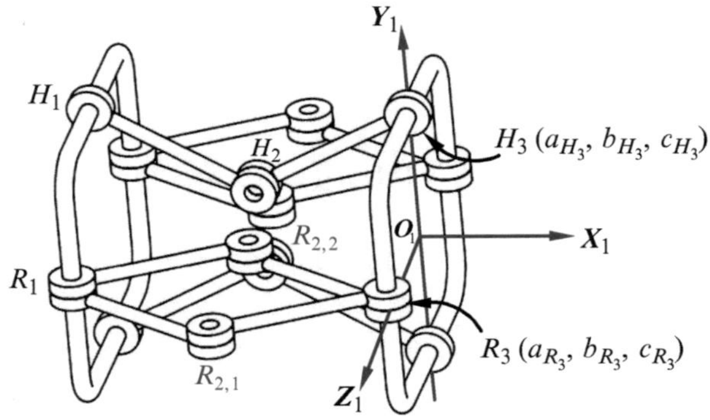


**Fig. 8.15** One quadrilateral element

Suppose the initial coordinates of the joint $R_3$ in the rectangle frame coordinate system are denoted by $R_3\left(a_{R_3}\ b_{R_3}\ c_{R_3}\right)$. So, the state vector of joint $R_3$ should satisfy that

$$\begin{bmatrix} X \\ Y \\ Z \\ \Theta_x \\ \Theta_y \\ \Theta_z \end{bmatrix}_{R_3}^{F_1} = \begin{bmatrix} X \\ Y \\ Z \\ \Theta_x \\ \Theta_y \\ \Theta_z \end{bmatrix}_{O_1}^{F_1} + \begin{bmatrix} \mathbf{R} & \mathbf{0} \\ \mathbf{0} & \mathbf{I} \end{bmatrix} \begin{bmatrix} a_{R_3} \\ b_{R_3} \\ c_{R_3} \\ 0 \\ 0 \\ 0 \end{bmatrix} \tag{8.133}$$

where $\mathbf{I} = \begin{bmatrix} 1 & 0 & 0 \\ 0 & 1 & 0 \\ 0 & 0 & 1 \end{bmatrix}$, and

$$\mathbf{R} = \begin{bmatrix} 1 & 0 & 0 \\ 0 & \cos\Theta_x & -\sin\Theta_x \\ 0 & \sin\Theta_x & \cos\Theta_x \end{bmatrix} \begin{bmatrix} \cos\Theta_y & 0 & \sin\Theta_y \\ 0 & 1 & 0 \\ -\sin\Theta_y & 0 & \cos\Theta_y \end{bmatrix} \begin{bmatrix} \cos\Theta_z & -\sin\Theta_z & 0 \\ \sin\Theta_z & \cos\Theta_z & 0 \\ 0 & 0 & 1 \end{bmatrix}$$

$$= \begin{bmatrix} \cos\Theta_y\cos\Theta_z & -\cos\Theta_y\sin\Theta_z & \sin\Theta_y \\ \sin\Theta_x\sin\Theta_y\cos\Theta_z + \cos\Theta_x\sin\Theta_z & -\sin\Theta_x\sin\Theta_y\sin\Theta_z + \cos\Theta_x\cos\Theta_z & -\sin\Theta_x\cos\Theta_y \\ -\cos\Theta_x\sin\Theta_y\cos\Theta_z + \sin\Theta_x\sin\Theta_z & \cos\Theta_x\sin\Theta_y\sin\Theta_z + \sin\Theta_x\cos\Theta_z & \cos\Theta_x\cos\Theta_y \end{bmatrix}$$

Because $\Theta_x$, $\Theta_y$, and $\Theta_z$ are all tiny values, **R** can be approximately replaced by

$$\mathbf{R}=\begin{bmatrix}1 & -\Theta_z & \Theta_y\\ \Theta_z & 1 & -\Theta_x\\ -\Theta_y & \Theta_x & 1\end{bmatrix} \tag{8.134}$$

From Fig. 8.15 and Eq. (8.131), we obtain

$$\begin{bmatrix}X\\Y\\Z\\\Theta_x\\\Theta_y\\\Theta_z\end{bmatrix}_{R_3}^{F_1}=\mathbf{R}_{6\times 6}\left(-\frac{\beta_3}{2}\right)\mathbf{M}_{6\times 12}(1,2,3,4,5,6)\left[[\boldsymbol{b}_{12}]_{R_3}^{2,1}\right]_o \tag{8.135}$$

where $\mathbf{R}_{6\times 6}\left(-\frac{\beta_3}{2}\right)=\begin{bmatrix}\mathbf{R}_y\left(-\frac{\beta_3}{2}\right) & 0\\ 0 & \mathbf{R}_y\left(-\frac{\beta_3}{2}\right)\end{bmatrix}$, and $\mathbf{R}_y\left(-\frac{\beta_3}{2}\right)=\begin{bmatrix}\cos\frac{\beta_3}{2} & 0 & \sin\frac{\beta_3}{2}\\ 0 & 1 & 0\\ -\sin\frac{\beta_3}{2} & 0 & \cos\frac{\beta_3}{2}\end{bmatrix}$

Substituting Eq. (8.131) into (8.135) presents

$$\begin{bmatrix}X\\Y\\Z\\\Theta_x\\\Theta_y\\\Theta_z\end{bmatrix}_{R_3}^{F_1}=\mathbf{D}_1[\boldsymbol{b}_6]_{R_1}^{1,1} \tag{8.136}$$

where $\mathbf{D}_1=\mathbf{R}_{6\times 6}\left(-\frac{\beta_3}{2}\right)\mathbf{M}_{6\times 12}(1,2,3,4,5,6)\{\left[\mathbf{H}_{R_1R_{2,2}R_3}\right]_{12\times 6}\mathbf{E}^{1,2\to 1,1}+M_{6\times 12}$ $(7,8,9,10,11,12)\left[\mathbf{H}_{R_1R_{2,1}R_3}\right]_{12\times 6}\}[\boldsymbol{b}_6]_{R_1}^{1,1}$ which denotes the mapping operation from $[\boldsymbol{b}_6]_{R_1}^{1,1}$ to the deflection and slope vector of the $i$th rectangle frame and $\mathbf{D}_1$ is always invertible for every multiplier of it is full rank.

Associating Eqs. (8.133) and (8.136), we find that

$$\begin{bmatrix}X\\Y\\Z\\\Theta_x\\\Theta_y\\\Theta_z\end{bmatrix}_{O_1}^{F_1}=\mathbf{D}_1[\boldsymbol{b}_6]_{R_1}^{1,1}-\begin{bmatrix}\mathbf{R} & \mathbf{0}\\ \mathbf{0} & \mathbf{I}\end{bmatrix}\begin{bmatrix}a_{R_3}\\b_{R_3}\\c_{R_3}\\0\\0\\0\end{bmatrix} \tag{8.137}$$

When considering another chain, $H_1H_2H_3$ shown in Fig. 8.15, for example, its state vector is similarly obtained from the above process:

$$[\boldsymbol{b}_{12}]^{1,1}_{H_{2,1}} = \mathbf{H}_{12\times12}(l_1)[\boldsymbol{b}_{12}]^{1,1}_{H_{1,1}} \tag{8.138}$$

In the $H_{2,1} - x_{2,1}y_{2,1}z_{2,1}$ frame, the state vector represented with Eq. (8.138) will be

$$[\boldsymbol{b}_{12}]^{2,1}_{H_{2,1}} = \mathbf{R}_{12\times12}(\chi_{2,1})[\boldsymbol{b}_{12}]^{1,1}_{H_{2,1}} \tag{8.139}$$

where

$$\mathbf{R}_{12\times12}(\chi_{2,1}) = \mathrm{diag}\left(\mathbf{R}_z(\chi_{2,1})\ \mathbf{R}_z(\chi_{2,1})\ \mathbf{R}_z(\chi_{2,1})\ \mathbf{R}_z(\chi_{2,1})\right) \tag{8.140}$$

and diag (·) represents a matrix whose diagonal elements are composed of "·", and

$$\mathbf{R}_y(\chi_{2,1}) = \begin{bmatrix} \cos\chi_{2,1} & -\sin\chi_{2,1} & 0 \\ \sin\chi_{2,1} & \cos\chi_{2,1} & 0 \\ 0 & 0 & 1 \end{bmatrix} \tag{8.141}$$

which denotes the transformation matrix from the coordinate frame $H_{1,1} - x_{1,1}y_{1,1}z_{1,1}$ to $H_{2,1} - x_{2,1}y_{2,1}z_{2,1}$ by rotating $\chi_{2,1}$ around the $z_{1,1}$-direction at joint $H_{2,1}$.

In the link $H_{2,1}H_{3,1}$, the state vector of joint $H_{2,1}$ can be obtained via Eq. (8.115):

$$[\boldsymbol{b}_{12}]^{2,1}_{H_{3,1}} = \mathbf{H}_{12\times12}(l_2)[\boldsymbol{b}_{12}]^{2,1}_{H_{2,1}} = \mathbf{H}_{H_{1,1}H_{2,1}H_{3,1}}[\boldsymbol{b}_{12}]^{1,1}_{H_{1,1}} \tag{8.142}$$

where $\mathbf{H}_{H_{1,1}H_{2,1}H_{3,1}} = \mathbf{H}_{12\times12}(l_2)\mathbf{R}_{12\times12}(\chi_{2,1})\mathbf{H}_{12\times12}(l_1)$. When this state vector is transformed into the first rectangle coordinate frame, we have

$$[\boldsymbol{b}_{12}]^{F_1}_{H_{3,1}} = R_{12\times12}(\chi_{2,1})[\boldsymbol{b}_{12}]^{3,1}_{H_{3,1}} = R_{12\times12}(\chi_{2,1})\mathbf{H}_{H_{1,1}H_{2,1}H_{3,1}}[\boldsymbol{b}_{12}]^{1,1}_{H_{1,1}} \tag{8.143}$$

The columns of $\mathbf{R}_{12\times12}(\chi_{2,1})\mathbf{H}_{H_{1,1}H_{2,1}H_{3,1}}$ corresponding to the zero rows of $[\boldsymbol{b}_{12}]^{1,1}_{H_{1,1}}$ can be removed in the calculation. Therefore, similar to Eq. (8.126), we suppose that

$$\begin{cases} \left[\mathbf{H}_{H_{1,1}H_{2,1}H_{3,1}}\right]_{12\times6} = \mathbf{R}_{12\times12}(\chi_{2,1})\mathbf{H}_{H_{1,1}H_{2,1}H_{3,1}}[\mathbf{M}_{6\times12}(5,7,9,10,11,12)]^{\mathrm{T}} \\ [\boldsymbol{b}_6]^{1,1}_{H_{1,1}} = \mathbf{M}_{6\times12}(5,7,9,10,11,12)[\boldsymbol{b}_{12}]^{1,1}_{H_{1,1}} \end{cases} \tag{8.144}$$

Equation (8.143) can be simplified as

$$[\boldsymbol{b}_{12}]^{F_1}_{H_{3,1}} = \left[\mathbf{H}_{H_{1,1}H_{2,1}H_{3,1}}\right]_{12\times6}[\boldsymbol{b}_6]^{1,1}_{H_{1,1}} \tag{8.145}$$

The deflection/displacement and the slope of joint $H_{3,1}$ can be expressed as

$$\begin{bmatrix} X \\ Y \\ Z \\ \Theta_x \\ \Theta_y \\ \Theta_z \end{bmatrix}_{H_{3,1}}^{F_1} = \mathbf{D}_{H_{3,1}} [\boldsymbol{b}_6]_{H_{1,1}}^{1,1} \tag{8.146}$$

where $\mathbf{D}_{H_{3,1}} = [\mathbf{M}_{6\times 12}(1,2,3,4,5,6)] \left[\mathbf{H}_{H_{1,1}H_{2,1}H_{3,1}}\right]_{12\times 6}$ and therefore $\left[\mathbf{H}_{H_{1,1}H_{2,1}H_{3,1}}\right]_{12\times 6}$ can be expressed as

$$\left[\mathbf{H}_{H_{1,1}H_{2,1}H_{3,1}}\right]_{12\times 6} = \begin{bmatrix} \left[\mathbf{D}_{H_{3,1}}\right]_{6\times 6} \\ \hline \left[\mathbf{D}_{H_{3,1}}^{C}\right]_{6\times 6} \end{bmatrix} \tag{8.147}$$

where $\mathbf{D}_{H_{3,1}}^{C} = [\mathbf{M}_{6\times 12}(7,8,9,10,11,12)] \left[\mathbf{H}_{H_{1,1}H_{2,1}H_{3,1}}\right]_{12\times 6}$

While $\left[\left[X\ Y\ Z\ \Theta_x\ \Theta_y\ \Theta_z\right]_{H_{3,1}}^{F_1}\right]^{\mathrm{T}}$ can also be expressed with the state vector of center point of the first rectangle frame in accordance to Eq. (8.133):

$$\begin{bmatrix} X \\ Y \\ Z \\ \Theta_x \\ \Theta_y \\ \Theta_z \end{bmatrix}_{H_{3,1}}^{F_1} = \begin{bmatrix} X \\ Y \\ Z \\ \Theta_x \\ \Theta_y \\ \Theta_z \end{bmatrix}_{O_1}^{F_1} + \begin{bmatrix} \mathbf{R} & \mathbf{0} \\ \mathbf{0} & \mathbf{I} \end{bmatrix} \begin{bmatrix} a_{H_{3,1}} \\ b_{H_{3,1}} \\ c_{H_{3,1}} \\ 0 \\ 0 \\ 0 \end{bmatrix} \tag{8.148}$$

Substituting Eq. (8.137) into Eq. (8.148) yields

$$\begin{bmatrix} X \\ Y \\ Z \\ \Theta_x \\ \Theta_y \\ \Theta_z \end{bmatrix}_{H_{3,1}}^{F_1} = \mathbf{D}_1 [\boldsymbol{b}_6]_{R_1}^{1,1} + \begin{bmatrix} \mathbf{R} & \mathbf{0} \\ \mathbf{0} & \mathbf{I} \end{bmatrix} \begin{bmatrix} a_{H_{3,1}} - a_{R_3} \\ b_{H_{3,1}} - b_{R_3} \\ c_{H_{3,1}} - c_{R_3} \\ 0 \\ 0 \\ 0 \end{bmatrix} \tag{8.149}$$

Therefore, associating Eqs. (8.146) and (8.149), we obtain

$$[\boldsymbol{b}_6]_{H_{1,1}}^{1,1} = \mathbf{D}_{H_{3,1}}^{-1} \mathbf{D}_1 [\boldsymbol{b}_6]_{R_1}^{1,1} + \mathbf{D}_{H_{3,1}}^{-1} \begin{bmatrix} \mathbf{R} & \mathbf{0} \\ \mathbf{0} & \mathbf{I} \end{bmatrix} \begin{bmatrix} a_{H_{3,1}} - a_{R_3} \\ b_{H_{3,1}} - b_{R_3} \\ c_{H_{3,1}} - c_{R_3} \\ 0 \\ 0 \\ 0 \end{bmatrix} \tag{8.150}$$

However, the state vector of the starting point of the chain $H_{1,1}H_{2,1}H_{3,1}$ in Eq. (8.150) is expressed in the coordinate frame $H_{1,1} - x_{1,1}y_{1,1}z_{1,1}$. Therefore, the coordinates of joint $H_{1,1}$ in Eq. (8.150) should be transformed into the $R_{1,1} - x_{1,1}y_{1,1}z_{1,1}$. This can be obtained from Eqs. (8.131) and (8.150):

$$[\boldsymbol{b}_6]_{H_{1,1}}^{1,1} = \mathbf{D}_{H_{3,1}}^{-1}\mathbf{D}_1[\boldsymbol{b}_6]_{R_1}^{1,1} + \mathbf{D}_{H_{3,1}}^{-1}\begin{bmatrix}\mathbf{R} & \mathbf{0}\\ \mathbf{0} & \mathbf{I}\end{bmatrix}\left(\begin{bmatrix} a_{H_{3,1}} - a_{R_3}\\ b_{H_{3,1}} - b_{R_3}\\ c_{H_{3,1}} - c_{R_3}\\ 0\\ 0\\ 0\end{bmatrix} - \begin{bmatrix} a_{H_{1,1}} - a_{R_1}\\ b_{H_{1,1}} - b_{R_1}\\ c_{H_{1,1}} - c_{R_1}\\ 0\\ 0\\ 0\end{bmatrix}\right) \tag{8.151}$$

Equation (8.151) indicates that the initial conditions of any other chain structure shown in Fig. 8.6 can be completely expressed by those of the first joint $R_1$ and the relative position parameters of joints $R_3$ and $H_3$ (or $R_1$ and $H_1$). Substituting Eqs. (8.147) and (8.151) into (8.145) yields

$$\begin{aligned}
[\boldsymbol{b}_{12}]_{H_{3,1}}^{F_1} &= \begin{bmatrix}\dfrac{[\mathbf{D}_{H_{3,1}}]_{6\times 6}}{[\mathbf{D}_{H_{3,1}}^{C}]_{6\times 6}}\end{bmatrix}\mathbf{D}_{H_{3,1}}^{-1}\mathbf{D}_1[\boldsymbol{b}_6]_{R_1}^{1,1} + \begin{bmatrix}\dfrac{[\mathbf{D}_{H_{3,1}}]_{6\times 6}}{[\mathbf{D}_{H_{3,1}}^{C}]_{6\times 6}}\end{bmatrix}\mathbf{D}_{H_{3,1}}^{-1}\begin{bmatrix}\mathbf{R} & \mathbf{0}\\ \mathbf{0} & \mathbf{I}\end{bmatrix}\\
&\quad\times\begin{bmatrix} a_{H_{3,1}} - a_{R_3}\\ b_{H_{3,1}} - b_{R_3}\\ c_{H_{3,1}} - c_{R_3}\\ 0\\ 0\\ 0\end{bmatrix} - \begin{bmatrix}\dfrac{[D_{H_{3,1}}]_{6\times 6}}{[\mathbf{D}_{H_{3,1}}^{C}]_{6\times 6}}\end{bmatrix}\begin{bmatrix} a_{H_{1,1}} - a_{R_1}\\ b_{H_{1,1}} - b_{R_1}\\ c_{H_{1,1}} - c_{R_1}\\ 0\\ 0\\ 0\end{bmatrix}\\
&= \begin{bmatrix}\dfrac{[\mathbf{I}]_{6\times 6}}{[\mathbf{D}_{H_{3,1}}^{C}]_{6\times 6}\mathbf{D}_{H_{3,1}}^{-1}}\end{bmatrix}\mathbf{D}_1[\boldsymbol{b}_6]_{R_1}^{1,1} + \begin{bmatrix}\dfrac{[I]_{6\times 6}}{[\mathbf{D}_{H_{3,1}}^{C}]_{6\times 6}\mathbf{D}_{H_{3,1}}^{-1}}\end{bmatrix}\\
&\quad\times\begin{bmatrix}\mathbf{R} - \mathbf{I} & \mathbf{0}\\ \mathbf{0} & \mathbf{I}\end{bmatrix}\begin{bmatrix} a_{H_{3,1}} - a_{R_3}\\ b_{H_{3,1}} - b_{R_3}\\ c_{H_{3,1}} - c_{R_3}\\ 0\\ 0\\ 0\end{bmatrix}
\end{aligned} \tag{8.152}$$

where $\left[\mathbf{D}_{H3,1}\right]_{6\times6}\begin{bmatrix} a_{H_{1,1}}-a_{R_1} \\ b_{H_{1,1}}-b_{R_1} \\ c_{H_{1,1}}-c_{R_1} \\ 0 \\ 0 \\ 0 \end{bmatrix}=\begin{bmatrix} a_{H3,1}-a_{R3} \\ b_{H3,1}-b_{R3} \\ c_{H3,1}-c_{R3} \\ 0 \\ 0 \\ 0 \end{bmatrix}$ and $\left[\mathbf{D}^C_{H3,1}\right]_{6\times6}\begin{bmatrix} a_{H_{1,1}}-a_{R_1} \\ b_{H_{1,1}}-b_{R_1} \\ c_{H_{1,1}}-c_{R_1} \\ 0 \\ 0 \\ 0 \end{bmatrix}$ $=\begin{bmatrix} 0 \\ 0 \\ 0 \\ 0 \\ 0 \\ 0 \end{bmatrix}$ which are specified by the geometry of Fig. 8.15.

The conditions of the released end of the cantilever are that the angles in **R** are all zero; therefore, Eq. (8.151) is simplified as

$$[\boldsymbol{b}_{12}]^{F_1}_{H3,1}=\left[\frac{[\mathbf{I}]_{6\times6}}{\left[\mathbf{D}^C_{H3,1}\right]_{6\times6}\mathbf{D}^{-1}_{H3,1}}\right]\mathbf{D}_1[\boldsymbol{b}_6]^{1,1}_{R_1} \tag{8.153}$$

Hereto, the state vector of the rectangle frame can be stated as that the deflection and slope are expressed by Eq. (8.136), while the shearing forces and the moments are the sum of all nodes connected with the frame.

$$\begin{bmatrix} M_x \\ M_y \\ M_z \\ Q_x \\ Q_y \\ Q_z \end{bmatrix}^{F_1}_{O_1}=\mathbf{M}_{6\times12}(7,8,9,10,11,12)\left\{\left[[\boldsymbol{b}_{12}]^{2,1}_{R_3}\right]_o+[\boldsymbol{b}_{12}]^{F_1}_{H3,1}\right\} \tag{8.154}$$

Expanding Eq. (8.154) yields

$$\begin{bmatrix} M_x \\ M_y \\ M_z \\ Q_x \\ Q_y \\ Q_z \end{bmatrix}^{F_1}_{O_1}=\mathbf{M}_{6\times12}(7,8,9,10,11,12)\left\{\begin{matrix}\left[\mathbf{H}_{R_1R_{2,2}R_3}\right]_{12\times6}\mathbf{E}^{1,2\to1,1}+\left[\mathbf{H}_{R_1R_{2,1}R_3}\right]_{12\times6} \\ +\left[\dfrac{[\mathbf{I}]_{6\times6}}{\left[\mathbf{D}^C_{H3,1}\right]_{6\times6}\mathbf{D}^{-1}_{H3,1}}\right]\mathbf{D}_1\end{matrix}\right\}[\boldsymbol{b}_6]^{1,1}_{R_1} \tag{8.155}$$

The released end of the cantilever requires that

$$\begin{bmatrix} M_x \\ M_y \\ M_z \\ Q_x \\ Q_y \\ Q_z \end{bmatrix}_{F_1} = 0$$

As a result, we gain that

$$\mathbf{A}[\boldsymbol{b}_6]_{R_1}^{1,1} = 0 \tag{8.156}$$

where $\left([\boldsymbol{b}_6]_{R_1}^{1,1}\right)_0$ represents the initial state vector of joint $R_1$,

$$\mathbf{A} = \mathbf{M}_{6\times12}\,(7,8,9,10,11,12)\left\{\begin{array}{l} \left[\mathbf{H}_{R_1R_{2,2}R_3}\right]_{12\times6}\mathbf{E}^{1,2\to1,1} + \left[\mathbf{H}_{R_1R_{2,1}R_3}\right]_{12\times6} \\ + \left[\dfrac{[\mathbf{I}]_{6\times6}}{\left[\mathbf{D}_{H_{3,1}}^{C}\right]_{6\times6}\mathbf{D}_{H_{3,1}}^{-1}}\right]\mathbf{D}_1 \end{array}\right\}.$$

Because $[\boldsymbol{b}_6]_{R_1}^{1,1}$ might be any nonzero vectors, the criteria for Eq. (8.153) should be

$$|\mathbf{A}| = 0 \tag{8.157}$$

Solving Eq. (8.157), we obtain the basic frequency of the deployable unit mechanism shown in Fig. 8.15. When one or more chain like $H_{1,1}H_{2,1}H_{3,1}$ or quadrilateral chain $R_1\left\langle\begin{matrix} R_{22} \\ R_{21}\end{matrix}\right\rangle R_3$ connect the unit, a similar process can be used to obtain the determinant represented by Eq. (8.157).

Based on the above analysis, the whole deployable frame can be treated as serial connection of a number of deployable units shown in Fig. 8.16. In a similar way, we can investigate the transfer matrix of state vector of the whole deployable frame of the metamorphic wing.

## 8.6 Conclusions

This chapter focuses on the structural dynamics of complex planar linkages. The revolute-jointed foldable linkage allows large displacements which surely introduces a lot of additional constraint equations when building the structural

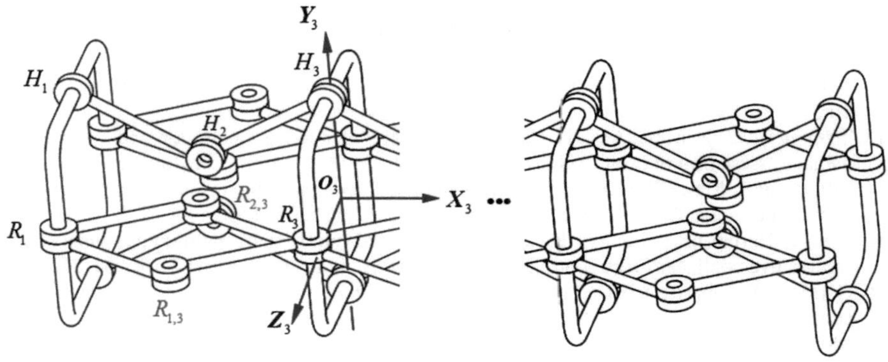


**Fig. 8.16** More units in series

dynamics by transfer-matrix method. Therefore, this chapter takes advantage of the fact that each unit is made up of identical units and establishes the structural dynamics through assembling the transfer matrices. The foldable stair is simplified as one planar linkage and the deployable wing is proposed as a double planar linkage. The structural dynamics of these linkages can be similarly modeled by transfer-matrix method. The frequency determinant of the whole structure at any position only subjects to the boundary conditions of the joints, and therefore it utilizes the least number of variables.

## References

1. Mirats Tur JM, Juan SH (2009) Tensegrity frameworks: dynamic analysis review and open problems. Mech Mach Theory 44(1):1–18
2. Escrig F, Valcarcel P (1993) Geometry of expandable space structures. Int J Space Struct 8(1-2):71–84
3. Kaveh A, Davaran A (1996) Analysis of pantograph foldable structures. Comput Struct 59(1):131–140
4. Chen Y, You Z, Tarnai T (2005) Threefold-symmetric Bricard linkages for deployable structures. Int J Solids Struct 42(8):2288–2301
5. Seffen KA, You Z, Pellegrino S (2000) Folding and deployment of curved tape springs. Int J Mech Sci 42(10):2055–2073
6. Mirats Tur JM, Juan SH (2008) Tensegrity frameworks: static analysis review. Mech Mach Theory 43(7):859–881
7. Xu LJ, Tian GY, Duan Y, Yang SX (2001) Inverse kinematic analysis for triple-octahedron variable-geometry truss manipulators. Proc Inst Mech Eng C J Mech Eng Sci 215(2):248–251
8. Gan WW, Pellegrino S (2006) Numerical approach to the kinematic analysis of deployable structures forming a closed loop. Proc Inst Mech Eng C J Mech Eng Sci 220(7):1045–1056
9. Park SW (2001) Analytical modelling of viscoelastic dampers for structural and vibration control. Int J Solids Struct 38(44-45):8065–8092
10. Tarnai T (2003) Zero stiffness elastic structures. Int J Mech Sci 45(3):425–431

11. Impollonia N (2006) A method to derive approximate explicit solutions for structural mechanics problems. Int J Solids Struct 43(22-23):7082–7098
12. Nohmi M, Matsumoto K, Ueno H, Yoshida T (2001) Deployable truss operation by ETS-VII robot arm using force accommodation control. Comput Aided Civ Infrastruct Eng 16(3):169–179
13. Talebinejad I, Fischer C, Ansari F (2011) Numerical evaluation of vibration-based methods for damage assessment of cable-stayed bridges. Comput Aided Civ Infrastruct Eng 26(3):239–251
14. Jafarkhani R, Masri SF (2011) Finite element model updating using evolutionary strategy for damage detection. Comput Aided Civ Infrastruct Eng 26(3):208–224
15. Kim Y, Hurlebaus S, Langari R (2010) Model-based multi-input, multi-output supervisory semi-active nonlinear fuzzy controller. Comput Aided Civ Infrastruct Eng 25(5):387–393
16. Cortés F, Elejabarrieta MJ (2008) Structural vibration of flexural beams with thick unconstrained layer damping. Int J Solids Struct 45(22-23):5805–5813
17. Mohan Rao AR, Shyju PP (2010) A meta-heuristic algorithm for multi-objective optimal design of hybrid laminate composite structures. Comput Aided Civ Infrastruct Eng 25(3): 149–170
18. Liu Y, Xu L, Grierson DE (2010) Influence of semi-rigid connections and local joint damage on progressive collapse of steel frameworks. Comput Aided Civ Infrastruct Eng 25(3):184–204
19. Chang SY (2010) A new family of explicit methods for linear structural dynamics. Comput Struct 88(11-12):755–772
20. Aldemir U (2010) A simple active control algorithm for earthquake excited structures. Comput Aided Civ Infrastruct Eng 25(3):218–225
21. Koh CG, Chen YF, Liaw CY (2003) A hybrid computational strategy for identification of structural parameters. Comput Struct 81(2):108–118
22. Correa FN, Jacob BP, Mansur WJ (2010) Formulation of an efficient hybrid time-frequency domain solution procedure for linear structural dynamic problems. Comput Struct 88(5-6):331–346
23. Langbecker T (1999) Kinematic analysis of deployable scissor structures. Int J Space Struct 14(1):1–15
24. Chen Y, You Z (2007) Spatial 6R linkages based on the combination of two Goldberg 5R linkages. Mech Mach Theory 42(11):1484–1489
25. Liu SY, Chen Y (2009) Myard linkage and its mobile assemblies. Mech Mach Theory 44(10):1950–1963
26. Kang SC, Miranda E (2009) Numerical methods to simulate and visualize detailed crane activities. Comput Aided Civ Infrastruct Eng 24(3):169–185
27. Graf W, Freitag S, Kaliske M, Sickert JU (2010) Recurrent neural networks for uncertain time-dependent structural behaviour. Comput Aided Civ Infrastruct Eng 25(5):322–333
28. Lin CC, Chen CL, Wang JF (2010) Vibration control of structures with initially accelerated passive tuned mass dampers under near-fault earthquake excitation. Comput Aided Civ Infrastruct Eng 25(1):69–75
29. Schoefs F, Yáñez-Godoy H, Lanata F (2011) Polynomial chaos representation for identification of mechanical characteristics of instrumented structures. Comput Aided Civ Infrastruct Eng 26(3):173–189
30. Shigley JE, Uicher JJ (1980) Theory of machines and mechanisms. McGraw-Hill Companies, Inc, New York
31. Gantes CJ, Konitopoulou E (2004) Geometric design of arbitrarily curved bi-stable deployable arches with discrete joint size. Int J Solids Struct 41(20):5517–5540
32. Adeli H (1988) Interactive microcomputer-aided structural steel design. Prentice Hall, Englewood Cliffs
33. Kaveh A, Shojaee S (2007) Optimal design of scissor-link foldable structures using ant colony optimization algorithm. Comput Aided Civ Infrastruct Eng 22(1):56–64
34. Adeli H (1994) Advances in design optimization. Chapman & Hall, London
35. Adeli H, Soegiarso R (1999) High-performance computing in structural engineering. CRC Press, London

36. Hachem C, Karni E, Hanaor A (2005) Evaluation of biological deployable systems. Int J Space Struct 20(4):189–200
37. Ragavan V, Made AM (2001) An algorithm for nonlinear stability analysis of an expandable self-erecting structure. Comput Struct 79(29-30):2588–2593
38. Ando K, Mitsugi J, Senbokuya Y (2000) Analyses of cable-membrane structure combined with deployable truss. Comput Struct 74(1):21–39
39. Chen WJ, Luo LY, Fu GY, Gong JH, Dong SL (2001) A study on space masts based on octahedral truss family. Int J Space Struct 16(1):75–82
40. Hanaor A, Levy R (2001) Evaluation of deployable structures for space enclosures. Int J Space Struct 16(4):211–229
41. Bolander JE, Hong GS, Yoshitake K (2000) Structural concrete analysis using rigid-body-spring networks. Comput Aided Civ Infrastruct Eng 15(2):120–133
42. Niu H, Wu Z (2005) Numerical analysis of debonding mechanisms in FRP-strengthened RC beams. Comput Aided Civ Infrastruct Eng 20(5):354–368
43. Adeli H, Sierakowski RL (1991) Mechanics computing in 1990's and beyond, vol 1. American Society of Civil Engineers, New York
44. Adeli H, Sierakowski RL (1991) Mechanics computing in 1990's and beyond, vol 2. American Society of Civil Engineers, New York
45. Pestel EC, Leckie FA (1963) Matrix method in elastomechanics. McGraw-Hill Book Company, New York
46. Rui X, Yun L, Lu Y, He B, Wang G (2008) Transfer matrix method of multibody system and its applications. Chinese Science Publishing House, Beijing
47. Ying L, Salari E (2010) Beamlet transform-based technique for pavement crack detection and classification. Comput Aided Civ Infrastruct Eng 25(8):572–580

# Chapter 9
# Foldable Tower with Multi-planar Linkages

This chapter proposes a foldable tower for huge electrical equipment, which should be used as the support structure of wind turbine. The advantage of the flexibility to fold and unfold can be utilized for protecting the wind turbine when destructive weather occurs. The method is developed to establish the dynamic stiffness of the whole structure. The equivalent stiffness of the foldable tower is investigated as the function of deployment angle of the actuator. To expand the applications of the foldable structures, multi-planar linkages and the topology changeable linkages are discussed at the end of this chapter.

## 9.1 Foldable Structures for Wind Turbine Tower

With the shortage of fuel in the world, wind power becomes one of the most potential energy today. For the demand for more energy from wind turbine, the tower tends to be designed much higher than before to capture the wind with larger speed. However, the higher the tower is, the more possible the catastrophe of whole wind turbine structure will occur when hurricanes come. The tower can be folded when hurricanes come and can be deployed fully to acquire more wind power in high sky in normal cases. Firstly, the geometry design of a deployable structure is presented in details by discussing the deployable ratio. Then, it focuses on the statics, stiffness analysis of this structure. The equivalent stiffness of the foldable tower is expressed as a function of the primary design parameters and the deployable angle. What must be pointed out is that the result comprehensively expresses the dependence of the stiffness on the deployable angle and the stiffness of the fully deployed case is the highest. This is particularly useful for the applications in wind turbine.

Wind turbine, which is also called wind generator set, is a kind of equipment that transforms wind energy into electricity. Wind turbine is both an ancient and modern machine. In terms of its age, windmill has been used for more than 3,000 years, and from as early as the thirteenth century, horizontal-axis wind mill has been an

J. Zhao et al., *Design of Special Planar Linkages*, Springer Tracts in Mechanical Engineering, DOI 10.1007/978-3-642-38448-6_9,

integral part of rural economy [1]. However, with the advent of cheap fossil-fuelled engines, wind energy and wind turbine fell into disuse for a very long time. During this time until the 1970s, only a few scientists such as Professor Hunter in German and corporations such as Electricite De France had made some finite advances in the development of wind turbine. But during the latest decades of the twentieth century, things have completely changed. The forecast of the fuel shortage in the near future combined with the negative environmental impacts caused by the use of the traditional electricity production methods forced all those involved in the energy production field to start exploring new directions in energy production [2]. In 1997, the *Commission of Europe Union* published its white paper calling for 12 % of the gross energy demand of Europe union to be contributed from renewable energy in 2010 [1]. Wind energy is the widely acknowledged renewable and clear energy source. The demand for sustainable energy rekindles the enthusiasm of the investigation of wind energy and wind turbine. As a result, wind turbine technology has developed a lot during these years, with a continuous increase in the rotor diameter and a doubling speed of the wind it makes use of, both contributing to an increase in energy output [1]. That is why wind energy is called a new kind of energy even though it has a long history of several thousand years.

Now that wind energy is in great demand in the world today, technological improvement should be done to increase its energy output. The power law $V_2/V_1 = (Z_2/Z_1)^n$ is often used for height projection of wind profiles, with $V_1$ and $V_2$ in m/s, $Z_1$ and $Z_2$ in m, and the exponent $n$ sometimes taken as depending on surface conditions or on atmospheric stability. (The power law profile for wind speed is shown here to be consistent with the observed height variation of Weibull wind speed probability distribution functions which have been found to fit the observed wind speed distributions, at least the above relevant threshold wind speeds [3].) It can be deduced that the wind turbine can obtain higher wind speed with the increase of height. In that case, the design of the higher wind turbine tower shows its importance.

Wind generator tower is used for carrying cabin and rotor, the primary components of a wind turbine. According to the discussion above, high tower has its advantages because the higher the tower is from the ground, the larger the wind speed will be on its top. Modern million-watt wind turbine has a tower as high as from 40 to 80 m. The tallest wind turbine in the world is 210 m (its axis is 160 m high from the ground) [4]. There are two kinds of wind turbine towers: one is tubular tower and the other is lattice tower. The highest wind turbine mentioned has a lattice tower, but the most popular tower is a single tubular column [5]. Tubular tower ensures personal safety, because they can reach the top through an internal stair or lifter.

However, until now there is no foldable wind generator tower in the world. When strong wind comes, such as the hurricanes from the sea and the tornadoes, a tower is very easily to be ruptured, bended, or even collapsed. Therefore, the foldable tower is particularly necessary to be synthesized to settle these problems. The tower can be fully deployed to capture the wind energy when the wind is not strong enough. When the wind grows to be a destroyable hurricane, the tower can be folded completely and covered by a shelter to avoid the disaster.

Deployable structures are prefabricated space frames consisting of straight bars linked together in the factory as a compact bundle, which can then be unfolded into large-span, load-bearing structural shapes by simple articulation [6]. Deployable structures often offer significant advantage compared to conventional fixed structure for a wide spectrum of applications mainly because of their feature of transforming and adapting to changing needs [6]. The investigation of deployable structure has started since 1960s, and many kinds of deployable structures have been designed to be applied in all kinds of fields. For example, in aerospace industry, Christiansen, Ker, Fuentes, and Schneider created the TransHab shields which contain a positive attribute of flexible and deployable structures. The foam is compressed during launch to minimize volume and released on-orbit to deploy the bumpers of the shield [7]. In biology area, the deployable structures are studied, and these structures are classified into four groups: planar, cylindrical, stiff, and complaint [8]. What is more, some new methods have been worked out to design new foldable structures. For example, Wei, Yao, Tian, and Fang proposed a method to create expandable structure for spatial objects [9]. Gan and Pellegrino presented a numerical approach to the kinematic analysis of deployable structures forming a closed loop [10]. According to the discussion above, deployable structure is a good choice for the wind turbine tower, if it can be designed properly.

A good foldable tower should have artful structural design and a proper foldable ratio. What is more important, it should have enough strength and stiffness. Namely, it should have the ability to prevent failure [11]. This chapter is trying to achieve these goals.

## 9.2 Geometry Design of Foldable Towers

### 9.2.1 Fundamental Structure Analysis

This section investigates an equivalent prismatic joint linkage. On the one hand, it provides the passive rectilinear motion requirement even under a certain transversal load, and on the other hand, it has good strength and stiffness for transversal loads. The linkage consists of 3-RRR kinematic chains which are shown in Fig. 9.1.

This is a rectilinear 3-RRR parallel planar linkage whose kinematic chains locate in different planes and can make full planar motions within their individual planes. As shown in Fig. 9.1, the kinematic chain, $R_{11}R_{12}R_{13}$, locates in plane $\pi_1$, and $R_{21}R_{22}R_{23}$ and $R_{31}R_{32}R_{33}$ locate in planes $\pi_2$ and $\pi_3$, respectively. The three planes have rotational symmetry about their common intersection, line $A_1B_1$, and therefore, the subtended angle of any two planes is 120°. For the sake of analysis below, it is assumed that the $y$-axis is superimposed with the intersection line and $z$-axis is within plane $\pi_1$, so the $x$-axis is parallel to the axes of the three revolute joints, $R_{11}$, $R_{12}$, and $R_{13}$, in accordance with the right-hand rule. What follows will prove that the relative motion between $R_{11}R_{21}R_{31}$ and $R_{13}R_{23}R_{33}$ is a pure translation parallel to the $y$-axis.

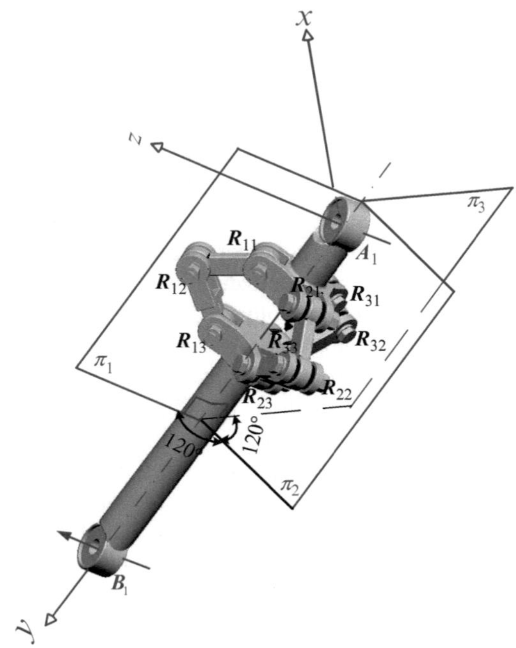


**Fig. 9.1** A 3-RRR parallel linkage

Suppose that $O_1R_{11} = O_1R_{21} = O_1R_{31} = r_0$, $O_2R_{13} = O_2R_{23} = O_2R_{33} = r_0$ and the lengths of $R_{i1}R_{i2}(i = 1, 2, 3)$ and $R_{i2}R_{i3}$ are $l_1$ and $l_2$, respectively. The subtended angle between $R_{i1}R_{i2}$ and $y$-axis is $\alpha$, and the subtended angle between $R_{i3}R_{i2}$ and $y$-axis is $\beta$ which are shown in Fig. 9.2. So the coordinates of the joints of the 3-RRR linkage are

$$R_{11}(0\ 0\ r_0),\quad R_{21}\left(-\frac{\sqrt{3}r_0}{2}\ 0\ -\frac{r_0}{2}\right),\quad R_{31}\left(\frac{\sqrt{3}r_0}{2}\ 0\ -\frac{r_0}{2}\right);$$

$$R_{12}(0\ l_1\cos\alpha\ r_0{+}l_1\sin\alpha),\quad R_{22}\left(-\frac{\sqrt{3}}{2}(r_0{+}l_1\sin\alpha)\ l_1\cos\alpha\ -\frac{1}{2}(r_0{+}l_1\sin\alpha)\right);$$

$$R_{32}\left(\frac{\sqrt{3}}{2}(r_0 + l_1\sin\alpha)\ l_1\cos\alpha\ -\frac{1}{2}(r_0 + l_1\sin\alpha)\right);$$

$$R_{13}(0\ l_1\cos\alpha + l_2\cos\beta\ r_0),\quad R_{23}\left(-\frac{\sqrt{3}}{2}r_0\ l_1\cos\alpha + l_2\cos\beta\ -\frac{r_0}{2}\right),$$

$$R_{33}\left(\frac{\sqrt{3}}{2}r_0\ l_1\cos\alpha + l_2\cos\beta\ -\frac{r_0}{2}\right).$$

With the joint coordinates above, we can write the screws of each kinematic chain. For the chain $R_{11}R_{12}R_{13}$, the kinematic screws can be expressed with

$$\mathbf{\$}_{R_{11}R_{12}R_{13}} = \left[\mathbf{\$}_{R_{11}}\quad \mathbf{\$}_{R_{12}}\quad \mathbf{\$}_{R_{13}}\right] \tag{9.1}$$

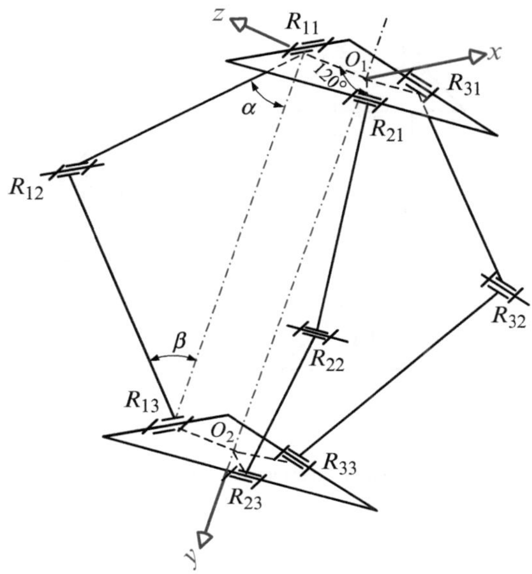


**Fig. 9.2** A simplified 3-RRR linkage

where $\$_{R_{11}} = (1 \quad 0 \quad 0 \quad 0 \quad r_0 \quad 0)^T$,

$$\$_{R_{12}} = (1 \quad 0 \quad 0 \quad 0 \quad r_0 + l_1 \sin\alpha \quad -l_1 \cos\alpha)^T,$$

$$\$_{R_{13}} = (1 \quad 0 \quad 0 \quad 0 \quad r_0 \quad -l_1 \cos\alpha - l_2 \cos\beta)^T.$$

Obviously, the degenerated condition for screw set $\$_{R_{11}R_{12}R_{13}}$ is

$$\begin{vmatrix} 1 & r_0 & 0 \\ 1 & r_0 + l_1 \sin\alpha & -l_1 \cos\alpha \\ 1 & r_0 & -l_1 \cos\alpha - l_2 \cos\beta \end{vmatrix} = 0 \tag{9.2}$$

Equation (9.2) can be simplified as

$$-l_1 \sin\alpha (l_1 \cos\alpha + l_2 \cos\beta) = 0 \tag{9.3}$$

The solution of Eq. (9.3) is $\alpha = 0 (\alpha = \pi)$ or $l_1 = l_2$ and $\alpha = \pi - \beta$ which means that the kinematic chain $R_{11}R_{12}R_{13}$ is kept in a straight line or the link $R_{11}R_{12}$ coincides with link $R_{12}R_{13}$ which is shown in Fig. 9.3. The case that $\alpha = 0 (\alpha = \pi)$ is the dead point position of the 3-RRR mechanism, and there is finite self-motion (FSM) of the 3-RRR mechanism when the links coincide with each other.

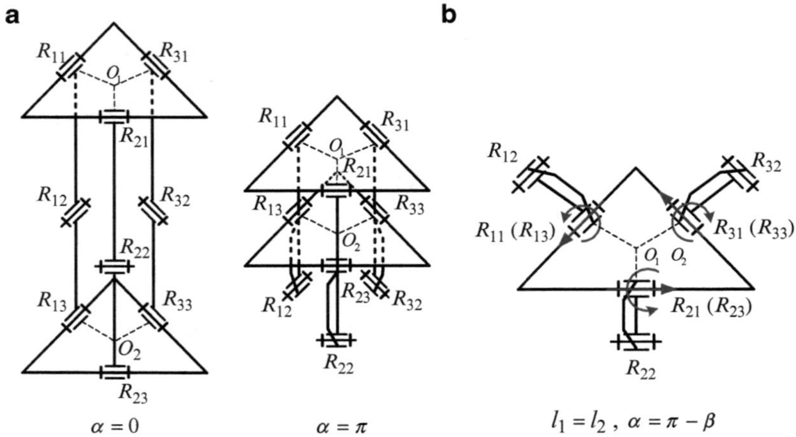


**Fig. 9.3** Singularity characteristic of 3-RRR mechanism. (**a**) Dead point position and (b) Finite self-motion (FSM)

In applications, those cases can be easily avoided by proper design. Therefore, $\mathbf{\$}_{R_{11}R_{12}R_{13}}$ does not degenerate in reality. As a result, the reciprocal screws of $\mathbf{\$}_{R_{11}R_{12}R_{13}}$ can be obtained through the following equation:

$$\$^{T}\Delta\$^{r}=0 \tag{9.4}$$

Solving Eq. (9.4) presents

$$\mathbf{\$}^{r}_{R_{11}R_{12}R_{13}}=\begin{bmatrix}1&0&0&0&0&0\\0&0&0&0&1&0\\0&0&0&0&0&1\end{bmatrix}^{T} \tag{9.5}$$

Similarly, the terminal constraints of $R_{21}R_{22}R_{23}$ and $R_{31}R_{32}R_{33}$ can be obtained via solving Eq. (9.4):

$$\mathbf{\$}^{r}_{R_{21}R_{22}R_{23}}=\begin{bmatrix}-\frac{1}{2}&0&\frac{\sqrt{3}}{2}&0&0&0\\0&0&0&0&1&0\\0&0&0&-\frac{\sqrt{3}}{2}&0&-\frac{1}{2}\end{bmatrix}^{T} \tag{9.6}$$

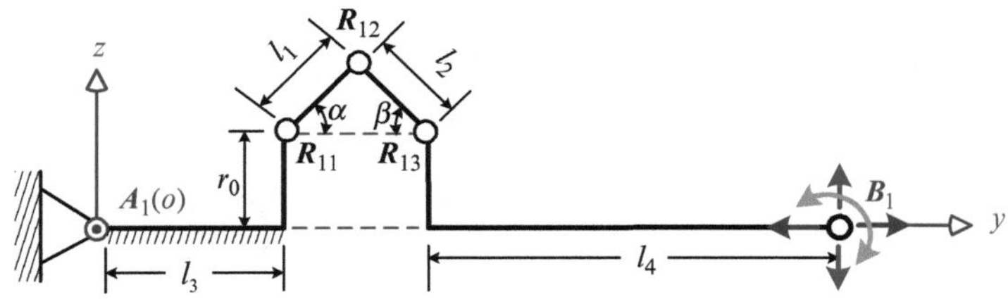


**Fig. 9.4** Possible motions of a planar RRR kinematic chain

$$\boldsymbol{\$}^r_{R_{31}R_{32}R_{33}} = \begin{bmatrix} \frac{1}{2} & 0 & \frac{\sqrt{3}}{2} & 0 & 0 & 0 \\ 0 & 0 & 0 & 0 & 1 & 0 \\ 0 & 0 & 0 & \frac{\sqrt{3}}{2} & 0 & -\frac{1}{2} \end{bmatrix}^T \tag{9.7}$$

Hence, the constraints exerted to the platform $R_{13}R_{23}R_{33}$ by the platform $R_{11}R_{21}R_{31}$ via the three kinematic chains are

$$\boldsymbol{\$}^r_{R_{13}R_{23}R_{33}} = \begin{bmatrix} \boldsymbol{\$}^r_{R_{11}R_{12}R_{13}} & \boldsymbol{\$}^r_{R_{21}R_{22}R_{23}} & \boldsymbol{\$}^r_{R_{31}R_{32}R_{33}} \end{bmatrix} \tag{9.8}$$

Substituting Eq. (9.8) into (9.4) again yields

$$\boldsymbol{\$}_{R_{13}R_{23}R_{33}} = \begin{pmatrix} 0 & 0 & 0 & 0 & 1 & 0 \end{pmatrix}^T \tag{9.9}$$

Equation (9.9) indicates that the relative motion of platform $R_{13}R_{23}R_{33}$ to platform $R_{11}R_{21}R_{31}$ is one translation along the $y$-axis. Therefore, this 3-RRR parallel linkage allows the axial stretch or compression motion just like a prismatic joint shown in Fig. 9.1 does, but avoids the resistance due to the limited contacting area of revolute joints in the 3-RRR parallel linkage.

On the other hand, the 3-RRR mechanism is an overconstrained mechanism with four redundant constraints, and those cases that the RRR kinematic chain kept in a straight line and the link $R_{i1}R_{i2}(i = 1, 2, 3)$ coincides with link $R_{i2}R_{i3}$ must be avoided by proper design. That is to say, the 3-RRR mechanism is non-singularity in practice.

As a matter of fact, the free motions of the RRR kinematic chain in plane $\pi_1$ can also be analyzed from the viewpoint of geometry. To simplify the analysis, this chain is equivalently represented by the one shown in Fig. 9.4.

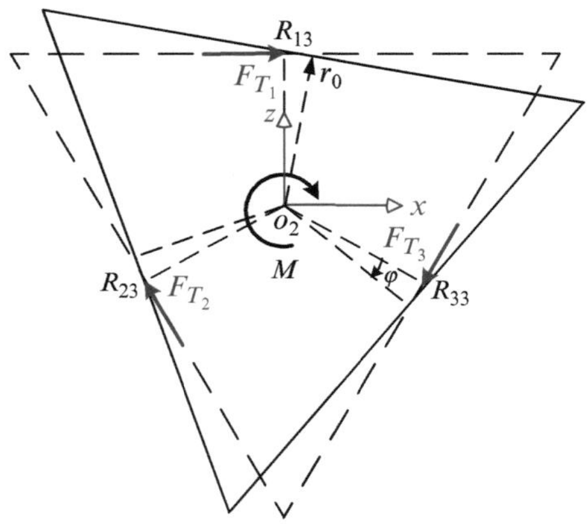


**Fig. 9.5** Terminal constraints of kinematic chain $A_1B_1$

The end point $B_1$ can translate left or right, up or down and rotate about any normal of the plane. Therefore, the end point $B_1$ could reach any point with any pose within a certain region in $\pi_1$-plane which is determined by the sizes of its links. Similarly, the other two chains, $R_{21}R_{22}R_{23}$ and $R_{31}R_{32}R_{33}$, also have the capacity to reach any point with any poses within their corresponding regions, determined by the lengths of the links, in $\pi_2$- and $\pi_3$-planes, respectively. As a result, their common distal platform $R_{13}R_{23}R_{33}$ can only move along the intersection of $\pi_1$-, $\pi_2$-, and $\pi_3$-planes, which is line $A_1B_1$ in Fig. 9.1 or line $o_1o_2$ in Fig. 9.2. This also proves that platforms $R_{11}R_{21}R_{31}$ and $R_{13}R_{23}R_{33}$ only have one relative degree of freedom of translation.

### 9.2.2 Constraint and Equivalent Stiffness Analysis of the Triple Rectilinear Planar Linkage

In this section, the constraint type of each compound kinematic chain of $A_iB_i(i = 1, 2, 3)$ and its equivalent stiffness will be analyzed. Suppose that a torque $M$ about $y$-axis is exerted to the platform $R_{11}R_{21}R_{31}$ shown in Fig. 9.5. This torque will be transformed into the lateral forces of the three RRR kinematic chains. Assume the force transferred to the $R_{i1}R_{i2}R_{i3}(i = 1, 2, 3)$ is $F_{T_i}$ and the platform $R_{11}R_{21}R_{31}$ rotates a small angle $\varphi$ with respect to the platform $R_{13}R_{23}R_{33}$.

Immediately, the equilibrium equation of the system can be expressed as

$$(F_{T_1} + F_{T_2} + F_{T_3})\, r_0 = M \tag{9.10}$$

For the sake of simplicity, the deflection of the RRR kinematic chain shown in Fig. 9.6 can be analyzed in two steps. To calculate the deflection of link $R_{12}R_{13}$,

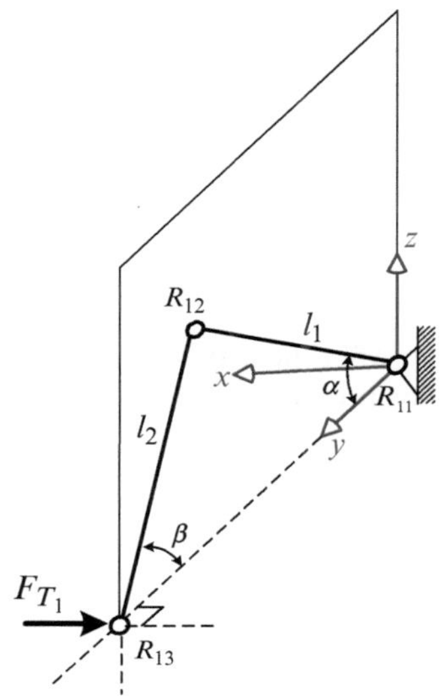


**Fig. 9.6** Deflection of a RRR kinematic chain

we can temporarily assume that $R_{11}R_{12}$ is a rigid link. Under the action of $F_{T_1}$, the deflection of $R_{13}$ with respect to $R_{12}$ is

$$\delta_1 = \frac{F_{T_1} l_2^3}{3EI} \tag{9.11}$$

where $\delta_1$ denotes the deflection of $R_{13}$ with respect to $R_{12}$ and $F_{T_1}$ denotes the lateral force exerted to the distal of kinematic chain $R_{11}R_{12}R_{13}$.

The second step is to transfer force $F_{T_1}$ to point $R_{12}$ accompanied by a bending moment of $F_{T_1} l_2$ which can be decomposed into a torque about the axis of link $R_{11}R_{12}$, $F_{T_1} l_2 \sin(\alpha + \beta)$, and a bending moment, $F_{T_1} l_2 \cos(\alpha + \beta)$. The increased deflection of point $R_{13}$ under the action of $F_{T_1}$ on the link $R_{11}R_{12}$ is

$$\delta_{21} = \frac{F_{T_1} l_1^3}{3EI} + \frac{F_{T_1} l_1^2 l_2 \cos(\alpha + \beta)}{2EI} \tag{9.12}$$

The deflection resulting from the torque $F_{T_1} l_2 \sin(\alpha + \beta)$ is

$$\delta_{22} = \frac{F_{T_1} l_1 l_2^2 \sin^2(\alpha + \beta)}{G I_{\mathrm{P}}} \tag{9.13}$$

where $G$ represents the shear modulus of link $R_{11}R_{12}$ and $I_{\mathrm{P}}$ stands for the polar inertia moment of link $R_{11}R_{12}$ about its axis.

The deflection resulting from the bending moment component, $F_{T_1} l_2 \cos(\alpha + \beta)$, is

$$\delta_{23} = \frac{F_{T_1} l_1^2 l_2 \cos(\alpha + \beta)}{2EI} + \frac{F_{T_1} l_1 l_2^2 \cos^2(\alpha + \beta)}{EI} \tag{9.14}$$

Therefore, the deflection of point $R_{13}$ resulting from the elasticity of link $R_{11}R_{12}$ under the transferred force $F_{T_1}$ is

$$\delta_2 = \delta_{21} + \delta_{22} + \delta_{23} = \frac{F_{T_1} l_2^3}{3EI} + \frac{F_{T_1} l_1^2 l_2 \cos(\alpha+\beta)}{EI} + \frac{F_{T_1} l_1 l_2^2 \cos^2(\alpha+\beta)}{EI} + \frac{F_{T_1} l_1 l_2^2 \sin^2(\alpha+\beta)}{GI_{\mathrm{P}}} \tag{9.15}$$

After the above two steps, the deflection of point $R_{13}$ under the action of $F_{T_1}$ at position $\varphi$ shown in Fig. 9.6 can be expressed as the sum of Eqs. (9.11) and (9.15)

$$\delta_{R_{11}R_{12}R_{13}} = \delta_1 + \delta_2 = \left[\frac{l_1^3}{3EI} + \frac{l_2^3}{3EI} + \frac{l_1^2 l_2 \cos(\alpha+\beta)}{EI} + \frac{l_1 l_2^2 \cos^2(\alpha+\beta)}{EI} + \frac{l_1 l_2^2 \sin^2(\alpha+\beta)}{GI_{\mathrm{P}}}\right] F_{T_1} \tag{9.16a}$$

In the same way, there are

$$\delta_{R_{21}R_{22}R_{23}} = \left[\frac{l_1^3}{3EI} + \frac{l_2^3}{3EI} + \frac{l_1^2 l_2 \cos(\alpha+\beta)}{EI} + \frac{l_1 l_2^2 \cos^2(\alpha+\beta)}{EI} + \frac{l_1 l_2^2 \sin^2(\alpha+\beta)}{GI_{\mathrm{P}}}\right] F_{T_2} \tag{9.16b}$$

and

$$\delta_{R_{31}R_{32}R_{33}} = \left[\frac{l_1^3}{3EI} + \frac{l_2^3}{3EI} + \frac{l_1^2 l_2 \cos(\alpha+\beta)}{EI} + \frac{l_1 l_2^2 \cos^2(\alpha+\beta)}{EI} + \frac{l_1 l_2^2 \sin^2(\alpha+\beta)}{GI_{\mathrm{P}}}\right] F_{T_3} \tag{9.16c}$$

Equations (9.16a), (9.16b), and (9.16c) indicate that the deflections of the three RRR kinematic chains are linear functions of the distal forces. Because the deflections are very small in reality, the rotated angle $\varphi$ is very tiny in value. So, the deflections of the kinematic chains can also be approximately expressed as

$$\delta = r_0 \varphi \tag{9.17}$$

where $r_0$ stands for the distance from the reference point to the rotational center. Because the distances from the distal points of the three RRR kinematic chains to the platform center are identically equal to $r_0$, the real deflections of them should satisfy

$$\delta_{R_{11}R_{12}R_{13}} = \delta_{R_{21}R_{22}R_{23}} = \delta_{R_{31}R_{32}R_{33}} = r_0 \varphi \tag{9.18}$$

Substituting Eq. (9.18) into (9.16) yields

$$\begin{cases} F_{T_1} = \dfrac{r_0\varphi}{\frac{l_1^3}{3EI}+\frac{l_2^3}{3EI}+\frac{l_1^2 l_2\cos(\alpha+\beta)}{EI}+\frac{l_1 l_2^2\cos^2(\alpha+\beta)}{EI}+\frac{l_1 l_2^2\sin^2(\alpha+\beta)}{GI_{\mathrm{P}}}} \\ F_{T_2} = \dfrac{r_0\varphi}{\frac{l_1^3}{3EI}+\frac{l_2^3}{3EI}+\frac{l_1^2 l_2\cos(\alpha+\beta)}{EI}+\frac{l_1 l_2^2\cos^2(\alpha+\beta)}{EI}+\frac{l_1 l_2^2\sin^2(\alpha+\beta)}{GI_{\mathrm{P}}}} \\ F_{T_3} = \dfrac{r_0\varphi}{\frac{l_1^3}{3EI}+\frac{l_2^3}{3EI}+\frac{l_1^2 l_2\cos(\alpha+\beta)}{EI}+\frac{l_1 l_2^2\cos^2(\alpha+\beta)}{EI}+\frac{l_1 l_2^2\sin^2(\alpha+\beta)}{GI_{\mathrm{P}}}} \end{cases} \tag{9.19}$$

Equation (9.19) indicates that the terminal forces of the three RRR kinematic chains are identical. Consequently, the solution of Eq. (9.10) can be gained:

$$F_{T_1} = F_{T_2} = F_{T_3} = \frac{M}{3r_0} \tag{9.20}$$

Substituting Eq. (9.19) into Eq. (9.20) yields

$$K_{T(3-\mathrm{RRR})} = \frac{M}{\varphi} = \frac{3r_0^2}{\frac{l_1^3}{3EI}+\frac{l_2^3}{3EI}+\frac{l_1^2 l_2\cos(\alpha+\beta)}{EI}+\frac{l_1 l_2^2\cos^2(\alpha+\beta)}{EI}+\frac{l_1 l_2^2\sin^2(\alpha+\beta)}{GI_{\mathrm{P}}}} \tag{9.21}$$

where $K_{T(3-\mathrm{RRR})}$ represents the equivalent torsional stiffness of the parallel 3-RRR linkage shown in Fig. 9.2. Generally, the equivalent torsional stiffness of a parallel $n$–RRR linkage whose kinematic chains are rotationally symmetric about the geometry center of the platform will be

$$K_{T(n-\mathrm{RRR})} = \frac{M}{\varphi} = \frac{nr_0^2}{\frac{l_1^3}{3EI}+\frac{l_2^3}{3EI}+\frac{l_1^2 l_2\cos(\alpha+\beta)}{EI}+\frac{l_1 l_2^2\cos^2(\alpha+\beta)}{EI}+\frac{l_1 l_2^2\sin^2(\alpha+\beta)}{GI_{\mathrm{P}}}} \tag{9.22}$$

Assuming that $l_1 = l_2 = l$, so $\alpha = \beta$, and Eq. (9.22) can be simplified as

$$K_{T(n-\mathrm{RRR})} = \frac{M}{\varphi} = \frac{n\zeta^2}{\frac{l}{3EI}\left[2+3\cos^2(2\alpha)+3\cos(2\alpha)\right]+\frac{l\sin^2(2\alpha)}{GI_{\mathrm{P}}}} \tag{9.23}$$

where $\zeta = \frac{r_0}{l}$ stands for the ratio of distance from the joint on the platform to the platform center and the length of the link.

Equation (9.23) shows that the torsional stiffness of the parallel linkage is proportional to the number of kinematic chains. Consequently, the more kinematic chains one use, the higher the torsional stiffness of the n-RRR mechanism is. The equivalent bending stiffness of the 3-RRR mechanism will be investigated next.

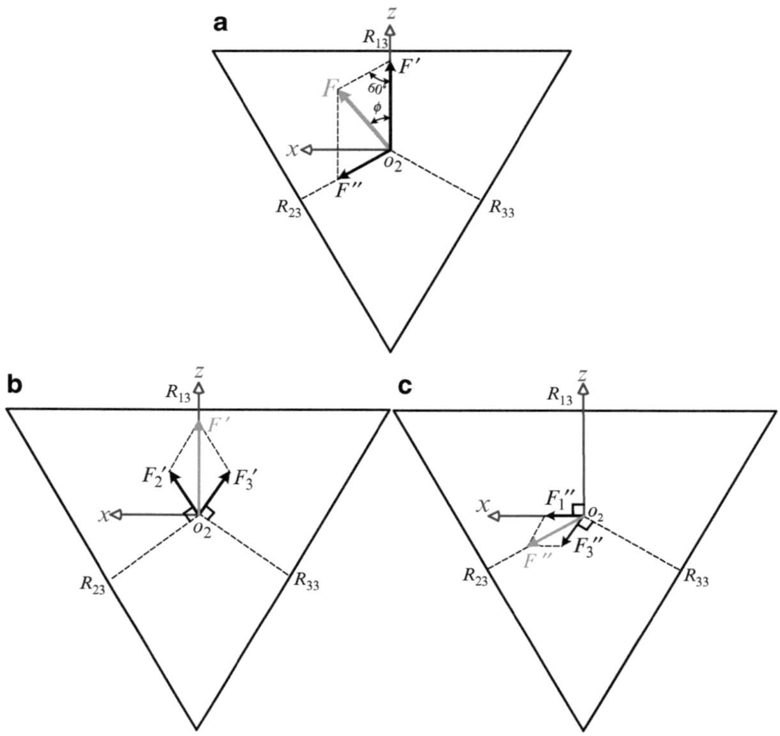


**Fig. 9.7** Decomposition of a force. (**a**) Decomposition of force $F$, (**b**) Decomposition of the component force $F'$, and (**c**) Decomposition of the component force $F''$

Suppose that there is a lateral force $F$ in the plane of $R_{13}R_{23}R_{33}$ and the subtended angle with the $z$-axis is $\phi$. For the sake of simplicity, the force can be decomposed into any two planes of the kinematic chains. For example, the planes determined by $R_{11}R_{12}R_{13}$ and $R_{21}R_{22}R_{23}$ kinematic chains are selected here.

So, $F$ can be replaced by the forces $F'$ and $F''$ as shown in Fig. 9.7a:

$$\begin{cases} F' = F\cos\phi + \dfrac{\sqrt{3}}{3}F\sin\phi \\ F'' = \dfrac{2\sqrt{3}}{3}F\sin\phi \end{cases} \tag{9.24}$$

Because every RRR kinematic chain can only provide the forces in the normal direction of its plane, $F'$ cannot be supported by the kinematic chain $R_{11}R_{12}R_{13}$ and $F''$ cannot be supported by $R_{21}R_{22}R_{23}$ either. $\boldsymbol{F}'$ is therefore decomposed as

$$\boldsymbol{F}' = \boldsymbol{F}'_2 + \boldsymbol{F}'_3 \tag{9.25}$$

where $F'_2 = F'_3 = \frac{\sqrt{3}}{3}F'$ which is shown in Fig. 9.7b. Similarly, a decomposition of $F''$ is expressed as

$$\boldsymbol{F''} = \boldsymbol{F''}_1 + \boldsymbol{F''}_3 \tag{9.26}$$

where $F''_1 = F''_3 = \frac{\sqrt{3}}{3}F''$ which is shown in Fig. 9.7c. As a result, the following relations hold

$$\begin{cases} F_1 = \dfrac{\sqrt{3}}{3}F'' = \dfrac{2}{3}F\sin\phi \\ F_2 = \dfrac{\sqrt{3}}{3}F' = \dfrac{\sqrt{3}}{3}F\cos\phi + \dfrac{1}{3}F\sin\phi \\ F_3 = \dfrac{\sqrt{3}}{3}(F' - F'') = \dfrac{\sqrt{3}}{3}F\cos\phi - \dfrac{1}{3}F\sin\phi \end{cases} \tag{9.27}$$

where $F_i\,(i = 1, 2, 3)$ denotes the resultant force exerted to the $i$th chain.

From Eq. (9.16a), it can be found that the shear stiffness of one RRR kinematic chain should be

$$K_{S(\mathrm{RRR})} = \frac{F_{T_1}}{\delta_{R_{11}R_{12}R_{13}}} = \frac{1}{\frac{l_1^3}{3EI} + \frac{l_2^3}{3EI} + \frac{l_1^2 l_2\cos(\alpha+\beta)}{EI} + \frac{l_1 l_2^2\cos^2(\alpha+\beta)}{EI} + \frac{l_1 l_2^2\sin^2(\alpha+\beta)}{GI_{\mathrm{P}}}} \tag{9.28}$$

where $K_{S(\mathrm{RRR})}$ stands for the lateral shear stiffness of the RRR kinematic chain.

Therefore, the deflections of the three kinematic chains, $R_{11}R_{12}R_{13}$, $R_{21}R_{22}R_{23}$, and $R_{31}R_{32}R_{33}$, will be

$$\begin{cases} \delta_1 = \dfrac{F_1}{K_{S(\mathrm{RRR})}} = \dfrac{2F}{3K_{S(\mathrm{RRR})}}\sin\phi \\ \delta_2 = \dfrac{F_2}{K_{S(\mathrm{RRR})}} = \left(\dfrac{\sqrt{3}}{3}\cos\phi + \dfrac{1}{3}\sin\phi\right)\dfrac{F}{K_{S(\mathrm{RRR})}} \\ \delta_3 = \dfrac{F_3}{K_{S(\mathrm{RRR})}} = \left(\dfrac{\sqrt{3}}{3}\cos\phi - \dfrac{1}{3}\sin\phi\right)\dfrac{F}{K_{S(\mathrm{RRR})}} \end{cases} \tag{9.29}$$

The resultant deflection of the platform $R_{11}R_{21}R_{31}$ under the shearing force is shown in Fig. 9.8.

This can be expressed as

$$\delta = \frac{F}{K_{S(\mathrm{RRR})}} \tag{9.30}$$

**Fig. 9.8** Resultant deflection

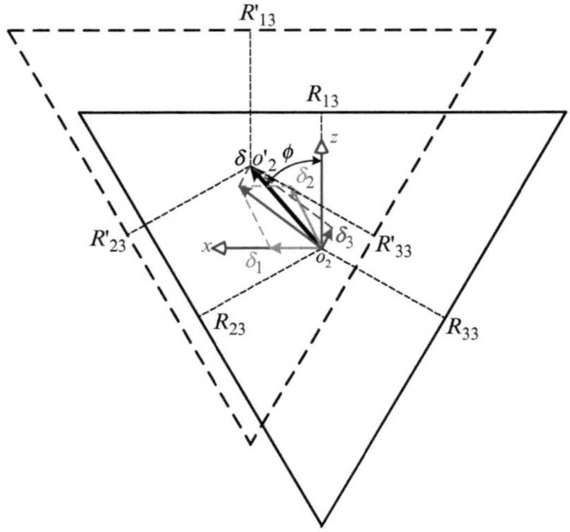


Therefore, the lateral shear stiffness of the 3-RRR mechanism is

$$K = \frac{F}{\delta} = K_{S(\text{RRR})} \tag{9.31}$$

Comparing Eq. (9.31) with (9.28), it can be found that the lateral shear stiffness of the 3-RRR mechanism equals that of one RRR chain. The bending stiffness of a RRR linkage can be approximately analyzed below.

The deflection of an RRR linkage shown in Fig. 9.6 where the force is substituted by a pure bending moment, denoted by $M$, is

$$\delta = \frac{Ml_1^2\cos\alpha}{2EI} + \frac{Ml_2^2\cos\beta}{2EI} + \frac{Ml_1l_2\cos\alpha\cos(\alpha+\beta)}{EI} + \frac{Ml_1l_2\sin\alpha\sin(\alpha+\beta)}{GI_{\text{P}}} \tag{9.32}$$

The distance between $R_{11}$ and $R_{13}$ is

$$s = l_1\cos\alpha + l_2\cos\beta \tag{9.33}$$

The slope of the RRR linkage can be approximately expressed as

$$\varphi_{B(\text{RRR})} \approx \frac{\delta}{s} = \frac{\frac{Ml_1^2\cos\alpha}{2EI} + \frac{Ml_2^2\cos\beta}{2EI} + \frac{Ml_1l_2\cos\alpha\cos(\alpha+\beta)}{EI} + \frac{Ml_1l_2\sin\alpha\sin(\alpha+\beta)}{GI_{\text{P}}}}{l_1\cos\alpha + l_2\cos\beta} \tag{9.34}$$

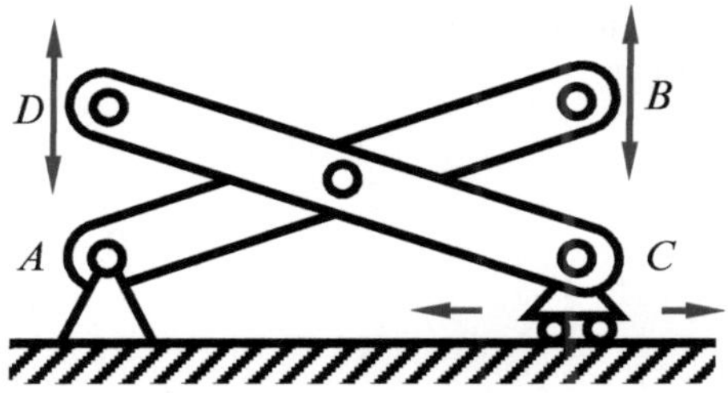


**Fig. 9.9** A foldable SLE structure

The bending stiffness can be approximately expressed as

$$K_{B(\mathrm{RRR})} \approx \frac{M}{\varphi_{B(\mathrm{RRR})}} = \frac{l_1 \cos\alpha + l_2 \cos\beta}{\frac{l_1^2 \cos\alpha}{2EI} + \frac{l_2^2 \cos\beta}{2EI} + \frac{l_1 l_2 \cos\alpha \cos(\alpha+\beta)}{EI} + \frac{l_1 l_2 \sin\alpha \sin(\alpha+\beta)}{GI_{\mathrm{P}}}} \tag{9.35a}$$

Assuming that $l_1 = l_2 = l$, so $\alpha = \beta$, and Eq. (9.35a) can be simplified as

$$K_{B(\mathrm{RRR})} \approx \frac{M}{\varphi_{B(\mathrm{RRR})}} = \frac{2}{\frac{l(1+\cos 2\alpha)}{EI} + \frac{l \tan\alpha \sin 2\alpha}{GI_{\mathrm{P}}}} \tag{9.35b}$$

For the linkage is used in the conditions when the torsional torque, the lateral load, and the bending moment coexist, 3-RRR structure will be the best scheme because the 2-RRR structure cannot realize the rectilinear guidance when they are in rotational symmetry about the central axis of the linkage while more kinematic chains should decrease the strength of each chain.

### *9.2.3 Structures for Foldable Towers*

The simplest pantograph structure is a pair of links, $AB$ and $CD$ shown in Fig. 9.9, which are pivot-jointed at their middle points. When one end of this structure, $A$, is pivot jointed with the fixed base and a second end, $D$, can make a determinate translation, this structure forms a foldable scissorlike-element (SLE) structure shown in Fig. 9.9.

When a number of SLEs shown in Fig. 9.9 are connected one by one in series shown in Fig. 9.10, a foldable structure is obtained. The top of the structure will move up and down when joint $C_1$ translates left and right along the guide on the fixed base.

Fig. 9.10 A foldable structure

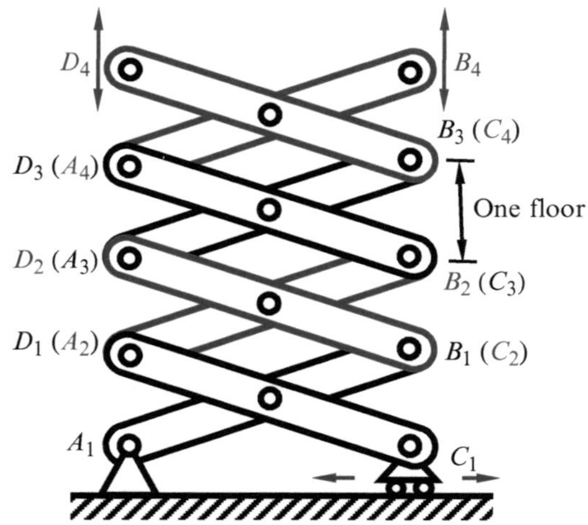


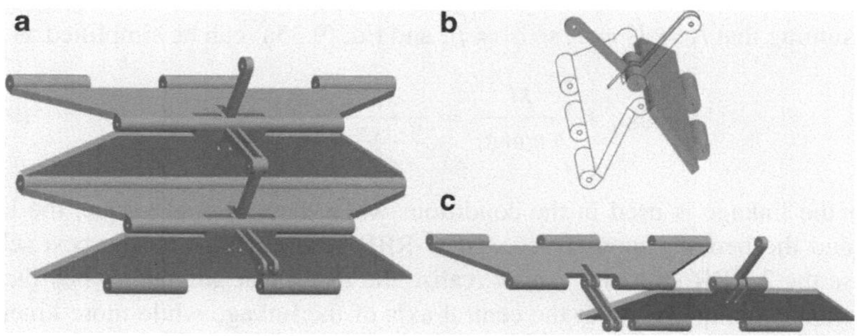


**Fig. 9.11** Assembly of a branch. (**a**) Assembly of two levels, (**b**) assembly of one level, and (**c**) a pair of members

However, the strength and stiffness of the foldable structure shown in Fig. 9.10 might not satisfy the needs of tower engineering. If half of each member is particularly designed to be thick plate, the lateral strength and stiffness of the whole structure will be improved greatly which is shown in Fig. 9.11 where the subfigure (a) shows the assembly of the structure and subfigures (b) and (c) illustrate the assembly of a pair of members.

Simultaneously, if two or more such planar structures are connected in parallel, the strength and stiffness will be increased a lot more. As shown in Figs. 9.12 and 9.13, the whole structure will become stronger if three such structures are connected through an equilateral triangular in each level correspondingly.

Theoretically, the tower can be completely folded. Figure 9.14 shows the folded tower.

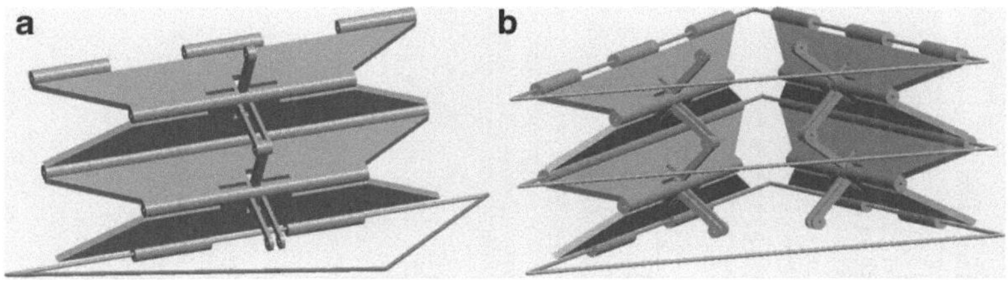


**Fig. 9.12** Parallel structure scheme. (**a**) One foldable structure and (**b**) two branches connected by an equilateral triangle in each level

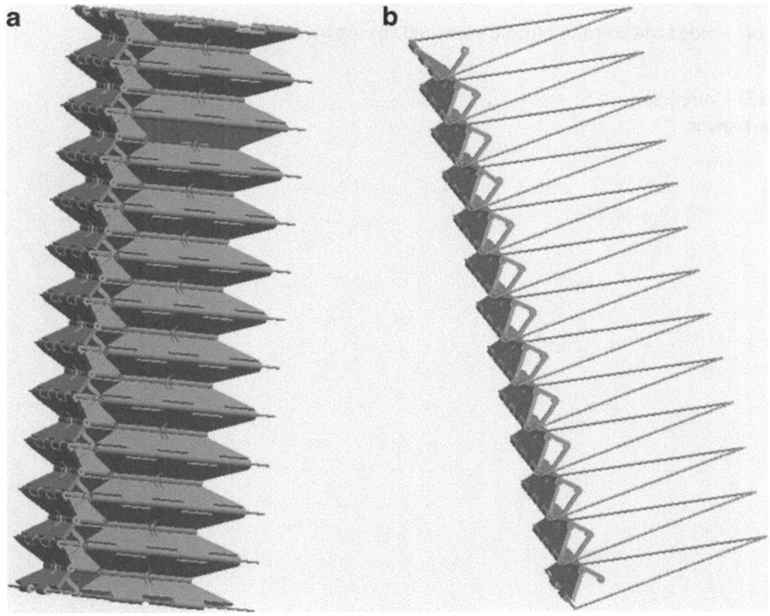


**Fig. 9.13** Assembly of a foldable tower and its one branch. (**a**) Assembly of the foldable structure and (**b**) one side of the structure

## *9.2.4 Calculation of Foldable Ratio*

Foldable ratio is defined as the quotient of the maximum length to the minimum length of a foldable structure. It is an important quality for any foldable structures to have larger foldable ratio that can save more material to reach the same height and save the required space when completely folded. But the foldable ratio should not be very large because of the limitations of the structure design.

Figure 9.15 shows the case when the tower stretches to the maximum, and Fig. 9.14 shows the case when it is completely folded. Assume the maximum

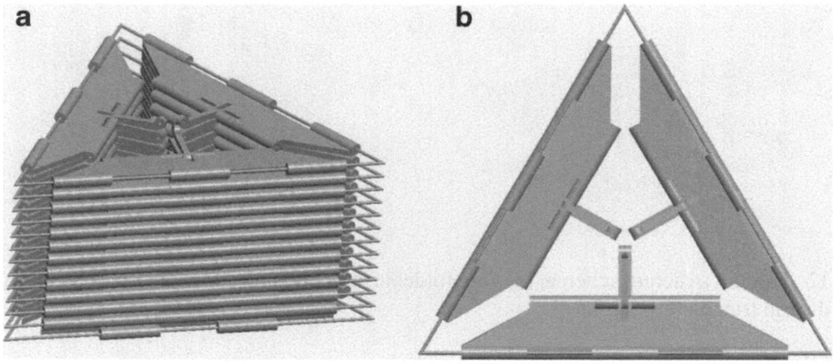


**Fig. 9.14** Folded tower. (**a**) Isotropic view and (**b**) top view of the tower

**Fig. 9.15** Completely deployed tower

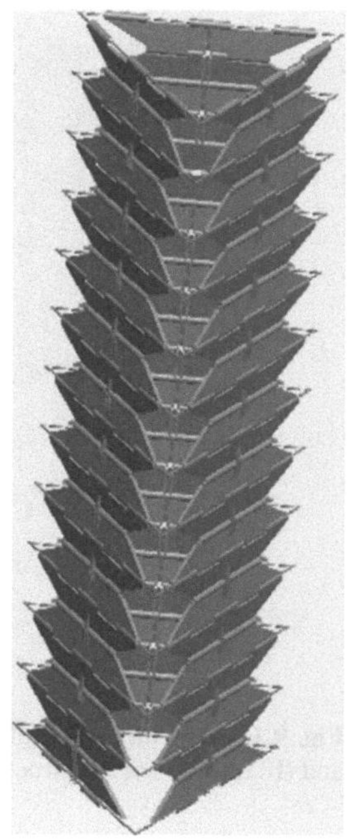

angle is $2\theta_{\max}$ and the minimum subtended angle of the two planes is $2\theta_{\min} = 0$ in Fig. 9.16. According to the geometry relationship, $2\theta_{\max}$ corresponds to the minimum of $\alpha$.

Theoretically,

$$\alpha_{\min} = 2\arcsin\frac{r}{l} \tag{9.36}$$

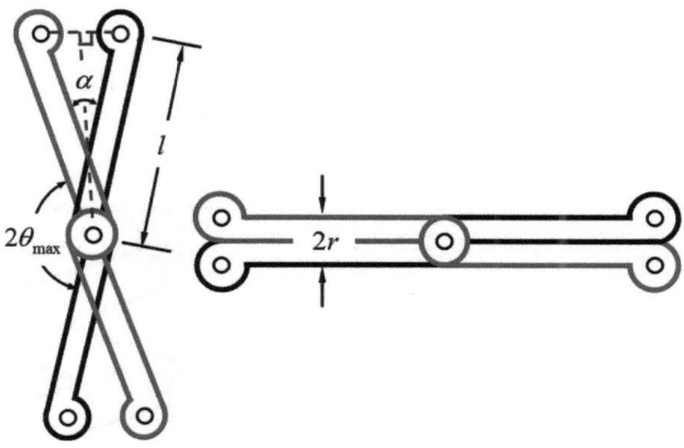


**Fig. 9.16** Geometry position when the tower is completely deployed

where $r$ denotes the radius of the revolute joint and $l$ represents the half length of each link.

Therefore,

$$2\theta_{\max} = \pi - \alpha_{\min} = \pi - 2\arcsin\frac{r}{l} \tag{9.37}$$

As a result, the deployable ratio of the foldable tower is

$$q = \frac{2l\sin\theta_{\max}}{2r} = \sqrt{\left(\frac{l}{r}\right)^2 - 1} \tag{9.38}$$

Equation (9.38) indicates that the deployable ratio of the tower is uniquely determined by the ratio of $l$ to $r$. The deployable ratio $q$ will be as high as 19.97 if we select $\frac{l}{r} = \frac{2{,}000\,\text{mm}}{100\,\text{mm}} = 20$ which is a very high ratio in engineering. That is to say, a folded tower of 4 m high can be expanded to 79.88 m. This kind of foldable tower can be effectively used in the wind turbine of offshore field for both providing the working condition for wind turbine and avoiding the hurricane's destroy by folding and covering a protective shelter.

## 9.3 Stiffness Analysis of the Foldable Tower

### *9.3.1 Statics of the Foldable Tower*

When the foldable tower works, it is deployed fully just as if it is a kind of normal fixed tower, staying in a state without moving. But the tower is subjected to the gravity of the wind turbine and the gravity of itself, so how can it be in a state of

Fig. 9.17 Statics model the foldable structure

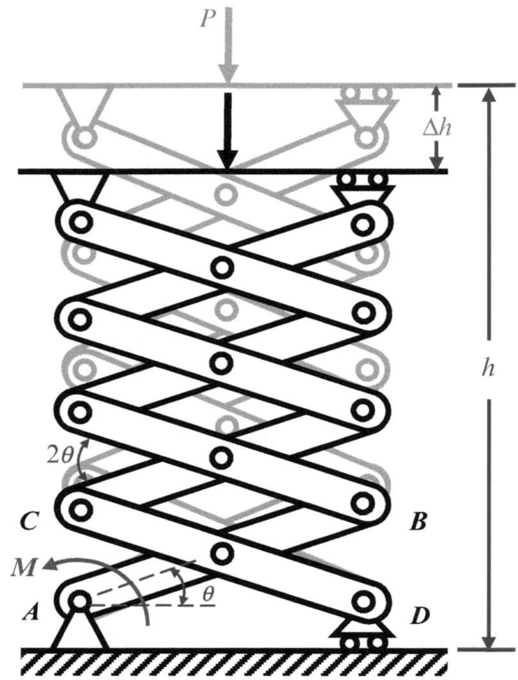


equilibrium at any poses including the pose, $\theta_{\max}$, shown in Fig. 9.18? Therefore, the equivalent stiffness of the tower should be investigated.

Suppose that the bending moment, $M$, which is used to make the foldable tower keep in a state of equilibrium, when the subtended angle between $AB$ and the horizontal plane is $\theta$. Assume that the mass of every link is $m$, the length of each link is $2l$ as shown in Fig. 9.16. Suppose the bending moment $M$ is exerted to joint $A$ shown in Fig. 9.17. According to the principle of virtual power, we obtain

$$M\delta\theta - \sum_{i=1}^{n}\{(2mg)\delta\left[(2i-1)l\sin\theta\right] - P\delta(2nl\sin\theta)\} = 0 \tag{9.39}$$

where $\delta(\cdot)$ represents the variation of "$\cdot$".

Expanding Eq. (9.39) yields

$$M\delta\theta - 2mg\sum_{i=1}^{n}[(2i-1)l\cos\theta]\delta\theta - 2nlP\cos\theta\delta\theta = 0$$

Considering the arbitrary of variation $\delta\theta$, we immediately gain that

$$M = 2(n^2mg + nP)l\cos\theta \tag{9.40}$$

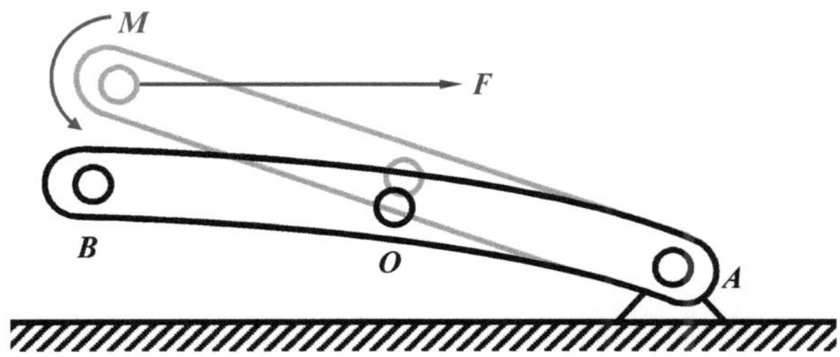


**Fig. 9.18** Deformations of a link under a bending moment and a force

The start value of moment needed to unfold the tower is $M = 2(n^2mg + nP)l\cos\theta_{\min}$, and then the tower is deployed in a constant velocity. When $\theta$ changes from $\theta_{\min}$ to $\theta_{\max}$, the whole work done by the torque $M$ is

$$\begin{aligned} W &= \int_{\theta_{\min}}^{\theta_{\max}} M\,\mathrm{d}\theta \\ &= l(6n^2mg + 2nP)\int_{\theta_{\min}}^{\theta_{\max}} \cos\theta\,\mathrm{d}\theta \\ &= l(6n^2mg + 2nP)(\sin\theta_{\max} - \sin\theta_{\min}) \end{aligned} \tag{9.41}$$

where $W$ denotes the work done to the foldable tower by $M$.

### 9.3.2 *Equivalent Stiffness of the Foldable Tower*

When a load, $P$ shown in Fig. 9.17, is exerted on the top of the foldable tower, the whole structure is subjecting to a compression, and therefore, its height will surely reduce a distance, denoted by $\Delta h$. Each link will be deformed correspondingly, compressed, and bended when it subjects to the bending moment and axial force, which is shown in Fig. 9.18.

When a given force, $P$, is exerted on the top of the tower, the whole deformation $\Delta h$ can be calculated. According to Hooke's law, the equivalent linear elastic coefficient $k$ can be equivalently expressed as

$$k = \frac{P}{\Delta h} \tag{9.42}$$

To simplify the analysis of the deformations, all links in the structure are treated as slender and long straight beams. The subtended angle between $AB$ and the horizontal plane shown in Fig. 9.19 is $\theta$ which indicates the deployment angle of the whole structure. Also suppose $l_{AB} = l_{CD} = 2l$. The external force, $P$,

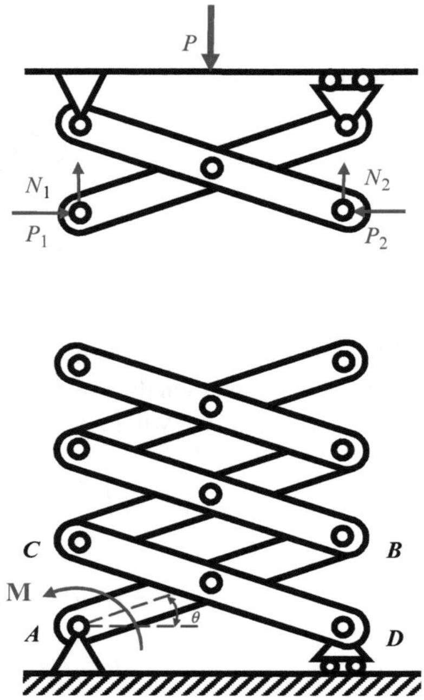


**Fig. 9.19** Statics analysis of the first level

is exerted on the geometry center of the tower's top plane. Assume there are $n$ layers in the tower. Let the top layer be the 1st layer, and the bottom layer is the $n$th one.

First of all, the statics of the 1st layer in Fig. 9.19 can be analyzed as the free body system under static constraints. According to symmetry, we immediately find that

$$\begin{cases} N_1 = N_2 = \dfrac{P}{2} + mg \\ P_1 = P_2 \end{cases} \tag{9.43}$$

According to the principle of virtual power, there is

$$-P_2\delta(2l\cos\theta) - P\delta(2l\sin\theta) - 2mg\delta(l\sin\theta) = 0 \tag{9.44}$$

Expanding Eq. (9.44) yields

$$2P_2 l\sin\theta\delta\theta - 2Pl\cos\theta\delta\theta - 2mgl\cos\theta\delta\theta = 0$$

Associating Eq. (9.43), we have

$$P_1 = P_2 = (P + mg)\cot\theta \tag{9.45}$$

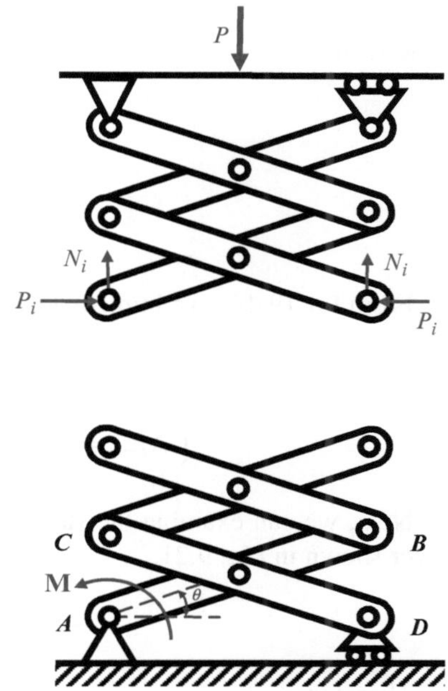


**Fig. 9.20** Statics of the tower

Next, for the $i$th layer shown in Fig. 9.20 where $i = 1, 2, \ldots, n-1$, there is

$$N_i = \frac{P}{2} + img \tag{9.46}$$

According to the principle of virtual power, we obtain that

$$-P_i\delta(2l\cos\theta) - P\delta(2il\sin\theta) - \sum_{j=1}^{i} 2mg\delta\left[(2j-1)l\sin\theta\right] = 0 \tag{9.47}$$

Expanding Eq. (9.47) yields

$$2lP_i\sin\theta\delta\theta - 2ilP\cos\theta\delta\theta - \sum_{j=1}^{i} 2mg(2j-1)l\cos\theta\delta\theta = 0$$

Therefore, there is

$$P_i = i(P + img)\cot\theta \tag{9.48}$$

Associating Eqs. (9.46) and (9.48) generates

$$\begin{cases} P_i = i(P + img)\cot\theta \\ N_i = \dfrac{P}{2} + img \end{cases} \tag{9.49}$$

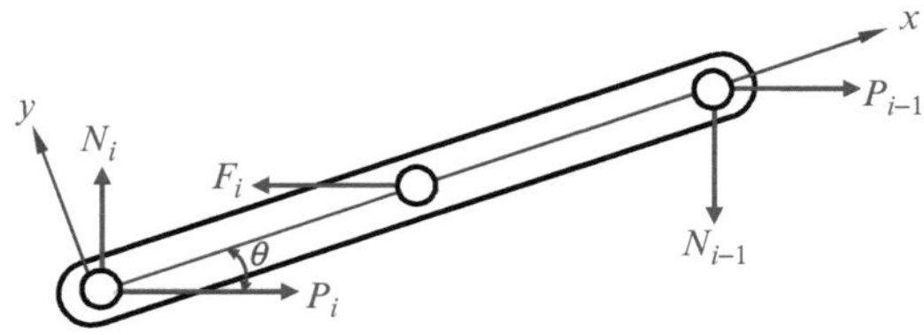


**Fig. 9.21** Statics of a bar in the *i* th floor

From Eq. (9.49), we immediately obtain that

$$\begin{cases} P_{i-1} = (i-1)\left[P + (i-1)mg\right]\cot\theta \\ N_{i-1} = \dfrac{P}{2} + (i-1)mg \end{cases} \tag{9.50}$$

Next, we can examine the statics of every link in the $i$th ($i = 1, 2, \ldots, n-1$) layer shown in Fig. 9.21.

$$-F_i + P_i + P_{i-1} = 0 \tag{9.51}$$

Therefore,

$$F_i = P_i + P_{i-1} = \left[(2i-1)P + (2i^2 - 2i + 1)mg\right]\cot\theta \tag{9.52}$$

When $0 \le x \le l$, the bending moment, the axial and shearing forces of the link are

$$\begin{cases} M_z(x) = P_i x\sin\theta - N_i x\cos\theta + \displaystyle\int_0^x \frac{mg}{2l}\xi\cos\theta\,\mathrm{d}\xi \\ \quad = \left[\left(i - \dfrac{1}{2}\right)P + (i^2 - i)mg + \dfrac{mg}{4l}x\right]x\cos\theta \\ N_x(x) = P_i\cos\theta + N_i\sin\theta - \dfrac{mg}{2l}x\sin\theta = \left(i\cot\theta\cos\theta + \dfrac{1}{2}\sin\theta\right)P \\ \quad + \left(i^2\cot\theta\cos\theta + i\sin\theta - \dfrac{x}{2l}\sin\theta\right)mg \\ Q_y(x) = -P_i\sin\theta + N_i\cos\theta - \dfrac{mg}{2l}x\cos\theta \\ \quad = -\left[\left(i - \dfrac{1}{2}\right)P - (i^2 - i)mg + \dfrac{mg}{2l}x\right]\cos\theta \end{cases} \tag{9.53a}$$

When $l \leq x \leq 2l$, the bending moment, the axial and shearing forces of the link are

$$\begin{cases} M_z(x) = P_{i-1}(2l-x)\sin\theta + N_{i-1}(2l-x)\cos\theta + \int_0^{2l-x}\frac{mg}{2l}\xi\cos\theta\,\mathrm{d}\xi \\ \quad = P\left(i-\frac{1}{2}\right)(2l-x)\cos\theta + \left[(i^2-i)+\frac{1}{4l}(2l-x)\right](2l-x)mg\cos\theta \\ N_x(x) = -P_{i-1}\cos\theta + N_i\sin\theta - \frac{mg}{2l}x\sin\theta \\ \quad = -(i-1)\left[P+(i-1)mg\right]\cot\theta\cos\theta + \left(\frac{P}{2}+img\right)\sin\theta - \frac{mg}{2l}x\sin\theta \\ Q_y(x) = P_{i-1}\sin\theta + N_i\cos\theta - \frac{mg}{2l}x\cos\theta \\ \quad = \left[\left(i-\frac{1}{2}\right)P + (i^2-i+1)mg\right]\cos\theta - \frac{mg}{2l}x\cos\theta \end{cases} \tag{9.53b}$$

Finally, for the $n$th layer (bottom layer), in line with the principle of virtual power, we have

$$M\delta\theta - 2N_{n-1}\delta(2l\sin\theta) + P_{n-1}\delta(2l\cos\theta) - 2mg\delta(l\sin\theta) = 0 \tag{9.54}$$

Substituting Eq. (9.50) into (9.54) yields

$$M = 2(nP + n^2mg)l\cos\theta \tag{9.55}$$

This again proves Eq. (9.40). In line with the static equilibrium, these are the following equations for the last layer:

$$\begin{cases} N_{n1} + N_{n2} - (P + 2nmg) = 0 \\ M + 2N_{n2}l\cos\theta - 2N_{n-1}l\cos\theta - 2mgl\cos\theta = 0 \end{cases} \tag{9.56}$$

Solving equation set (9.56) yields

$$\begin{cases} N_{n1} = \left(n+\frac{1}{2}\right)P + (n^2+n)mg \\ N_{n2} = -\left(n-\frac{1}{2}\right)P - (n^2-n)mg \end{cases} \tag{9.57}$$

where $n \geq 2$; therefore, $N_{n2} < 0$ which indicates that the direction of the force is inverse to the one shown in Fig. 9.22.

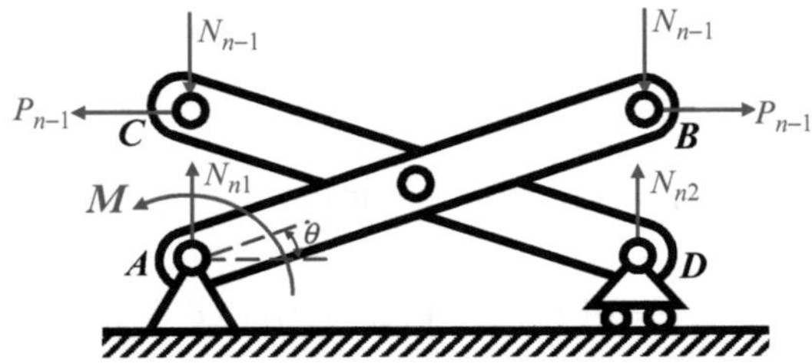


**Fig. 9.22** Statics for the bottom floor

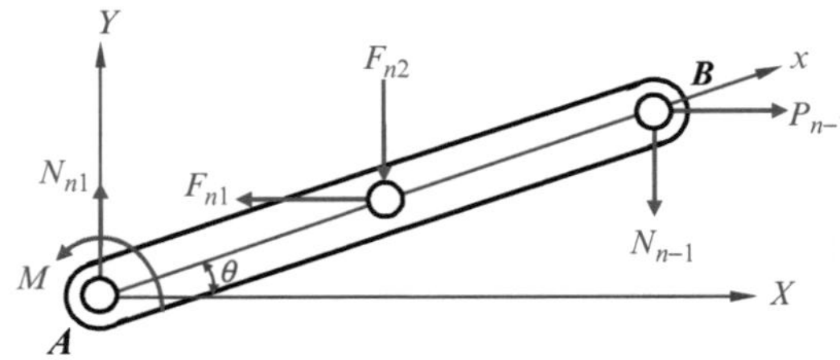


**Fig. 9.23** Statics of link $AB$

Next, we can investigate the statics of link $AB$. The equilibrium criteria of link $AB$ can be expressed as (Fig. 9.23)

$$\begin{cases} \sum F_X = 0, & -F_{n1} + P_{n-1} = 0 \\ \sum F_Y = 0, & N_{n1} - F_{n2} - N_{n-1} = 0 \end{cases} \tag{9.58}$$

Substituting Eqs. (9.52) and (9.57) into (9.58) yields

$$\begin{cases} F_{n1} = (n-1)\left[P + (n-1)mg\right]\cot\theta \\ F_{n2} = nP + (n^2+1)mg \end{cases} \tag{9.59}$$

When $0 \le x \le l$, the bending moment, the axial and shearing forces of the link are

$$\begin{cases} M_z(x) = M - N_{n1}x\cos\theta + \int_0^x \frac{mg}{2l}\xi\cos\theta\,\mathrm{d}\xi \\ \quad = 2(nP+n^2mg)l\cos\theta - \left[\left(n+\frac{1}{2}\right)P + (n^2+n)mg\right]x\cos\theta + \frac{mg}{4l}x^2\cos\theta \\ N_x(x) = N_{n1}\sin\theta - \frac{mg}{2l}x\sin\theta \\ \quad = \left[\left(n+\frac{1}{2}\right)P + (n^2+n)mg - \frac{mg}{2l}x\right]\sin\theta \\ Q_y(x) = N_{n1}\cos\theta - \frac{mg}{2l}x\cos\theta \\ \quad = \left[\left(n+\frac{1}{2}\right)P + (n^2+n)mg - \frac{mg}{2l}x\right]\cos\theta \end{cases} \tag{9.60a}$$

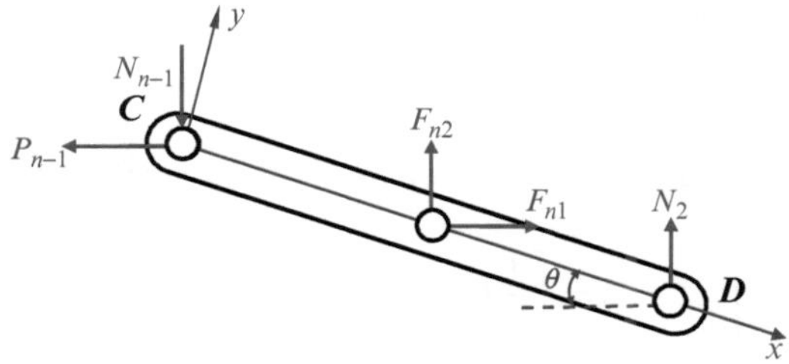


**Fig. 9.24** Statics of link $CD$

When $l \le x \le 2l$, the bending moment, the axial and shearing forces of the link are

$$\begin{cases} M_z(x) = N_{n-1}(2l-x)\cos\theta + P_{n-1}(2l-x)\sin\theta + \displaystyle\int_0^{2l-x}\frac{mg}{2l}\xi\cos\theta\,\mathrm{d}\xi \\ \quad = \left[\dfrac{1}{2}P + (n-1)mg\right](2l-x)\cos\theta \\ \qquad + (n-1)\left[P+(n-1)mg\right](2l-x)\cos\theta + \dfrac{mg}{4l}(2l-x)^2\cos\theta \\ \quad = \left[\dfrac{1}{2}P + (n-1)mg + (n-1)P + (n-1)^2 mg\right](2l-x)\cos\theta \\ \qquad + \dfrac{mg}{4l}(2l-x)^2\cos\theta \\ \quad = \left[\left(n-\dfrac{1}{2}\right)P + \left(n^2 - n + \dfrac{1}{2} - \dfrac{x}{4l}\right)mg\right](2l-x)\cos\theta \\ N_x(x) = -P_{n-1}\cos\theta + N_{n-1}\sin\theta + \dfrac{mg}{2l}(2l-x)\sin\theta \\ \quad = -\left[(n-1)P + (n-1)^2 mg\right]\cot\theta\cos\theta + \left(\dfrac{1}{2}P + nmg - \dfrac{mg}{2l}x\right)\sin\theta \\ Q_y(x) = -P_{n-1}\sin\theta + N_{n-1}\cos\theta - \dfrac{mg}{2l}(2l-x)\cos\theta \\ \quad = -\left[\dfrac{1}{2}P + \dfrac{1}{2}(n-1)mg\right]\cos\theta - (n-1)\left[P+(n-1)mg\right]\cos\theta \\ \qquad - \dfrac{mg}{2l}(2l-x)\cos\theta \\ \quad = -\left[\left(n-\dfrac{1}{2}\right)P + (n^2-n)mg + \dfrac{mg}{2l}(2l-x)\right]\cos\theta \end{cases} \tag{9.60b}$$

Because the equilibrium criteria of link $AB$ are different from those of link $CD$, the statics of link $CD$ should be established using the same method as well. The equilibrium conditions of link $CD$ can be expressed as (Fig. 9.24)

When $0 \leq x \leq l$, the bending moment, the axial and shearing forces of the link are

$$\begin{cases} M_z(x) = N_{n-1}x\cos\theta + P_{n-1}x\sin\theta + \int_0^x \frac{mg}{2l}\xi\cos\theta\,\mathrm{d}\xi \\ \quad = \left[\frac{1}{2}P + (n-1)mg\right]x\cos\theta \\ \quad + (n-1)\left[P + (n-1)mg\right]x\cot\theta\sin\theta + \frac{mg}{4l}x^2\cos\theta \\ \quad = \left[\left(n-\frac{1}{2}\right)P + (n^2-n)mg + \frac{mg}{4l}x\right]x\cos\theta \\ N_x(x) = N_{n-1}\sin\theta - P_{n-1}\cos\theta + \frac{mg}{2l}x\sin\theta \\ \quad = \left[\frac{1}{2}P + (n-1)mg\right]\sin\theta - (n-1)\left[P + (n-1)mg\right]\cot\theta\cos\theta \\ \quad + \frac{mg}{2l}x\sin\theta \\ Q_y(x) = -N_{n-1}\cos\theta - P_{n-1}\sin\theta - \frac{mg}{2l}x\cos\theta \\ \quad = -\left[\left(n-\frac{1}{2}\right)P + (n^2-n)mg\right]\cos\theta - \frac{mg}{2l}x\cos\theta \end{cases} \tag{9.61a}$$

When $l \leq x \leq 2l$, the bending moment, the axial and shearing forces of the link are

$$\begin{cases} M_z(x) = -N_{n2}(2l-x)\cos\theta - \int_0^{2l-x} \frac{mg}{2l}\xi\cos\theta\,\mathrm{d}\xi \\ \quad = \left[\left(n-\frac{1}{2}\right)P + \left(n^2-n+\frac{1}{2}-\frac{x}{4l}\right)mg\right](2l-x)\cos\theta \\ N_x(x) = N_{n2}\sin\theta - \frac{mg}{2l}(2l-x)\sin\theta \\ \quad = -\left[\left(n-\frac{1}{2}\right)P + (n^2-n+1)mg - \frac{mg}{2l}x\right]\sin\theta \\ Q_y(x) = N_{n2}\cos\theta - \frac{mg}{2l}(2l-x)\cos\theta \\ \quad = -\left[\left(n-\frac{1}{2}\right)P + (n^2-n+1)mg - \frac{mg}{2l}x\right]\cos\theta \end{cases} \tag{9.61b}$$

Now the bending moment, the axial and shearing forces of all links of the structure are known. Therefore, the whole deformation $\Delta h$ of the tower when the force $P$ is exerted on the top can be obtained in line with Mohr integral theory [11]:

$$\Delta h = \Delta h_1 + \Delta h_2 + \Delta h_3 \tag{9.62}$$

where
$$\begin{cases} \Delta h_1 = \sum\limits_{i=1}^{n-1}\left(\int\limits_{2l}\frac{N_x}{EA}\frac{\partial N_x}{\partial P}\mathrm{d}x + \int\limits_{2l}\frac{M_z}{EI}\frac{\partial M_z}{\partial P}\mathrm{d}x + \int\limits_{2l}\frac{KQ_y}{GA}\frac{\partial Q_y}{\partial P}\mathrm{d}x\right) \\ \Delta h_2 = \int\limits_{AB}\left(\frac{N_x}{EA}\frac{\partial N_x}{\partial P} + \frac{M_z}{EI}\frac{\partial M_z}{\partial P} + \frac{KQ_y}{GA}\frac{\partial Q_y}{\partial P}\right)\mathrm{d}x \\ \Delta h_3 = \int\limits_{CD}\left(\frac{N_x}{EA}\frac{\partial N_x}{\partial P} + \frac{M_z}{EI}\frac{\partial M_z}{\partial P} + \frac{KQ_y}{GA}\frac{\partial Q_y}{\partial P}\right)\mathrm{d}x \end{cases}$$
and

where $K = \frac{6}{5}$ for rectangular cross-section link, $\frac{\partial(\cdot)}{\partial P}$ indicates the partial derivative of "·" with respect to force $P$, $E$ represents the modulus of elasticity of the material, and $G$ represents shear modulus of elasticity of the material. And $I$ represents the inertia moment of the transverse section of the beam; $A$ represents the area of the transverse section of the link.

According to Mohr integral theory, we obtain

$$\begin{aligned}
\Delta h_1 =& \sum_{i=1}^{n-1}\left(\int\limits_{2l}\frac{N_x}{EA}\frac{\partial N_x}{\partial P}\mathrm{d}x + \int\limits_{2l}\frac{M_z}{EI}\frac{\partial M_z}{\partial P}\mathrm{d}x + \int\limits_{2l}\frac{KQ_y}{GA}\frac{\partial Q_y}{\partial P}\mathrm{d}x\right) \\
=& \frac{1}{EA}\sum_{i=1}^{n-1}\left(\begin{aligned} &\int_0^l\left[\left(i\cot\theta\cos\theta + \frac{1}{2}\sin\theta\right)P + \left(i^2\cot\theta\cos\theta + i\sin\theta - \frac{x}{2l}\sin\theta\right)mg\right] \\ &\left(i\cot\theta\cos\theta + \frac{1}{2}\sin\theta\right)\mathrm{d}x \\ &+\int_l^{2l}\left\{-(i-1)\left[P + (i-1)mg\right]\cot\theta\cos\theta + \left(\frac{P}{2} + img\right)\sin\theta \right. \\ &\left. -\frac{mg}{2l}x\sin\theta\right\}\left[-(i-1)\cot\theta\cos\theta + \frac{1}{2}\sin\theta\right]\mathrm{d}x \end{aligned}\right) \\
&+ \frac{1}{EI}\sum_{i=1}^{n-1}\left(\begin{aligned} &\int_0^l\left[\left(i-\frac{1}{2}\right)P + (i^2-i)mg + \frac{mg}{4l}x\right]\left(i-\frac{1}{2}\right)x^2\cos^2\theta\mathrm{d}x \\ &+\int_l^{2l}\left\{\left(i-\frac{1}{2}\right)P + \left[(i^2-i) + \frac{1}{4l}(2l-x)\right]mg\right\}\left(i-\frac{1}{2}\right)(2l-x)^2\cos^2\theta\mathrm{d}x \end{aligned}\right) \\
&+ \frac{K}{GA}\sum_{i=1}^{n-1}\left(\begin{aligned} &\int_0^l\left[\left(i-\frac{1}{2}\right)P - (i^2-i)mg + \frac{mg}{2l}x\right]\left(i-\frac{1}{2}\right)\cos^2\theta\,\mathrm{d}x \\ &+\int_l^{2l}\left[\left(i-\frac{1}{2}\right)P + (i^2-i+1)mg - \frac{mg}{2l}x\right]\left(i-\frac{1}{2}\right)\cos^2\theta\mathrm{d}x \end{aligned}\right)
\end{aligned}$$

$$
\begin{aligned}
&= \frac{2l^3}{3EI}\sum_{i=1}^{n-1}\left[\left(i-\frac{1}{2}\right)P+\left(i^2-i+\frac{3}{16}\right)mg\right]\left(i-\frac{1}{2}\right)\cos^2\theta \\
&+\frac{l}{EA}\sum_{i=1}^{n-1}\left[\left(i\cot\theta\cos\theta+\frac{1}{2}\sin\theta\right)P+\left(i^2\cot\theta\cos\theta+i\sin\theta-\frac{1}{4}\sin\theta\right)mg\right] \\
&\times\left(i\cos\theta\cot\theta+\frac{1}{2}\sin\theta\right) \\
&+\frac{l}{EA}\sum_{i=1}^{n-1}\left\{\left[(1-i)\cot\theta\cos\theta+\frac{1}{2}\sin\theta\right]P\right. \\
&\qquad\left.+\left[\left(i-\frac{3}{4}\right)\sin\theta-(i-1)^2\cot\theta\cos\theta\right]mg\right\}\left[(1-i)\cos\theta\cot\theta+\frac{1}{2}\sin\theta\right] \\
&+\frac{Kl}{GA}\sum_{i=1}^{n-1}\left[(2i-1)P+\frac{1}{2}mg\right]\left(i-\frac{1}{2}\right)\cos^2\theta \\
&= \frac{l^3}{EI}\left[\left(\frac{2}{9}n^3-\frac{2}{3}n^2+\frac{11}{18}n-\frac{1}{6}\right)P+\left(\frac{1}{6}n^4-\frac{2}{3}n^3+\frac{43}{48}n^2-\frac{11}{24}n+\frac{1}{16}\right)mg\right]\cos^2\theta \\
&+\frac{l}{EA}\left\{\begin{aligned}&\left[\left(\frac{2}{3}n^3-2n^2+\frac{7}{3}n-1\right)\cot^2\theta\cos^2\theta+(n-1)\cos^2\theta+\left(\frac{1}{2}n-\frac{1}{2}\right)\sin^2\theta\right]P \\ &+\left[\left(\frac{1}{2}n^4-2n^3+\frac{7}{2}n^2-3n+1\right)\cot^2\theta\cos^2\theta+\left(\frac{5}{4}n^2-\frac{5}{2}n+\frac{5}{4}\right)\cos^2\theta\right. \\ &\left.+\left(\frac{1}{2}n^2-n+\frac{1}{2}\right)\sin^2\theta\right]mg\end{aligned}\right\} \\
&+\frac{Kl}{GA}\left[\left(\frac{2}{3}n^3-2n^2+\frac{11}{6}n-\frac{1}{2}\right)P+\left(\frac{1}{4}n^2-\frac{1}{2}n+\frac{1}{4}\right)mg\right]\cos^2\theta
\end{aligned}
\tag{9.63a}
$$

$$
\begin{aligned}
\Delta h_2 &= \int_{AB}\left(\frac{N_x}{EA}\frac{\partial N_x}{\partial P}+\frac{M_z}{EI}\frac{\partial M_z}{\partial P}+\frac{KQ_y}{GA}\frac{\partial Q_y}{\partial P}\right)\mathrm{d}x \\
&= \frac{1}{EA}\int_0^l\left[\left(n+\frac{1}{2}\right)P+\left(n^2+n\right)mg-\frac{mg}{2l}x\right]\left(n+\frac{1}{2}\right)\sin^2\theta\mathrm{d}x \\
&\quad+\frac{1}{EA}\int_l^{2l}\left\{-\left[(n-1)P+(n-1)^2mg\right]\cot\theta\cos\theta\right. \\
&\qquad\left.+\left(\frac{1}{2}P+nmg-\frac{mg}{2l}x\right)\sin\theta\right\}\left[-(n-1)\cot\theta\cos\theta+\frac{1}{2}\sin\theta\right]\mathrm{d}x
\end{aligned}
$$

$$
\begin{aligned}
&+\frac{1}{EI}\int_0^l\left\{2\left(nP+n^2mg\right)l-\left[\left(n+\frac{1}{2}\right)P+\left(n^2+n\right)mg\right]x+\frac{mg}{4l}x^2\right\}\\
&\times\left[2nl-\left(n+\frac{1}{2}\right)x\right]\cos^2\theta\mathrm{d}x\\
&+\frac{1}{EI}\int_l^{2l}\left[\left(n-\frac{1}{2}\right)P+\left(n^2-n+\frac{1}{2}-\frac{x}{4l}\right)mg\right]\left(n-\frac{1}{2}\right)(2l-x)^2\cos^2\theta\mathrm{d}x\\
&+\frac{K}{GA}\int_0^l\left[\left(n+\frac{1}{2}\right)P+\left(n^2+n\right)mg-\frac{mg}{2l}x\right]\left(n+\frac{1}{2}\right)\cos^2\theta\mathrm{d}x\\
&+\frac{K}{GA}\int_l^{2l}\left[\left(n-\frac{1}{2}\right)P+\left(n^2-n\right)mg+\frac{mg}{2l}(2l-x)\right]\left(n-\frac{1}{2}\right)\cos^2\theta\mathrm{d}x\\
=&\frac{l}{EA}\left[\left(n+\frac{1}{2}\right)P+\left(n^2+n-\frac{1}{4}\right)mg\right]\left(n+\frac{1}{2}\right)\sin^2\theta\\
&+\frac{l}{EA}\left\{\left[(1-n)\cot\theta\cos\theta+\frac{1}{2}\sin\theta\right]P\right.\\
&\qquad\left.+\left[\left(n-\frac{3}{4}\right)\sin\theta-(n-1)^2\cot\theta\cos\theta\right]mg\right\}\left[(1-n)\cot\theta\cos\theta+\frac{1}{2}\sin\theta\right]\\
&+\frac{l^3}{EI}\left[\left(\frac{7}{3}n^2-\frac{2}{3}n+\frac{1}{12}\right)P+\left(\frac{7}{3}n^3-n^2+\frac{13}{48}n-\frac{1}{32}\right)mg\right]\cos^2\theta\\
&+\frac{l^3}{EI}\left[\left(\frac{1}{3}n^2-\frac{1}{3}n+\frac{1}{12}\right)P+\left(\frac{1}{3}n^3-\frac{1}{2}n^2+\frac{11}{48}n-\frac{1}{32}\right)mg\right]\cos^2\theta\\
&+\frac{Kl}{GA}\left[\left(2n^2+\frac{1}{2}\right)P+\left(2n^3+n-\frac{1}{4}\right)mg\right]\cos^2\theta\\
=&\frac{l}{EA}\left\{\begin{array}{l}\left[(n^2-2n+1)\cot^2\theta\cos^2\theta+(1-n)\cos^2\theta+\left(n^2+n+\frac{1}{2}\right)\sin^2\theta\right]P\\
+\left[(n^3-3n^2+3n-1)\cot^2\theta\cos^2\theta+\left(-\frac{3}{2}n^2+\frac{11}{4}n-\frac{5}{4}\right)\cos^2\theta\right.\\
\quad\left.+\left(n^3+\frac{3}{2}n^2+\frac{3}{4}n-\frac{1}{2}\right)\sin^2\theta\right]mg\end{array}\right\}\\
&+\frac{l^3}{EI}\left[\left(\frac{8}{3}n^2-n+\frac{1}{6}\right)P+\left(\frac{8}{3}n^3-\frac{3}{2}n^2+\frac{1}{2}n-\frac{1}{16}\right)mg\right]\cos^2\theta\\
&+\frac{Kl}{GA}\left[\left(2n^2+\frac{1}{2}\right)P+\left(2n^3+n-\frac{1}{4}\right)mg\right]\cos^2\theta \qquad (9.63b)
\end{aligned}
$$

$$
\begin{aligned}
\Delta h_3 &= \int\limits_{CD} \left( \frac{N_x}{EA}\frac{\partial N_x}{\partial P} + \frac{M_z}{EI}\frac{\partial M_z}{\partial P} + \frac{KQ_y}{GA}\frac{\partial Q_y}{\partial P} \right) \mathrm{d}x \\
&= \frac{1}{EA}\int_0^l \left\{ \left[ \frac{1}{2}P + (n-1)mg \right] \sin\theta - (n-1)\left[P + (n-1)mg\right] \cot\theta\cos\theta \right. \\
&\quad \left. + \frac{mg}{2l}x\sin\theta \right\} \left[ \frac{1}{2}\sin\theta - (n-1)\cot\theta\cos\theta \right] \mathrm{d}x \\
&\quad + \frac{1}{EA}\int_l^{2l} \left[ \left(n-\frac{1}{2}\right)P + (n^2-n+1)mg - \frac{mg}{2l}x \right] \left(n-\frac{1}{2}\right) \sin^2\theta \mathrm{d}x \\
&\quad + \frac{1}{EI}\int_0^l \left[ \left(n-\frac{1}{2}\right)P + (n^2-n)mg + \frac{mg}{4l}x \right] \left(n-\frac{1}{2}\right) x^2\cos^2\theta \mathrm{d}x \\
&\quad + \frac{1}{EI}\int_l^{2l} \left[ \left(n-\frac{1}{2}\right)P + \left(n^2-n+\frac{1}{2}-\frac{x}{4l}\right)mg \right] \left(n-\frac{1}{2}\right) (2l-x)^2\cos^2\theta \mathrm{d}x \\
&\quad + \frac{K}{GA}\int_0^l \left\{ \left[ \left(n-\frac{1}{2}\right)P + (n^2-n)mg \right] + \frac{mg}{2l}x \right\} \left(n-\frac{1}{2}\right) \cos^2\theta \mathrm{d}x \\
&\quad + \frac{K}{GA}\int_l^{2l} \left[ \left(n-\frac{1}{2}\right)P + (n^2-n+1)mg - \frac{mg}{2l}x \right] \left(n-\frac{1}{2}\right) \cos^2\theta \mathrm{d}x \\
&= \frac{l}{EA}\left\{ \left[ \frac{1}{2}\sin\theta - (n-1)\cot\theta\cos\theta \right] P \right. \\
&\quad \left. + \left[ \left(n-\frac{3}{4}\right)\sin\theta - (n-1)^2\cot\theta\cos\theta \right] mg \right\} \left[ \frac{1}{2}\sin\theta - (n-1)\cot\theta\cos\theta \right] \\
&\quad + \frac{l}{EA}\left[ \left(n-\frac{1}{2}\right)P + \left(n^2-n+\frac{1}{4}\right)mg \right] \left(n-\frac{1}{2}\right)\sin^2\theta \\
&\quad + \frac{l^3}{EI}\left[ \left(\frac{2}{3}n-\frac{1}{3}\right)P + \left(\frac{2}{3}n^2-\frac{2}{3}n+\frac{1}{8}\right)mg \right] \left(n-\frac{1}{2}\right)\cos^2\theta \\
&\quad + \frac{Kl}{GA}\left[ (2n-1)P + \left(2n^2-2n+\frac{1}{2}\right)mg \right] \left(n-\frac{1}{2}\right)\cos^2\theta \\
&= \frac{l}{EA}\left\{ \begin{array}{l} \left[ (n^2-2n+1)\cot^2\theta\cos^2\theta + (1-n)\cos^2\theta + \left(n^2-n+\frac{1}{2}\right)\sin^2\theta \right] P \\ + \left[ (n^3-3n^2+3n-1)\cot^2\theta\cos^2\theta + \left(-\frac{3}{2}n^2+\frac{11}{4}n-\frac{5}{4}\right)\cos^2\theta \right. \\ \quad \left. + \left(n^3-\frac{3}{2}n^2+\frac{5}{4}n-\frac{1}{2}\right)\sin^2\theta \right] mg \end{array} \right\} \\
&\quad + \frac{l^3}{EI}\left[ \left(\frac{2}{3}n^2-\frac{2}{3}n+\frac{1}{6}\right)P + \left(\frac{2}{3}n^3-n^2+\frac{11}{24}n-\frac{1}{16}\right)mg \right]\cos^2\theta \\
&\quad + \frac{Kl}{GA}\left[ \left(2n^2-2n+\frac{1}{2}\right)P + \left(2n^3-3n^2+\frac{3}{2}n-\frac{1}{4}\right)mg \right]\cos^2\theta
\end{aligned}
\tag{9.63c}
$$

Substituting Eqs. (9.63a), (9.63b), and (9.63c) into (9.62) yields

$$\begin{aligned}
\Delta h &= \Delta h_1 + \Delta h_2 + \Delta h_3 \\
&= \frac{l^3}{EI}\left[\left(\frac{2}{9}n^3 + \frac{8}{3}n^2 - \frac{19}{18}n + \frac{1}{6}\right)P \right.\\
&\qquad \left. + \left(\frac{1}{6}n^4 + \frac{8}{3}n^3 - \frac{77}{48}n^2 + \frac{1}{2}n - \frac{1}{16}\right)mg\right]\cos^2\theta \\
&\quad + \frac{l}{EA}\left\{\begin{aligned}
&\left[\left(\frac{2}{3}n^3 - \frac{5}{3}n^2 + 1\right)\cot^2\theta\cos^2\theta - (n-1)\cos^2\theta\right. \\
&\qquad \left. + \left(2n^2 + \frac{1}{2}n + \frac{1}{2}\right)\sin^2\theta\right]P \\
&+ \left[\left(\frac{1}{2}n^4 - \frac{5}{2}n^3 + 3n - 1\right)\cot^2\theta\cos^2\theta + \left(-\frac{7}{4}n^2 + 3n - \frac{5}{4}\right)\right. \\
&\qquad \left. \cos^2\theta + \left(2n^3 + \frac{1}{2}n^2 + n - \frac{1}{2}\right)\sin^2\theta\right]mg
\end{aligned}\right\} \\
&\quad + \frac{6l}{5GA}\left[\left(\frac{2}{3}n^3 + 2n^2 - \frac{1}{6}n + \frac{1}{2}\right)P + \left(4n^3 - \frac{11}{4}n^2 + 2n - \frac{1}{4}\right)mg\right]\cos^2\theta
\end{aligned} \tag{9.64}$$

The equivalent elastic coefficient of the structure shown in Fig. 9.9 is

$$\begin{aligned}
k_1(\theta) &= \frac{P}{\Delta h} \\
&= \frac{1}{\left\{\begin{aligned}
&\frac{l^3}{EI}\left[\left(\frac{2}{9}n^3 + \frac{8}{3}n^2 - \frac{19}{18}n + \frac{1}{6}\right)\right. \\
&\qquad \left. + \left(\frac{1}{6}n^4 + \frac{8}{3}n^3 - \frac{77}{48}n^2 + \frac{1}{2}n - \frac{1}{16}\right)\frac{mg}{P}\right]\cos^2\theta \\
&+ \frac{l}{EA}\left\{\begin{aligned}
&\left[\left(\frac{2}{3}n^3 - \frac{5}{3}n^2 + 1\right)\cot^2\theta\cos^2\theta - (n-1)\cos^2\theta\right. \\
&\qquad \left. + \left(2n^2 + \frac{1}{2}n + \frac{1}{2}\right)\sin^2\theta\right] \\
&+ \left[\left(\frac{1}{2}n^4 - \frac{5}{2}n^3 + 3n - 1\right)\cot^2\theta\cos^2\theta + \left(-\frac{7}{4}n^2 + 3n - \frac{5}{4}\right)\right. \\
&\qquad \left. \cos^2\theta + \left(2n^3 + \frac{1}{2}n^2 + n - \frac{1}{2}\right)\sin^2\theta\right]\frac{mg}{P}
\end{aligned}\right\} \\
&+ \frac{6l}{5GA}\left[\left(\frac{2}{3}n^3 + 2n^2 - \frac{1}{6}n + \frac{1}{2}\right) + \left(4n^3 - \frac{11}{4}n^2 + 2n - \frac{1}{4}\right)\frac{mg}{P}\right]\cos^2\theta
\end{aligned}\right\}}
\end{aligned} \tag{9.65}$$

In most engineering cases, there is the following approximation:

$$\frac{mg}{P} = \frac{2}{n} \tag{9.66}$$

Substituting Eq. (9.66) into Eq. (9.65) obtains

$$k_1(\theta) = \frac{1}{\left\{\begin{aligned} &\frac{l^3}{EI}\left[\frac{5}{9}n^3 + 8n^2 - \frac{307}{72}n + \frac{7}{6} - \frac{1}{8n}\right]\cos^2\theta \\ &+ \frac{l}{EA}\left[\left(\frac{5}{3}n^3 - \frac{20}{3}n^2 + 7 - \frac{2}{n}\right)\cot^2\theta\cos^2\theta - \left(\frac{9}{2}n - 7 + \frac{5}{2n}\right)\right. \\ &\left.\cos^2\theta + \left(6n^2 + \frac{3}{2}n + \frac{5}{2} - \frac{1}{n}\right)\sin^2\theta\right] \\ &+ \frac{6l}{5GA}\left(\frac{2}{3}n^3 + 10n^2 - \frac{17}{3}n + \frac{9}{2} - \frac{1}{2n}\right)\cos^2\theta \end{aligned}\right\}} \tag{9.67}$$

Considering the structure of the foldable tower shown in Fig. 9.13 or 9.14, we can find that there are 3 structures of that in Fig. 9.25. Therefore, the ultimate equivalent stiffness of the foldable tower will be immediately obtained from Eq. (9.67):

$$k(\theta) = \frac{3}{\left\{\begin{aligned} &\frac{l^3}{EI}\left[\frac{5}{9}n^3 + 8n^2 - \frac{307}{72}n + \frac{7}{6} - \frac{1}{8n}\right]\cos^2\theta \\ &+ \frac{l}{EA}\left[\left(\frac{5}{3}n^3 - \frac{20}{3}n^2 + 7 - \frac{2}{n}\right)\cot^2\theta\cos^2\theta - \left(\frac{9}{2}n - 7 + \frac{5}{2n}\right)\right. \\ &\left.\cos^2\theta + \left(6n^2 + \frac{3}{2}n + \frac{5}{2} - \frac{1}{n}\right)\sin^2\theta\right] \\ &+ \frac{6l}{5GA}\left(\frac{2}{3}n^3 + 10n^2 - \frac{17}{3}n + \frac{9}{2} - \frac{1}{2n}\right)\cos^2\theta \end{aligned}\right\}} \tag{9.68}$$

Because

$$k'(\theta) = \frac{3\left[Q(1 + \sin^2\theta)2\cot^3\theta + R\sin 2\theta\right]}{S^2} \tag{9.69}$$

**Table 9.1** Primary parameters of the foldable tower

| Set series | $E$(Pa) | $G$(Pa) | $I$(m$^4$) | $A$(m$^2$) | $l$(m) | $n$ |
|---|---|---|---|---|---|---|
| a | $2.06 \times 10^{11}$ | $7.9 \times 10^{10}$ | $8.33 \times 10^{-6}$ | $10^{-2}$ | 1 | 20 |
| b | $2.06 \times 10^{11}$ | $7.9 \times 10^{10}$ | $8.33 \times 10^{-6}$ | $10^{-2}$ | 1 | 10 |
| c | $2.06 \times 10^{11}$ | $7.9 \times 10^{10}$ | $8.33 \times 10^{-6}$ | $10^{-2}$ | 1 | 6 |
| d | $2.06 \times 10^{11}$ | $7.9 \times 10^{10}$ | $5.21 \times 10^{-7}$ | $2.5 \times 10^{-3}$ | 1 | 6 |

where

$$
\begin{aligned}
S = {} & \frac{l^3}{EI}\left[\frac{5}{9}n^3 + 8n^2 - \frac{307}{72}n + \frac{7}{6} - \frac{1}{8n}\right]\cos^2\theta \\
& + \frac{l}{EA}\left[\left(\frac{5}{3}n^3 - \frac{20}{3}n^2 + 7 - \frac{2}{n}\right)\cot^2\theta\cos^2\theta - \left(\frac{9}{2}n - 7 + \frac{5}{2n}\right)\cos^2\theta \right. \\
& \left. + \left(6n^2 + \frac{3}{2}n + \frac{5}{2} - \frac{1}{n}\right)\sin^2\theta\right] \\
& + \frac{6l}{5GA}\left(\frac{2}{3}n^3 + 10n^2 - \frac{17}{3}n + \frac{9}{2} - \frac{1}{2n}\right)\cos^2\theta \\
Q = {} & \frac{2l}{EA}\left(\frac{5}{3}n^3 - \frac{20}{3}n^2 + 7 - \frac{2}{n}\right) \\
R = {} & \frac{l^3}{EI}\left(\frac{5}{9}n^3 + 8n^2 - \frac{307}{72}n + \frac{7}{6} - \frac{1}{8n}\right) - \frac{l}{EA}\left(6n^2 + 6n - \frac{9}{2} + \frac{3}{2n}\right) \\
& + \frac{6l}{5GA}\left(\frac{2}{3}n^3 + 10n^2 - \frac{17}{3}n + \frac{9}{2} - \frac{1}{2n}\right)
\end{aligned}
$$

Obviously, the relationships of $\begin{cases} S^2 > 0 \\ Q > 0 \\ R > 0 \end{cases}$ always hold, and $k'(\theta) = \frac{3\left[Q(1+\sin^2\theta)2\cot^3\theta + R\sin 2\theta\right]}{S^2} > 0$ for $0 \le \theta < \frac{\pi}{2}$. Therefore, function $k(\theta)$ increases with the increasing of $\theta$. Letting $k'(\theta) = 0$, we can find that $\theta = \frac{\pi}{2}$. So

$$
[k(\theta)]_{\max} = k\left(\frac{\pi}{2}\right) = \frac{3EA}{l\left(6n^2 + \frac{3}{2}n + \frac{5}{2} - \frac{1}{n}\right)} \tag{9.70}
$$

Suppose there are four sets of parameters of the foldable tower shown in Table 9.1. Next we can examine the effects of the parameters on the stiffness of the foldable tower.

With the parameters in Table 9.1 and Eq. (9.67), we can gain the curves of the stiffness of the foldable tower with respect to $\theta$. Figure 9.25 shows the different effects of the parameters of the foldable tower.

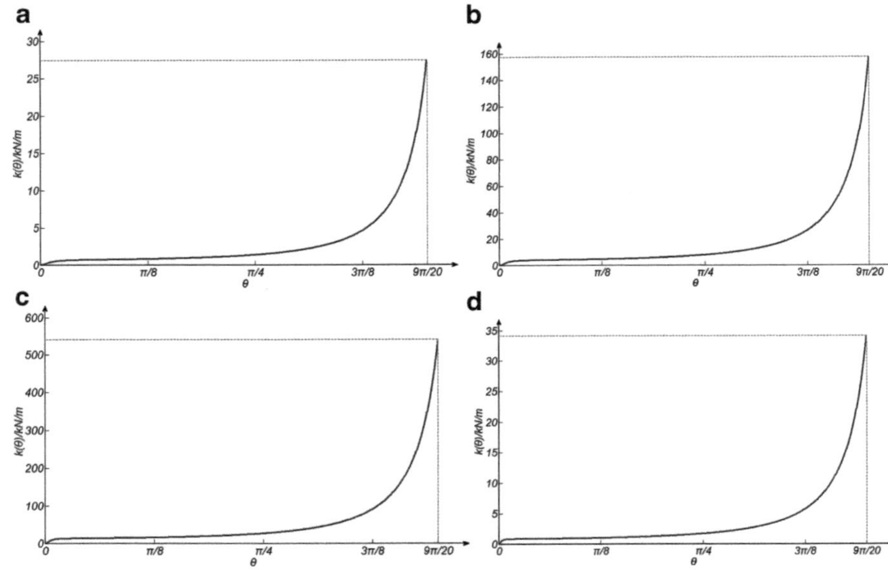


**Fig. 9.25** Stiffness of the foldable tower of different parameters. (**a**) First tower, (**b**) Second tower, (**c**) Third tower, and (**d**) Fourth tower

Figure 9.25 shows that the stiffness of the foldable tower changes with the change of parameters and the deployable angle $\theta$. The result that the stiffness of the foldable tower increases much faster than the increasing of the deployable angle $\theta$ indicates the most invaluable possibility in the applications of wind turbine.

## 9.4 Foldable Tower Consisting of Multi-planar Linkages

Section 9.3 discusses the statics and stiffness of one planar scissorlike-element structure of a foldable tower made up of SLEs. From Eq. (9.35), we can find that the stiffness of the tower will be parallel to the number of such linkages. In addition to triangle form shown in Fig. 9.14, every unit of the structure can also be designed to be a square, pentagon, hexagon, and so on. Figure 9.26 illustrates a foldable tower that is made up of four planar linkages whose top view of every unit is a regular square. Figure 9.26a illustrates the assembly when completely folded, Fig. 9.26b represents the assembly when completely unfolded, and Fig. 9.26c shows the top view of a unit when completely folded.

Figure 9.27 shows another pentagon foldable tower that is made up of five planar linkages whose top view of every unit is a regular pentagon.

A hexagon foldable tower shown in Fig. 9.28 is made up of six planar linkages whose top view of every unit is a regular hexagon. Figure 9.28a illustrates the assembly when completely folded, Fig. 9.28b illustrates the assembly when completely unfolded, and Fig. 9.28c illustrates the top view of a unit when completely folded.

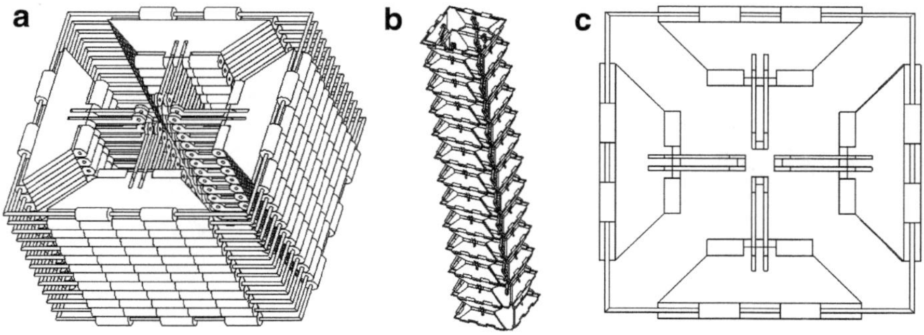

Fig. 9.26 A foldable tower made up of four planar linkages. (a) Assembly when completely folded, (b) assembly whencompletely deployed, and (c) top view of one unit structure

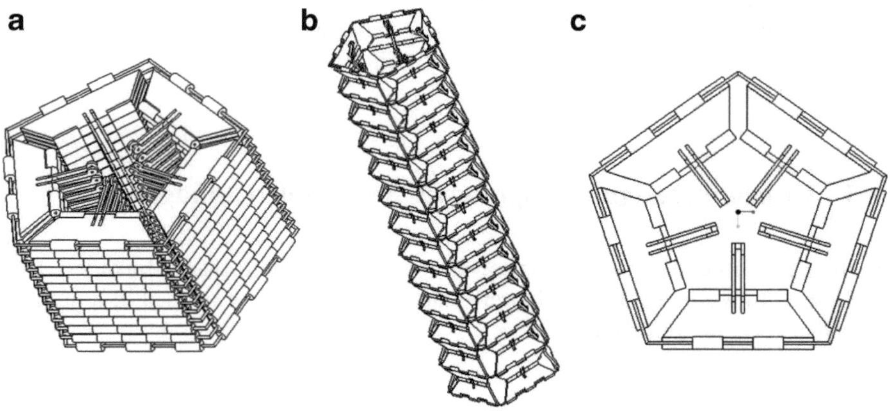

Fig. 9.27 A foldable tower made up of five planar linkages. (a) Assembly when completely folded, (b) assembly when completely deployed, and (c) top view of one unit structure

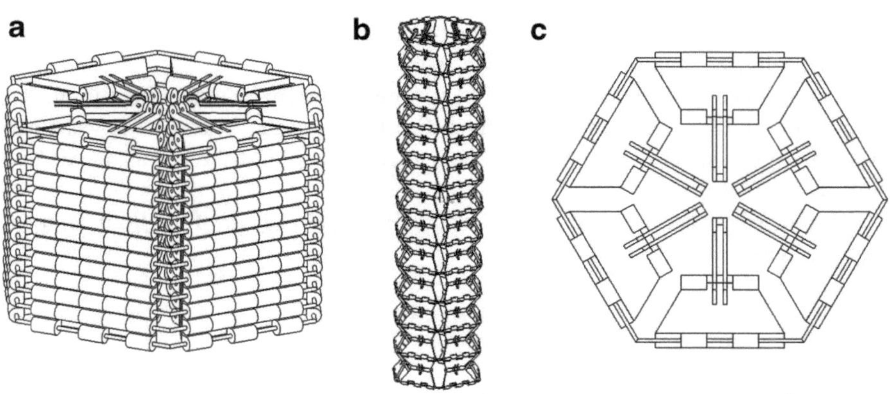

Fig. 9.28 A foldable tower made up of six planar linkages. (a) Assembly when completely folded, (b) assembly when completely deployed, and (c) top view of one unit structure

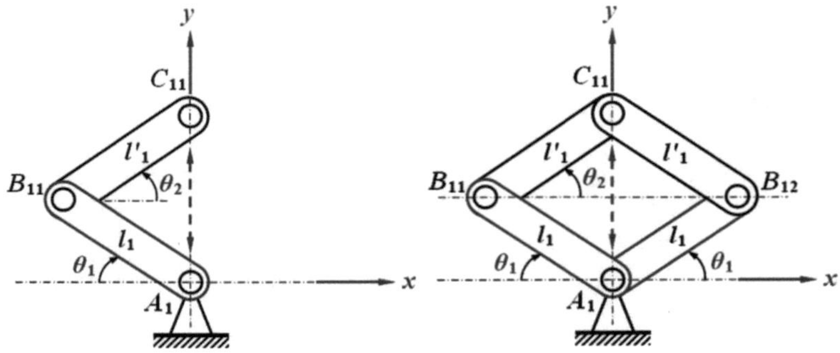


**Fig. 9.29** A planar quadrilateral linkage

The virtues of these three types of foldable tower are that the vertical stiffness and the transversal stiffness of the structure are higher than those of the triangle tower. Of course, the more sides the tower has, the more high the stiffness of the tower will be.

## 9.5 Concept Design of Topology Changeable Linkages

Section 9.5 discusses the foldable tower consisting of multi-planar linkages. This section will investigate a reconfigurable structure which is composed of a number of planar revolute-jointed linkages.

### *9.5.1 Synthesis of a Unit Reconfigurable Mechanism*

The planar linkage, $A_1B_{11}C_{11}$ shown in Fig. 9.29 (left), is an RRR open chain. The relative free motions at this instant of the revolute joint $C_{11}$ with respect to the fixed revolute joint, $A_1$, can be first examined as follows:

A coordinate system is established by setting the origin be at the fixed joint, $A_1$, $x$-axis be rightward along the horizontal direction, and the $y$-axis be upward and pass through joint $C_{11}$. Suppose that the subtended angle of the minus $x$-axis and link $A_1B_{11}$ is $\theta_1$.

The screw matrix of the kinematic chain, $A_1B_{11}C_{11}$ in Fig. 9.29, can be expressed as

$$\$_{A_1B_{11}C_{11}} = \begin{bmatrix} \$_{A_1} & \$_{B_{11}} & \$_{C_{11}} \end{bmatrix} \tag{9.71}$$

where $\$_{A_1}$, $\$_{B_{11}}$, and $\$_{C_{11}}$ denote the kinematic screws of the joints $A_1$, $B_{11}$, and $C_{11}$, respectively, and $\$_{A_1} = (0 \quad 0 \quad 1 \quad 0 \quad 0 \quad 0)^{\mathrm{T}}$ which represents a pure rotation around joint $A_1$, $\$_{B_{11}} = (0 \quad 0 \quad 1 \quad -l_1\sin\theta_1 \quad -l_1\cos\theta_1 \quad 0)^{\mathrm{T}}$ which indicates a rotation around joint $B_{11}$, and $\$_{C_{11}} = (0 \quad 0 \quad 1 \quad -l_1 \sin\theta_1 - l'_1 \sin\theta_2 \quad l_1 \cos\theta_1 - l'_1 \cos\theta_2 \quad 0)^{\mathrm{T}}$ which stands for a pure rotation around joint $C_{11}$, and where $l_1$ denotes the length of $A_1B_{11}$ and $l'_1$ denotes the length of $B_{11}C_{11}$. Therefore, the terminal constraint screw matrix of kinematic chain, $A_1B_{11}C_{11}$, can be obtained by solving the reciprocal screw equation

$$\mathbf{\$}^{\mathrm{T}} \mathbf{\Delta} \mathbf{\$}^{\tau} = \mathbf{0} \tag{9.72}$$

where $\mathbf{\$}$ is a kinematic screw matrix, $\mathbf{\$}^{\tau}$ is the terminal constraint, and $\mathbf{\Delta} = \begin{bmatrix} \mathbf{0} & \mathbf{I} \\ \mathbf{I} & \mathbf{0} \end{bmatrix}$, in which $\mathbf{0}$ and $\mathbf{I}$ are zero and unit matrices, respectively.

According to Eq. (9.2), we can obtain the terminal constraint screw matrix for the chain $A_1B_{11}C_{11}$.

$$\mathbf{\$}^{\tau}_{A_1B_{11}C_{11}} = \begin{bmatrix} \mathbf{\$}^{\tau}_1 & \mathbf{\$}^{\tau}_2 & \mathbf{\$}^{\tau}_3 \end{bmatrix} \tag{9.73}$$

where $\mathbf{\$}^{\tau}_1 = (0 \quad 0 \quad 1 \quad 0 \quad 0 \quad 0)^{\mathrm{T}}$ which denotes force constraint passing through the origin and being along the $z$-direction, $\mathbf{\$}^{\tau}_2 = (0 \quad 0 \quad 0 \quad 1 \quad 0 \quad 0)^{\mathrm{T}}$ which represents a torque constraint being along the $x$-direction, and $\mathbf{\$}^{\tau}_3 = (0 \quad 0 \quad 0 \quad 0 \quad 1 \quad 0)^{\mathrm{T}}$ which is a torque constraint being along the $y$-direction.

The constraint screws represented by Eq. (9.73) are therefore called planar constraints that are restricted by the $xA_1y$-plane shown in Fig. 9.29. Similarly, when another identical RRR kinematic chain is connected $A_1$ with $C_{11}$ in parallel shown in Fig. 9.29 (right), the screw matrix of the chain can also be expressed as

$$\mathbf{\$}_{A_1B_{12}C_1} = \begin{bmatrix} \mathbf{\$}_{A_1} & \mathbf{\$}_{B_{12}} & \mathbf{\$}_{C_{11}} \end{bmatrix} \tag{9.74}$$

where $\mathbf{\$}_{B_{12}} = (0 \quad 0 \quad 1 \quad -l_1 \sin\theta_1 \quad l_1 \cos\theta_1 \quad 0)^{\mathrm{T}}$ which denotes the kinematic screw of joint $B_{11}$. The terminal constraint screw matrix of kinematic chain, $A_1B_{12}C_{11}$, can be similarly obtained:

$$\mathbf{\$}^{\tau}_{A_1B_{12}C_{11}} = \begin{bmatrix} 0 & 0 & 1 & 0 & 0 & 0 \\ 0 & 0 & 0 & 1 & 0 & 0 \\ 0 & 0 & 0 & 0 & 1 & 0 \end{bmatrix}^{\mathrm{T}} \tag{9.75}$$

The constraint screw system exerted on the pin $C_{11}$ depends entirely on the union of Eqs. (9.73) and (9.75). Consequently, the constraint screw matrix for pin $C_{11}$ is expressed as

$$\mathbf{\$}^{\tau}_{C_{11}} = \begin{bmatrix} 0 & 0 & 1 & 0 & 0 & 0 \\ 0 & 0 & 0 & 1 & 0 & 0 \\ 0 & 0 & 0 & 0 & 1 & 0 \end{bmatrix}^{\mathrm{T}} \tag{9.76}$$

Comparing Eq. (9.76) with Eqs. (9.73) and (9.75), we can find that the two chains, $A_1B_{11}C_{11}$ and $A_1B_{12}C_{11}$, provide the same constraints for pin $C_{11}$, and therefore, they are two overconstraint kinematic chains. According to the reciprocal screw theory, the free motions the pin $C_{11}$ has and the constraint screw matrix for pin $C_{11}$ should satisfy that

$$\left[\boldsymbol{\$}^{\tau}\right]^{\mathrm{T}} \Delta \boldsymbol{\$}^{F} = 0 \tag{9.77}$$

where $\boldsymbol{\$}^F$ denotes the screw of a free motion of an object.

The free motion screw matrix of pin $C_{11}$ can be obtained by substituting Eq. (9.76) into Eq. (9.77):

$$\boldsymbol{\$}_{C_{11}}^{F} = \left[\boldsymbol{\$}_1^F \quad \boldsymbol{\$}_2^F \quad \boldsymbol{\$}_3^F\right] \tag{9.78}$$

where $\boldsymbol{\$}_1^F = (0 \quad 0 \quad 1 \quad 0 \quad 0 \quad 0)^{\mathrm{T}}$, which indicates a pure rotation around the joint $A_1$; $\boldsymbol{\$}_2^F = (0 \quad 0 \quad 0 \quad 1 \quad 0 \quad 0)^{\mathrm{T}}$, which represents a translation along the $x$-direction; and $\boldsymbol{\$}_3^F = (0 \quad 0 \quad 0 \quad 0 \quad 1 \quad 0)^{\mathrm{T}}$, which stands for a translation along the $y$-direction. The real motion of point $C_{11}$ will be any combination of screws $\boldsymbol{\$}_1^F$, $\boldsymbol{\$}_2^F$, and $\boldsymbol{\$}_3^F$ which can be expressed as

$$\boldsymbol{\$}_{C_{11}}^{F} = \boldsymbol{\$}_{C_{11}}^{F} \boldsymbol{k} \tag{9.79}$$

where $\boldsymbol{\$}_{C_{11}}^F$ represents the instantaneous screw of point $C_{11}$ and $\boldsymbol{k} = (k_1 \; k_2 \; k_3)^{\mathrm{T}}$ and $k_i (i = 1, 2, 3)$ could be any real numbers, the range of which is determined by the sizes of the links shown in Fig. 9.29.

Equations (9.78) and (9.79) indicate that the lengths of link $A_1B_{11}$ and $B_{11}C_1$ do not provide any effect on the motions of the distal link connected by joint $C_1$ but only determine the range of its movements. Therefore, the assumption $l'_1 = l_1$ is adopted in the following analysis.

Now it is safe to draw a conclusion that the tip pin of a planar equilateral quadrangle can make full degrees of freedom motions within the plane it locates. In other words, the equivalent screw matrix for the tip pin of the planar equilateral quadrangle shown in the right subfigure of Fig. 9.29 has the form of Eq. (9.78). The motion of an equilateral triangle frame when its three side centers are connected with three such planar quadrangles will be discussed next. The equilateral triangle, $C_{12}C_{22}C_{32}$ shown in Fig. 9.30, is a platform which outputs the motion of the three complex chains.

To facilitate the analysis of the free motion of the triangle frame coupled by three planar equilateral quadrangle chains that are jointed in the middle of every side, a coordinate system is set as Fig. 9.30 shows. The constraints exerted to the triangular frame by the first equilateral quadrangle are expressed by Eq. (9.76) in the coordinate system shown in Fig. 9.30. Assume that the side length of the equilateral

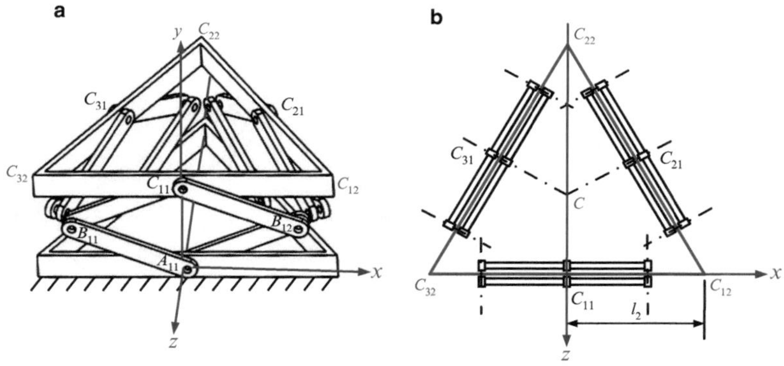


**Fig. 9.30** A triangular platform with three planar equilateral quadrangle chains. (**a**) Isometric view and (**b**) overview

triangle is denoted by $2l_2$, the constraints exerted to the triangular frame by the other two equilateral quadrangles can be similarly obtained:

$$\mathbf{\$}_{C_{21}}^{\tau} = \begin{bmatrix} \cos\alpha & 0 & -\sin\alpha & 2l_1\sin\theta\sin\alpha & -l_2\sin\alpha & 2l_1\sin\theta\cos\alpha \\ 0 & 0 & 0 & -\sin\alpha & 0 & -\cos\alpha \\ 0 & 0 & 0 & 0 & 1 & 0 \end{bmatrix}^{\mathrm{T}} \tag{9.80}$$

and

$$\mathbf{\$}_{C_{31}}^{\tau} = \begin{bmatrix} -\cos\alpha & 0 & -\sin\alpha & 2l_1\sin\theta\sin\alpha & l_2\sin\alpha & -2l_1\sin\theta\cos\alpha \\ 0 & 0 & 0 & -\sin\alpha & 0 & \cos\alpha \\ 0 & 0 & 0 & 0 & 1 & 0 \end{bmatrix}^{\mathrm{T}} \tag{9.81}$$

where $\alpha = \frac{\pi}{6}$ for the equilateral triangle and the geometry center of the triangular frame in the coordinate system shown in Fig. 9.30 are $C(0\ 2l_1\sin\theta\ -l_2\tan\alpha)$. Associating Eqs. (9.76), (9.80), and (9.81) and substituting them into Eq. (9.77) results in

$$\mathbf{\$}_{C_{11}C_{21}C_{31}}^{F} = k\begin{bmatrix} 0 & 0 & 0 & 0 & 1 & 0 \end{bmatrix}^{\mathrm{T}} \tag{9.82}$$

where $\mathbf{\$}_{C_{11}C_{21}C_{31}}^{F}$ represents the free motion of the triangular frame connected by the three equilateral quadrangles at points, $C_{11}$, $C_{21}$, and $C_{31}$. The instantaneous motion of the triangular frame can be represented by

$$\mathbf{\$}_{C_{11}C_{21}C_{31}}^{F} = k\begin{bmatrix} 0 & 0 & 0 & 0 & 1 & 0 \end{bmatrix}^{\mathrm{T}} \tag{9.83}$$

where $k$ denotes the instantaneous amplitude of the translation.

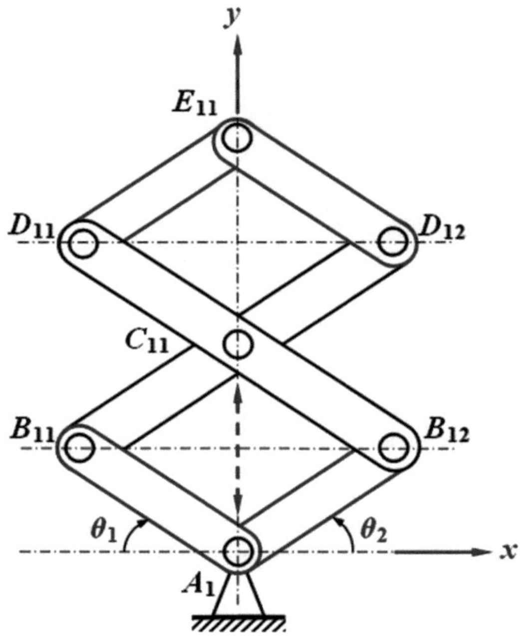


**Fig. 9.31** A two-level mechanism

Equation (9.83) indicates that the triangular frame $C_{12}C_{22}C_{32}$ can only make a translation along the $y$-axis. When the two links, $B_{11}C_{11}$ and $B_{12}C_{11}$, of parallelogram linkage shown in Fig. 9.29 extend in their own central line directions to form a pair of scissorlike links with the common center at joint $C_{11}$, another unit $C_{11}D_{12}E_{11}D_{11}$ comes into being. As shown in Fig. 9.31, $C_{11}D_{12}E_{11}D_{11}$ is also an identical rhombus to $A_1B_{12}C_{11}B_{11}$. The free motions of joint $E_{11}$ turn out to be expressed by Eq. (9.78) with a similar analysis process above. So is true to the triangular frame $E_{12}E_{22}E_{32}$, the free motion of which is also depicted by Eq. (9.82).

The three scissorlike linkages and the triangular frame connected by them shown in Fig. 9.32 are called a deployable unit of the lift mechanism.

What follows will discuss the case when the three vertexes of a triangular frame $C_{12}C_{22}C_{32}$ shown in Fig. 9.30 are split and connected by vertically placed pin joints. The overview of the mechanism is shown in Fig. 9.33.

To investigate the motion of the triangular frame pin jointed vertically at the vertexes, we can first analyze the motion of any side of the frame. For the sake of simplicity, the side $C_{32}C_{11}C_{12}$ of the frame is discussed here. As shown in Fig. 9.34, $C_{32}C_{11}C_{12}$ has three complex kinematic chains, $A_3\left\langle\begin{matrix}B_{31}\\B_{32}\end{matrix}\right\rangle C_{31}C_{32}$, $A_1\left\langle\begin{matrix}B_{11}\\B_{12}\end{matrix}\right\rangle C_{11}$, and $A_2\left\langle\begin{matrix}B_{21}\\B_{22}\end{matrix}\right\rangle C_{21}C_{12}$.

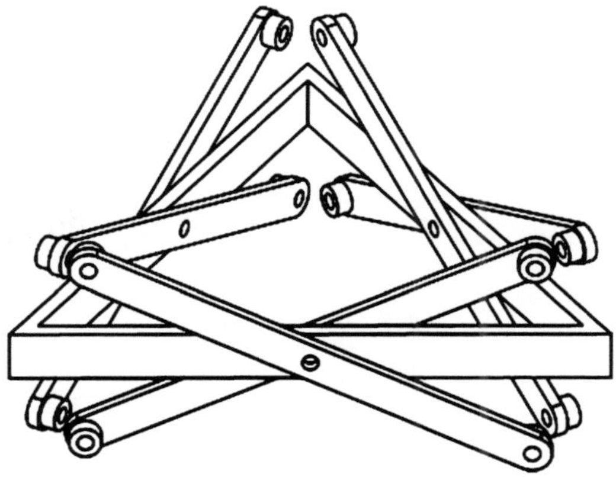

**Fig. 9.32** A deployable unit

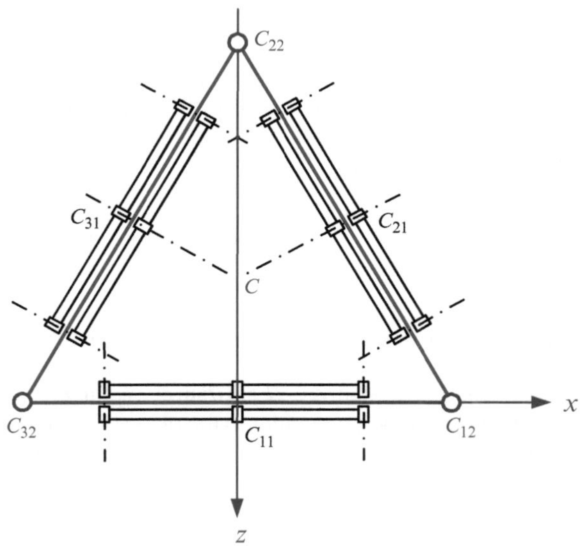


**Fig. 9.33** Triangular frame vertically pin jointed at the vertexes

Similar to Eq. (9.78), the screw matrix of the complex chain $A_3\begin{pmatrix} B_{31} \\ B_{32} \end{pmatrix} C_{31}C_{32}$ can be expressed as

$$\mathbf{\$}^F_{C_{32}} = \begin{bmatrix} \$^{C_{32}}_1 & \$^{C_{32}}_2 & \$^{C_{32}}_3 & \$^{C_{32}}_4 \end{bmatrix} \tag{9.84}$$

where $\mathbf{\$}^F_{C_{32}}$ represents the kinematic screw matrix of the chain $A_3\begin{pmatrix} B_{31} \\ B_{32} \end{pmatrix} C_{31}C_{32}$; $\$^{C_{32}}_1 = \left(-\cos\alpha \quad 0 \quad -\sin\alpha \quad 2l_1\sin\theta\sin\alpha \quad l_2\sin\alpha \quad -2l_1\sin\theta\cos\alpha\right)^{\mathrm{T}}$, which indicates

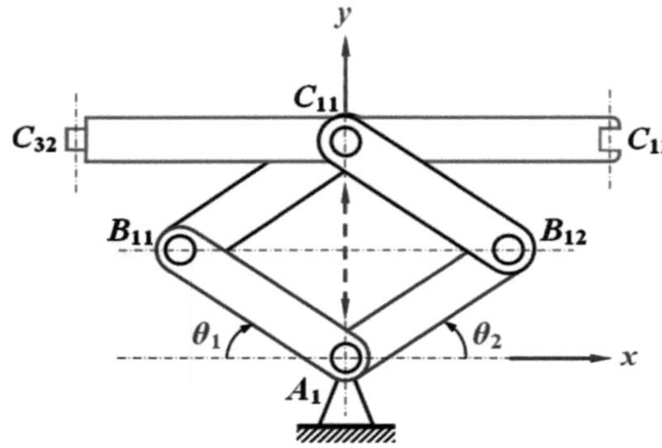


**Fig. 9.34** A branch of the quadrilateral chain

a pure rotation around the joint $C_{31}$; $\boldsymbol{\$}_2^{C_{32}} = (0 \quad 0 \quad 0 \quad -\sin\alpha \quad 0 \quad \cos\alpha)^{\mathrm{T}}$, which represents a translation along the $C_{22}C_{32}$-direction; $\boldsymbol{\$}_3^{C_{32}} = (0 \quad 0 \quad 0 \quad 0 \quad 1 \quad 0)^{\mathrm{T}}$, which stands for a translation along the $y$-direction; and $\boldsymbol{\$}_4^{C_{32}} = (0 \quad 1 \quad 0 \quad 0 \quad 0 \quad l_2)^{\mathrm{T}}$.

Substituting Eq. (9.84) into Eq. (9.72) produces the terminal constraints of the complex chain $A_3\left\langle\begin{matrix} B_{31} \\ B_{32} \end{matrix}\right\rangle C_{31}C_{32}$:

$$\boldsymbol{\$}^{\tau}_{C_{32}} = \begin{bmatrix} \boldsymbol{\$}_{\tau 1}^{C_{32}} & \boldsymbol{\$}_{2}^{C_{32}} \end{bmatrix} \tag{9.85}$$

where $\boldsymbol{\$}^{\tau}_{C_{32}}$ represents the terminal constraint matrix of the chain $A_3\left\langle\begin{matrix} B_{31} \\ B_{32} \end{matrix}\right\rangle C_{31}C_{32}$; $\boldsymbol{\$}_{\tau 1}^{C_{32}} = (-\cos\alpha \quad 0 \quad -\sin\alpha \quad 2l_1\sin\theta\sin\alpha \quad l_2\sin\alpha \quad -2l_1\sin\theta\cos\alpha)^{\mathrm{T}}$, which denotes a force constraint perpendicular to the plane of equilateral quadrangle $A_3B_{31}B_{32}C_{31}$; and $\boldsymbol{\$}_{\tau 2}^{C_{32}} = (0 \quad 0 \quad 0 \quad -\sin\alpha \quad 0 \quad \cos\alpha)^{\mathrm{T}}$, which denotes a torque constraint about the direction of side $C_{31}C_{32}$.

Similarly, the terminal constraints of the complex chain $A_2\left\langle\begin{matrix} B_{21} \\ B_{22} \end{matrix}\right\rangle C_{21}C_{12}$ can also be obtained with a process above:

$$\boldsymbol{\$}^{\tau}_{C_{12}} = \begin{bmatrix} \boldsymbol{\$}_{\tau 1}^{C_{12}} & \boldsymbol{\$}_{\tau 2}^{C_{12}} \end{bmatrix} \tag{9.86}$$

where $\boldsymbol{\$}^{\tau}_{C_{12}}$ represents the terminal constraint matrix of the chain $A_2\left\langle\begin{matrix} B_{21} \\ B_{22} \end{matrix}\right\rangle C_{21}C_{12}$, $\boldsymbol{\$}_{\tau 1}^{C_{12}} = (\cos\alpha \quad 0 \quad -\sin\alpha \quad 2l_1\sin\theta\sin\alpha \quad -l_2\sin\alpha \quad 2l_1\sin\theta\cos\alpha)^{\mathrm{T}}$ which denotes a force constraint perpendicular to the plane of equilateral quadrangle $A_2B_{21}B_{22}C_{12}$, $\boldsymbol{\$}_{\tau 2}^{C_{12}} = (0 \quad 0 \quad 0 \quad -\sin\alpha \quad 0 \quad -\cos\alpha)^{\mathrm{T}}$ which denotes a torque constraint about the direction of side $C_{21}C_{12}$.

The terminal constraints of $C_{32}C_{11}C_{12}$ can be expressed in the matrix form by associating Eqs. (9.76), (9.85), and (9.86):

$$\boldsymbol{\$}^{\tau}_{C_{32}C_{11}C_{12}} = \begin{bmatrix} 0 & 0 & 1 & 0 & 0 & 0 \\ 0 & 0 & 0 & 1 & 0 & 0 \\ 0 & 0 & 0 & 0 & 1 & 0 \\ -\cos\alpha & 0 & -\sin\alpha & 2l_1\sin\theta\sin\alpha & l_2\sin\alpha & -2l_1\sin\theta\cos\alpha \\ 0 & 0 & 0 & -\sin\alpha & 0 & \cos\alpha \\ \cos\alpha & 0 & -\sin\alpha & 2l_1\sin\theta\sin\alpha & -l_2\sin\alpha & 2l_1\sin\theta\cos\alpha \\ 0 & 0 & 0 & -\sin\alpha & 0 & -\cos\alpha \end{bmatrix}^{\mathrm{T}} \tag{9.87}$$

Substituting Eq. (9.87) into Eq. (9.77) yields

$$\boldsymbol{\$}^{F}_{C_{32}C_{11}C_{12}} = \begin{bmatrix} 0 & 0 & 0 & 0 & 1 & 0 \end{bmatrix}^{\mathrm{T}} \tag{9.88}$$

Equation (9.88) indicates that the frame side $C_{32}C_{11}C_{12}$ still has one translational degree of freedom along the $y$-axis shown in Fig. 9.34. In addition, it can be found from Eqs. (9.87) and (9.88) that the frame side, $C_{32}C_{11}C_{12}$, is subjected to overconstraints, the redundant number of which is 3.

In a similar way, the conclusion that the other two sides of the triangular frame also only have the same one translational degree of freedom along the $y$-axis shown in Fig. 9.34 can be drawn. Therefore, the rigid connection of the strengthened triangular frame can be released to vertically pin-jointed while not changing the motion of the frame which is still subjected to the overconstraints of three nonparallel planes. The conclusion will also be true when two or more deployable units are connected in series. This provides an outstanding advantage for the portable design of the mechanism.

### *9.5.2 Structure Synthesis of Completely Foldable Lift Mechanism*

When the unit mechanism shown in Fig. 9.34 compacts, any vertically pinned joint of the triangular frame can be unlocked, and therefore, the three equilateral quadrangles should be stretched out to form one planar mechanism. Figure 9.35 illustrates the case when the three complex chains are untied at joint $C_{22}$ and then deployed to one plane.

If the free motions of the outspread frame $C_{22}C_{31}C_{32}C_{11}C_{12}C_{21}C_{22}$ which forms one beam within the $xA_1y$-plane are further investigated, the terminal constraints contributed to this beam by the three complex planar kinematic chains, $A_1\left\langle \begin{matrix} B_{11} \\ B_{12} \end{matrix} \right\rangle C_{11}$, $A_2\left\langle \begin{matrix} B_{21} \\ B_{22} \end{matrix} \right\rangle C_{21}$, and $A_3\left\langle \begin{matrix} B_{31} \\ B_{32} \end{matrix} \right\rangle C_{31}$, can be analyzed in a like manner above:

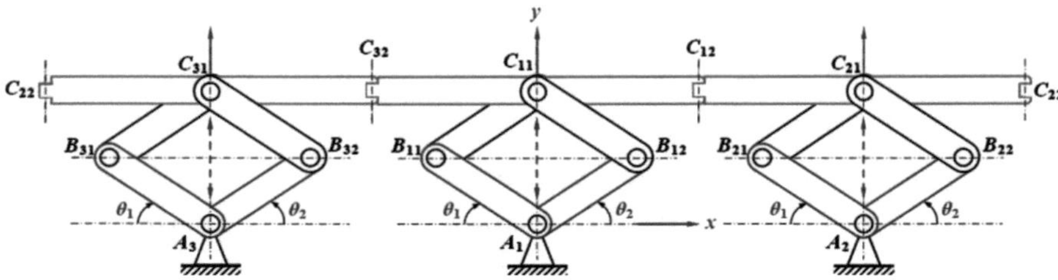


**Fig. 9.35** One level of the mechanism deployed to one planar linkage

$$\boldsymbol{\$}_{C_{22}-C_{22}}^{\tau} = \begin{bmatrix} 0 & 0 & 1 & 0 & 0 & 0 \\ 0 & 0 & 0 & 1 & 0 & 0 \\ 0 & 0 & 0 & 0 & 1 & 0 \end{bmatrix}^{\mathrm{T}} \tag{9.89}$$

where $\boldsymbol{\$}_{C_{22}-C_{22}}^{F}$ represents the terminal constraint matrix of the planar beam $C_{22}C_{31}C_{32}C_{11}C_{12}C_{21}C_{22}$.

Substituting Eq. (9.89) into Eq. (9.77) results

$$\boldsymbol{\$}_{C_{22}-C_{22}}^{\tau} = \begin{bmatrix} 0 & 0 & 1 & 0 & 0 & 0 \\ 0 & 0 & 0 & 1 & 0 & 0 \\ 0 & 0 & 0 & 0 & 1 & 0 \end{bmatrix}^{\mathrm{T}} \tag{9.90}$$

where $\boldsymbol{\$}_{C_{22}-C_{22}}^{F}$ represents the screw matrix of the full three planar free motions of the planar beam $C_{22}C_{31}C_{32}C_{11}C_{12}C_{21}C_{22}$. The instantaneous free motion of the outspread frame can be expressed as

$$\boldsymbol{\$}_{C_{22}-C_{22}}^{F} = \boldsymbol{\$}_{C_{22}-C_{22}}^{F} \boldsymbol{k}_{C_{22}-C_{22}} \tag{9.91}$$

where $\boldsymbol{k}_{C_{22}-C_{22}} = (k_1\ k_2\ k_3)^{\mathrm{T}}$ and $k_i (i = 1, 2, 3)$ can be any real numbers, the range of which is specified by the sizes of the links.

Comparing Eq. (9.90) with (9.88), we can find that the free motions of the frame $C_{32}C_{11}C_{12}$ has changed from one translational degree of freedom along the $y$-axis shown in Fig. 9.34 to the full three planar degrees of freedom. This indicates that the free motions of a mechanism might be shifted with the change of configurations.

To completely fold the structure shown in Fig. 9.35, the triangle frame can be modified as Fig. 9.8 shows. The final folding sequence is shown in the left of Fig. 9.36, and the completely folded case is shown in the right of Fig. 9.36.

The least length of the zigzag segment of each side shown in Fig. 9.34 can be calculated respectively. $C_{12}C_{22}C_{32}$ shown in the left subfigure of Fig. 9.36 is still an equilateral triangle, the side length of which is

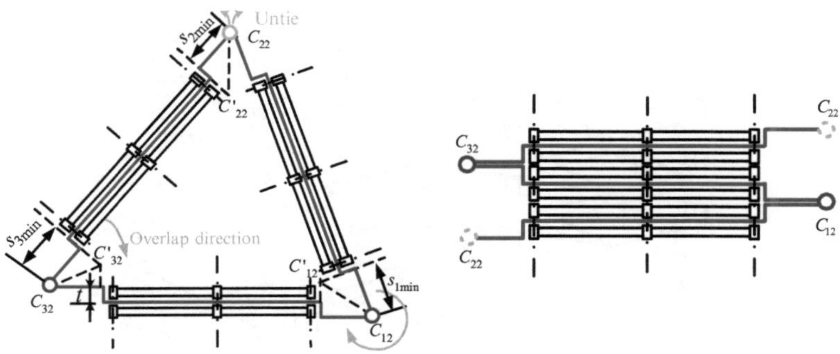


**Fig. 9.36** The final folding sequence

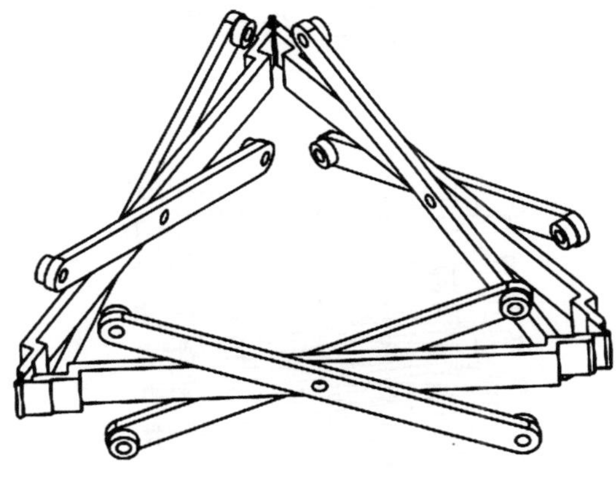

**Fig. 9.37** An unit mechanism jointed by a vertically pinned frame

$$l_{C_{12}C_{22}} = 2\sqrt{l_2^2 + t^2} \tag{9.92}$$

where $t$ is the thickness of the link.

The least length of the zigzag segment of each side should be

$$\begin{cases} s_{1\min} = 2t\cot\dfrac{\pi}{6} = 2\sqrt{3}t \\ s_{2\min} = 2t\cot\dfrac{\pi}{6} = 2\sqrt{3}t \\ s_{3\min} = 0 \end{cases} \tag{9.93}$$

Therefore, to keep the possibility of completely folding of the structure shown in the right subfigure of Fig. 9.36, the minimum length of the zigzag segment should be $2\sqrt{3}t$.

Corresponding to Fig. 9.32, the deployable unit mechanism connected by the pin-jointed triangular frame, which can also be untied to spread out in one plane, is shown in Fig. 9.37.

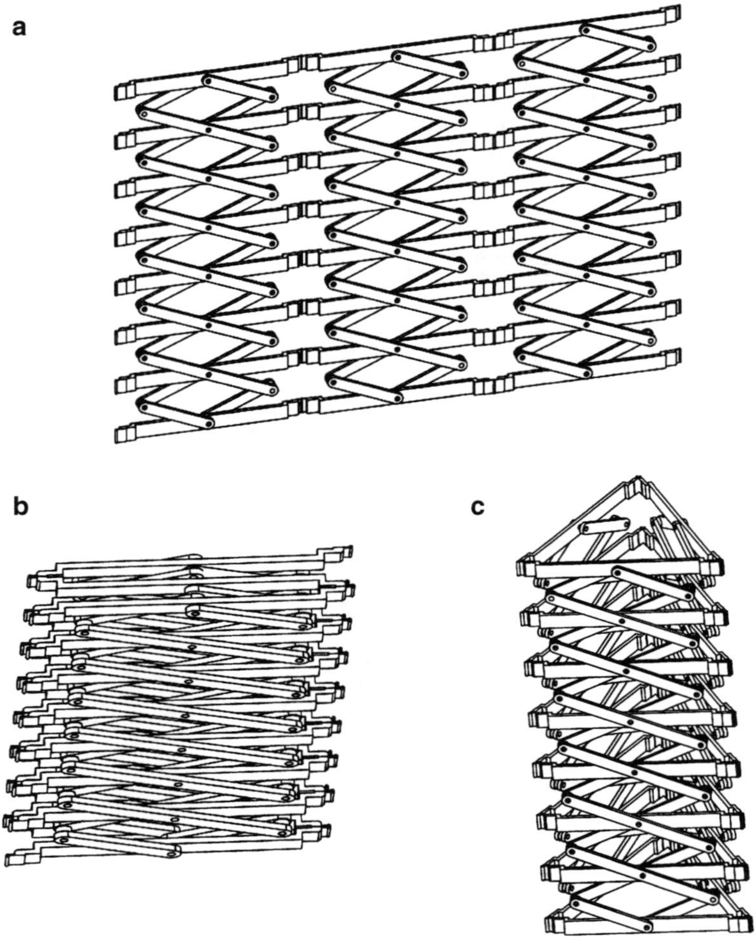


**Fig. 9.38** Reconfigurable lift mechanism. (**a**) Outspread structure of the mechanism in one plane, (**b**) a bundle of folded lift mechanism, and (**c**) deployed lift mechanism

When more deployable units shown in Fig. 9.37 are jointed one by one in series, a lift mechanism comes into existence. The outspread, folded, and deployed structures of the reconfigurable lift mechanism are presented in Fig. 9.38.

The deployed lift mechanism is a 3-plane overconstraint rectilinear moving platform. Therefore, the stiffness and stability of the platform should be higher than the existing counterparts. For the portable and reconfigurable requirements, the actuation unit should be particularly designed. Because every level of the unit shown in Fig. 9.38c only has linear relative motion, a hydraulic cylinder spherically jointed at both ends and some accessory parts are enough to keep the linear movement. Figure 9.39 illustrates this reconfigurable lift mechanism which is a

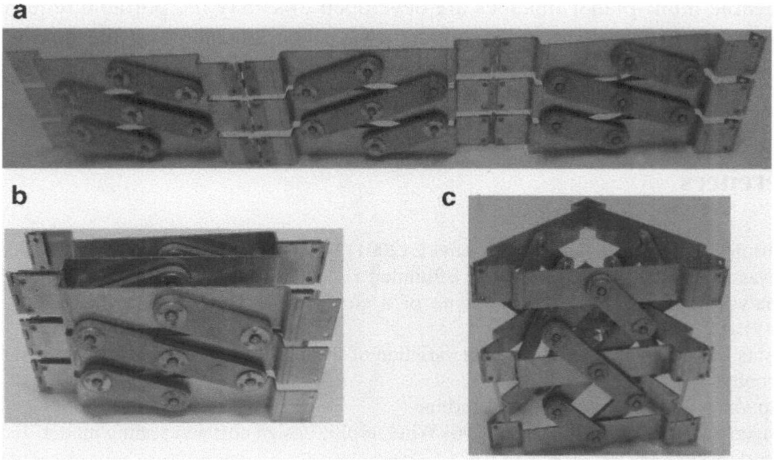


**Fig. 9.39** Prototype for the lift mechanism. (**a**) Outspread structure in one plane, (**b**) a bundle of folded mechanism, and (**c**) deployed mechanism

portable platform. The completely folded bundle shown in Fig. 9.39b can be taken away in a case or in the car boot.

This section investigates a reconfigurable linkage that can be used as the unit mechanism of a portable lift platform. The reconfigurable linkage is a planar revolute-jointed mechanism which could be compacted to one line and then be further folded to a bundle. This provides the possibility for the portable target. When using, the bundle of linkages should be first spread out and the two ends tied to form a closed triangular loop, and under the actuating of the actuation unit, the folded loop can be deployed to a lift mechanism. This kind of mechanism has a wide adaptation in engineering and can be surely used in the situations when a column structure must be enclosed by a lift mechanism for work. Because every reconfigurable unit is three-plane overconstraint mechanism connected by a strengthened triangular frame, the portable lift platform has very high stiffness and strength in applications. The fact that all the mechanism units only have linear relative movement provides the facility of actuator's setting. It can be mounted between any two adjacent levels.

## 9.6 Conclusions

This chapter has focused on the foldable tower which should be used as the support of huge equipment such as wind turbine and investigated the method to establish the equivalent stiffness expression of the whole structure. The stiffness of the foldable tower is especially high when it is deployed completely. Different multi-planar linkages are then discussed in the viewpoint of overconstraint structures. Topology changeable multi-planar linkages are developed to satisfy the portable requirement of nowadays lift mechanism or support tower. The advantage of the flexibility to fold and unfold should witness wider engineering applications in the future.

## References


1. Burton T, Sharp D, Jenkins N, Bossanyi E (2001) Wind energy: handbook. Wiley, Chichester
2. Lavassas I, Nikolaidis G, Zervas P, Efthimiou E, Doudoumis IN, Baniotopoulos CC (2003) Analysis and design of the prototype of a steel 1-MW wind turbine tower. Eng Struct 25(7):1097–1106
3. Justus CG, Mikhail A (1976) Height variation of wind speed and wind distributions statistics. Geophys Res Lett 3(5):261–264
4. http://en.wikipedia.org/wiki/Wind_turbine
5. Fingersh L, Hand M, Laxson A (2006) Wind turbine design cost and scaling model. Technical report, NREL/TP-500-40566
6. Gantes CJ, Konitopoulou E (2004) Geometric design of arbitrarily curved bi-stable deployable arches with discrete joint size. Int J Solids Struct 41(20):5517–5540
7. Christiansen EL, Kerr JH, Fuentes HM, Schneider WC (1999) Flexible and deployable meteoroid/debris shielding for spacecraft. Int J Impact Eng 23(1):125–136
8. Vincent JFV (2000) Deployable structures in nature: potential for biomimicking. Proc Inst Mech Eng 214(1):1–10
9. Wei XZ, Yao YA, Tian YB, Fang R (2006) A new method of creating expandable structure for spatial objects. Proc Inst Mech Eng Part C J Mech Eng Sci 220(12):1813–1818
10. Gan WW, Pellegrino S (2006) Numerical approach to the kinematic analysis of deployable structures forming a closed loop. Proc Inst Mech Eng Part C J Mech Eng Sci 220(7):1045–1056
11. Bedford A, Kenneth M (2000) Mechanics of materials. Prentice Hall, Upper Saddle River

# Chapter 10
# Extended Applications of Double Planar Linkages

This chapter discusses the design of foldable mounting equipment for solar cell panels. The equipment consists of a number of scissorlike elements. After assembly, each unit becomes a solar cell mounting bracket when it is deployed; the frame of each unit forms the space for fixing solar cell panels through fixtures. The merit of the structure is that the whole structure occupies less space when completely folded. So it is also very convenient for manufacturing and transporting because of its less occupied space when folded. Meanwhile, it is also available to form arc-shaped or any spatial curved space panels by changing the positions of the center joints of scissor units in assembly. The chapter first investigates the geometry of each foldable unit and then focuses on statics of the structure and simulation experiments.

## 10.1 Foldable Structures for Solar Cell Panels

The advancement of science and technology on the one hand propels the global economic development and on the other hand increases the demands of energy. The overconsumption of fossil fuels, such as oil, gas, and coal, renders a lot of global problems. According to the investigation of the American Petroleum Industry Association, the supply of fossil fuels will be dissipated in 2250 and the energy crisis is coming much faster than before [1]. Energy crisis will be one of major obstacles to sustainable development of the world. Therefore, contradictions between the limited resources and sustainable development will be a fierce challenge to the whole world.

Meanwhile, environmental pollution attracts more and more attention of humankind. The destruction of environmental pollution is an ignorable problem, and some contamination has a global influence. Renewable energy is one of the most promising sources. There are different kinds of renewable energy, such as hydrogen, solar, wind, tidal, biomass, and nuclear. Comparing with the traditional energy, the renewable one has outstanding advantages, especially solar energy. Solar power is a clean, safe, and widely distributed renewable energy [2].

J. Zhao et al., *Design of Special Planar Linkages*, Springer Tracts in Mechanical Engineering, DOI 10.1007/978-3-642-38448-6_10,

For its outstanding advantages, solar energy has attracted more and more attention of humankind. As a result, solar photovoltaic industry has obtained a rapid development. Meanwhile, a series of solar photovoltaic-related industries also develop subsequently. Among them, solar photovoltaic mounting equipment industry is one of the most promising industries. Among all cost of photovoltaic power generation system, solar cell components account for nearly 60 % [3] which is relatively difficult to reduce as it depends on the cell technology. However, the cost of solar photovoltaic mounting brackets accounts for about 10 %, which is convenient to deduce. Moreover, with the development of solar photovoltaic industry, the demand of solar photovoltaic mounting brackets is increasing. Therefore, it is of great significance to reasonably design photovoltaic mounting brackets and improve its production mode.

This chapter investigates the design of foldable solar photovoltaic mounting equipment, which includes a certain number of scissorlike elements. Each unit could be deployed to form a cell bracket after assembly with the scissorlike elements. When the structure is completely folded, the space that the whole structure takes is little which is very convenient to move and fabricate.

There are many deployable systems in biological organisms. Deployable structures are prefabricated space frames consisting of straight bars linked together in the factory as a compact bundle, which can then be unfolded into large-span, load-bearing structural shapes via simple articulations [4]. Deployable structures have significant advantages for their operational flexibility of a mechanism and the high strength and high rigidity of redundantly constraint structures. The deployable concept was initially applied for a moveable theater in the 1960s by an architect from Spain named Pienro [5, 6]. Then different kinds of deployable structures have been applied in different areas, such as small-scale structures [7], the covering of swimming pools [8], and aerospace structures [9, 10]. According to the discussion above, deployable structure is a good choice for the photovoltaic mounting bracket when it is designed properly.

## 10.2 Synthesis of Photovoltaic Mounting Bracket for Solar Panels

Among the deployable structures, which are mainly used in engineering, are based on scissorlike elements shown in Fig. 10.1. The scissorlike element (SLE) is a planar II-Assur group in the theory of mechanism. Links $A_1B_1$ and $A_2B_2$ are connected by their common middle point $C$ which is a revolute joint. It allows the links to rotate around the axis which is perpendicular to their common plane but limits all the other degrees of freedom. When two or more such elements jointed together through revolute joints as Fig. 10.1 shows, the degree of freedom will not change.

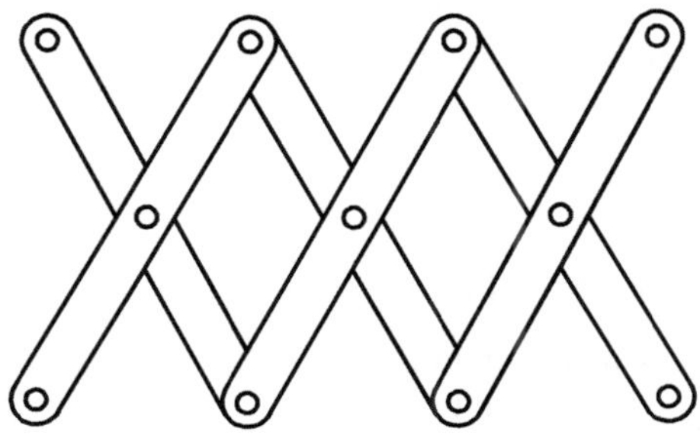

**Fig. 10.1** Assembly of three scissorlike elements

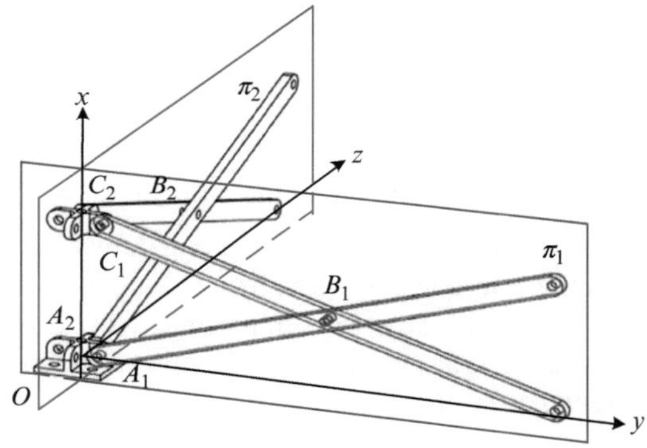


**Fig. 10.2** A structure consists of two scissorlike elements

The primary unit of the deployable structure consists of two scissorlike elements shown in Fig. 10.2. Suppose that the planes of the SLEs are named as planes $\pi_1$ and $\pi_2$. The angle between planes $\pi_1$ and $\pi_2$ is 90°. A coordinate system is established by setting the origin at the fixed joint $A_1$, the $x$-axis be upward and pass through joint $A_1$, the $y$-axis be rightward along the horizontal direction, and the $z$-axis can be confirmed by right-hand coordinate system.

The screw matrix of the kinematic chain, $A_1B_1C_1$ in Fig. 10.2, can be expressed as

$$\$_{A_1B_1C_1} = \begin{bmatrix} \$_{A_1} & \$_{B_1} & \$_{C_1} \end{bmatrix} \tag{10.1}$$

where $\$_{A_1}$, $\$_{B_1}$, and $\$_{C_1}$ denote the kinematic screws of the joints $A_1$, $B_1$, and $C_1$, which are

$$\begin{aligned} \$_{A_1} &= \begin{bmatrix} 0 & 0 & 1 & y_{A_1} & -x_{A_1} & 0 \end{bmatrix}^T \\ \$_{B_1} &= \begin{bmatrix} 0 & 0 & 1 & y_{B_1} & -x_{B_1} & 0 \end{bmatrix}^T \\ \$_{C_1} &= \begin{bmatrix} 0 & 0 & 1 & y_{C_1} & -x_{C_1} & 0 \end{bmatrix}^T \end{aligned} \tag{10.2}$$

Therefore, the terminal constraint screw matrix of the kinematic chain, $A_1B_1C_1$, can be obtained by solving the reciprocal screw equation

$$\$^T \Delta \$^r = 0 \tag{10.3}$$

where $\$$ is a kinematic screw matrix, $\$^r$ is the terminal constraint and $\Delta = \begin{bmatrix} 0 & I_3 \\ I_3 & 0 \end{bmatrix}$, in which 0 and $I_3$ are zero and unit matrices, respectively.

According to Eq. (10.2), one can obtain the terminal constraint screw matrix for the chain $A_1B_1C_1$.

$$\$^r = \begin{bmatrix} \$_{11}^r & \$_{12}^r & \$_{13}^r \end{bmatrix} \tag{10.4}$$

where $\$_{11}^r = \begin{bmatrix} 0 & 0 & 1 & 0 & 0 & 0 \end{bmatrix}^T$, which denotes force constraint passing through the origin and being along the $z$-direction; $\$_{12}^r = \begin{bmatrix} 0 & 0 & 0 & 1 & 0 & 0 \end{bmatrix}^T$, which represents a torque constraint being along the $x$-direction; and $\$_{13}^r = \begin{bmatrix} 0 & 0 & 0 & 0 & 1 & 0 \end{bmatrix}^T$, which is a torque constraint being along the $y$-direction.

The constraint screws represented by Eq. (10.4) are therefore called planar constraints that are restricted by the plane $\pi_1$ shown in Fig. 10.2. Similarly, the screw matrix of the chain $A_2B_2C_2$ can also be expressed as

$$\$_{A_2B_2C_2} = \begin{bmatrix} \$_{A_2} & \$_{B_2} & \$_{C_2} \end{bmatrix} \tag{10.5}$$

where $\$_{A_2}$, $\$_{B_2}$, and $\$_{C_2}$ denote the kinematic screws of the joints $A_2$, $B_2$, and $C_2$.

$$\begin{aligned} \$_{A_2} &= \begin{bmatrix} 0 & 1 & 0 & -z_{A_2} & 0 & x_{A_2} \end{bmatrix}^T \\ \$_{B_2} &= \begin{bmatrix} 0 & 1 & 0 & -z_{B_2} & 0 & x_{B_2} \end{bmatrix}^T \\ \$_{C_2} &= \begin{bmatrix} 0 & 1 & 0 & z_{C_2} & 0 & x_{C_2} \end{bmatrix}^T \end{aligned} \tag{10.6}$$

Therefore, the terminal constraint screw matrix of the kinematic chain, $A_2B_2C_2$, can be obtained by solving the reciprocal screw Eq. (10.3).

According to Eq. (10.6), one can obtain the terminal constraint screw matrix for the chain $A_1B_1C_1$.

$$\$^r = \begin{bmatrix} \$_{21}^r & \$_{22}^r & \$_{23}^r \end{bmatrix} \tag{10.7}$$

where $\$_{21}^r = \begin{bmatrix} 0 & 1 & 0 & 0 & 0 & 0 \end{bmatrix}^T$, which denotes force constraint passing through the origin and being along the $y$-direction; $\$_{22}^r = \begin{bmatrix} 0 & 0 & 0 & 1 & 0 & 0 \end{bmatrix}^T$, which represents a

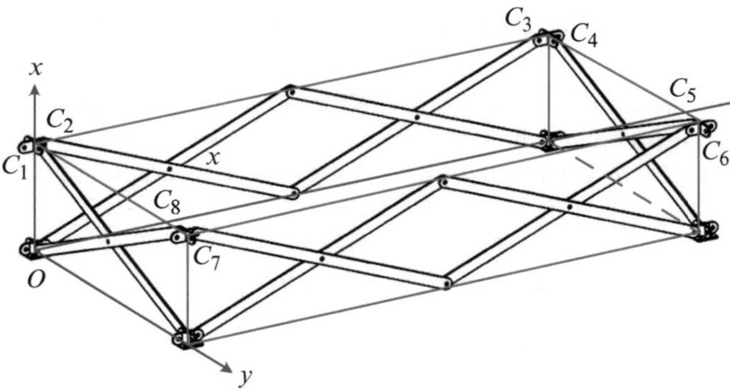


**Fig. 10.3** Basal frame for one solar photovoltaic cell

torque constraint being along the $x$-direction; and $\boldsymbol{\$}_{23}^{r} = [0 \quad 0 \quad 0 \quad 0 \quad 0 \quad 1]^{T}$, which is a torque constraint being along the $z$-direction. The constraint screws represented by Eq. (10.7) are therefore called planar constraints that are restricted by the plane $\pi_2$ shown in Fig. 10.2.

The constraint screw system exerted on $C_1C_2$ depends entirely on the union of Eqs. (10.4) and (10.7). Consequently, the constraint screw matrix for $C_1C_2$ is expressed as

$$\boldsymbol{\$}_{C_1C_2}^{r} = \begin{bmatrix} \boldsymbol{\$}_{11}^{r} & \boldsymbol{\$}_{12}^{r} & \boldsymbol{\$}_{13}^{r} & \boldsymbol{\$}_{21}^{r} & \boldsymbol{\$}_{22}^{r} & \boldsymbol{\$}_{23}^{r} \end{bmatrix} \tag{10.8}$$

According to the reciprocal screw theory, the free motions $C_1C_2$ has and the constraint screw matrix for $C_1C_2$ should satisfy

$$(\boldsymbol{\$}_{C_1C_2}^{r})^{T} \boldsymbol{\Delta} \boldsymbol{\$}_{C_1C_2} = 0 \tag{10.9}$$

where $\boldsymbol{\$}_{C_1C_2}$ denotes the screw of a free motion of an object.

The free motion screw matrix of $C_1C_2$ can be obtained by substituting Eq. (10.8) into Eq. (10.9):

$$\boldsymbol{\$}_{C_1C_2} = [0 \quad 0 \quad 0 \quad 1 \quad 0 \quad 0]^{T} \tag{10.10}$$

Equation (10.10) indicates that $C_1C_2$ can only make a translation along the $x$-axis. Similarly, one can also obtain the conclusion that $C_3C_4$, $C_5C_6$, and $C_7C_8$ have the same one translational degree of freedom along the $x$-axis shown in Fig. 10.3. Therefore, the whole frame only has the translational degree of freedom along the $x$-axis.

## 10.3 Geometry Design of the Photovoltaic Mounting Bracket

We can see the basal frame for one solar cell in Fig. 10.3. The long side and short side of the basal frame are shown in Fig. 10.4. Suppose that the length of $A_2A_3$, $A_3C_3$, $A_1A_8$ are $b$, $h$, $a$. The length of links $A_2C_{23}$ and $A_1C_8$ are $l_1$ and $l_2$.

$$\begin{cases} l_{A_2B_2} = pl_1 \\ l_{B_2C_{23}} = (1-p)l_1 \\ l_{A_1B_1} = 0.5l_2 \\ b = 2\sqrt{l_1^2 + h^2} \\ a = \sqrt{l_2^2 + h^2} \end{cases} \tag{10.11}$$

where $l_{A_2B_2}, l_{B_2C_{23}}, l_{A_1B_1}$ denote the length of $A_2B_2$, $B_2C_{23}$, and $A_1B_1$ separately. $p$ is a constant that satisfies the condition $0 < p < 1$. When the whole structure is completely folded, the length of the long side should be equal to the length of the short side as the following equation shows:

$$l_{A_2B_2} + l_{B_2C_2} = l_{A_1C_8} \tag{10.12}$$

where $l_{B_2C_2}, l_{A_1C_8}$ denote the length of $B_2C_2$ and $A_1C_8$ separately.

That is to say

$$pl_1 = 0.5l_2 \tag{10.13}$$

Under the prescribed conditions, the length of each frame can be easily figured out. The practical case is that the dimension of solar panels is 1580 mm × 808 mm × 30 mm. It requires bearing 0.6 kN/m$^2$ wind pressure and

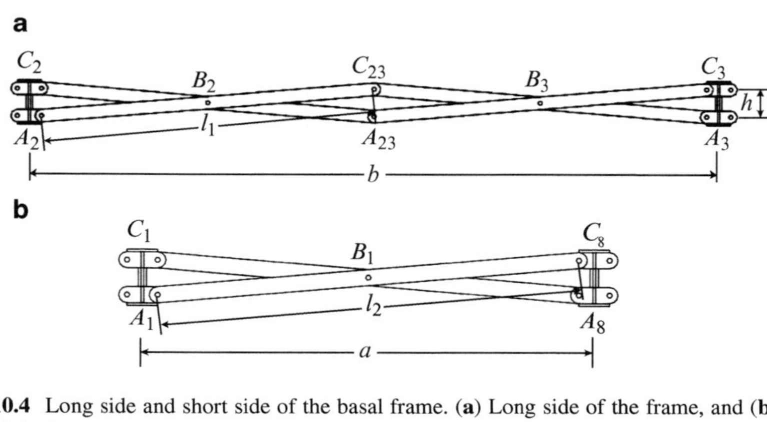


**Fig. 10.4** Long side and short side of the basal frame. (**a**) Long side of the frame, and (**b**) Short side of the frame

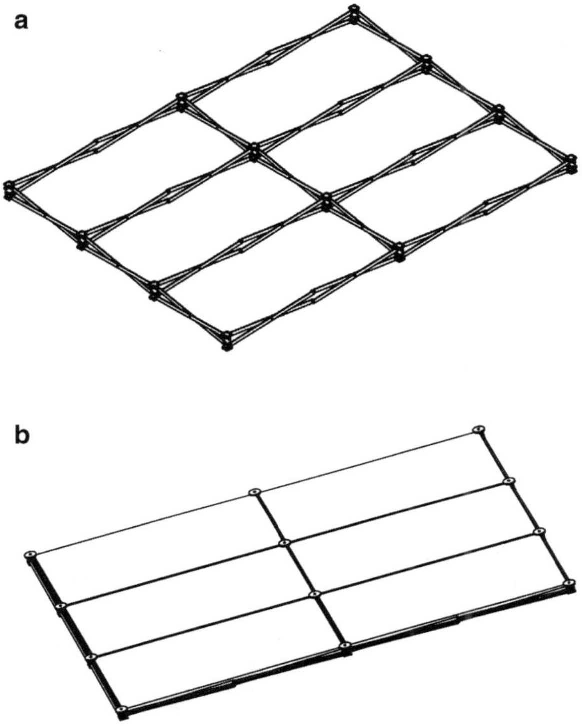

**Fig. 10.5** Deployed photovoltaic mounting bracket. (**a**) Completely deployed bracket and (**b**) deployed bracket assembled with solar panels

0.6 kN/m$^2$ snow load. Therefore, $h$ equals to 30 mm, $a$ equals to 808 mm, and $b$ equals to 1,580 mm. One can obtain

$$\begin{cases} l_1 = \sqrt{(b/2)^2 - h^2} = \sqrt{(1,580/2)^2 - 30^2}\,\text{mm} = 789.22\,\text{mm} \\ l_2 = \sqrt{a^2 - h^2} = \sqrt{808^2 - 30^2}\,\text{mm} = 807.24\,\text{mm} \\ p = \dfrac{0.5 l_2}{l_1} = \dfrac{0.5 \times 807.24}{789.22} = 0.51 \end{cases} \tag{10.14}$$

Simplifying the parameter generates

$$\begin{cases} l_1 = 790\,\text{mm} \\ l_2 = 810\,\text{mm} \\ p = 0.5127 \\ l_{A_2B_2} = pl_1 = 405\,\text{mm} \\ l_{B_2C_{23}} = 385\,\text{mm} \end{cases} \tag{10.15}$$

When six cells are connected together as shown in Fig. 10.5a, the whole structure only has one translational degree of freedom. Figure 10.5b shows the deployed bracket fixed with six solar panels. When folded, the structure possesses a small space shown in Fig. 10.6.

**Fig. 10.6** Folded photovoltaic mounting bracket

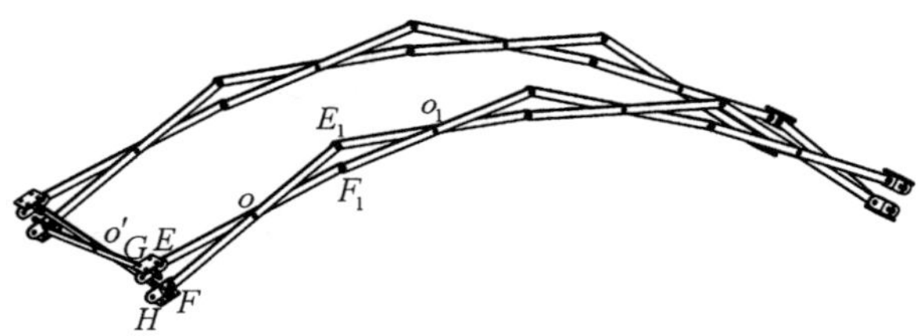


**Fig. 10.7** Folded bracket for arc-shaped solar panel

For further research, when the bracket is designed for arc-shaped or spherical-shaped solar panels, we can change the position of the rod hinge hole and scissor unit assembly to suit the shape as Fig. 10.7 shows. The shape of the long side of bracket is similar to an arc curve after the whole structure is being deployed. There are some requirements for geometry design. Suppose the arc radius is $r$ and the distance between hinge point $E_i$ ($i = 1,2,3,\ldots,n$) and hinge point $F_i$ ($i = 1,2,3\ldots,n$) is $h$ which is shown in Fig. 10.8. Based on geometry, we immediately obtain

$$\begin{cases} l_{l_k} = l_{l_{k+1}} = l_l, l_{s_k} = l_{s_{k+1}} = l_s \; (l_l > l_s) \\ l_s \sin\alpha + l_l \sin\beta = h \\ l_s \cos\alpha = l_l \cos\beta \\ l_l \cos\beta = r \sin\theta \\ \varphi = \frac{\pi}{2} - \beta \end{cases} \quad (k = 1, 2, \ldots, n) \tag{10.16}$$

where the definitions of parameters are shown in Fig. 10.8.

In the triangle $OO_2E_2$ of Fig. 10.9, $\gamma$ is its one external angle; therefore, there is

$$\gamma = \angle OE_2O_2 + \angle E_2OO_2 = \varphi + \theta \tag{10.17}$$

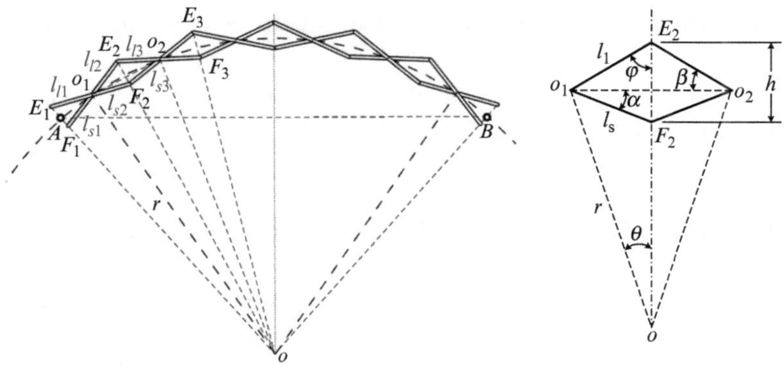


**Fig. 10.8** Geometry characteristics of arc-shaped solar panels

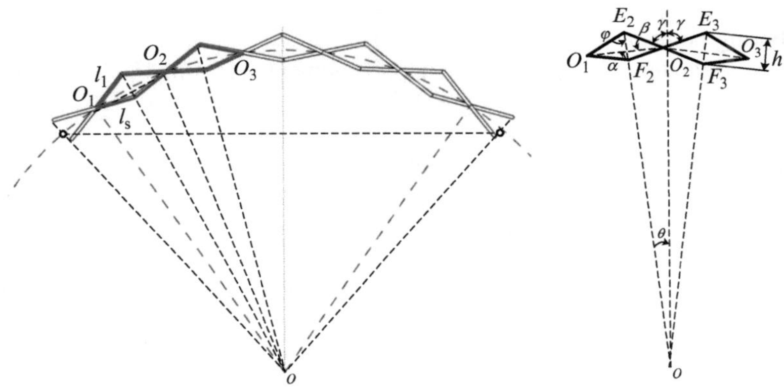


**Fig. 10.9** Angles relationship of arc-shaped solar panels

According to angle relationships in Fig. 10.9, we obtain that

$$\begin{cases} 2\gamma + \angle E_2O_2F_2 = \pi \\ \angle E_2O_2F_2 = \alpha + \beta \end{cases} \tag{10.18}$$

Associating Eqs. (10.17) and (10.18) yields

$$\alpha + \beta + 2(\theta + \varphi) = \pi \tag{10.19}$$

When $l_l$, $l_s$, and $h$ are given, we can immediately calculate the unknowns in Eq. (10.16). That is to say, we can design the corresponding structure shown in Fig. 10.8 based on the dimensions of the solar panels.

## 10.4 Stiffness Analysis of the Photovoltaic Mounting Bracket

After analyzing the geometry of a photovoltaic mounting bracket, the statics analysis should be investigated.

### 10.4.1 Statics Analysis of the Photovoltaic Mounting Bracket

When the photovoltaic mounting bracket works, it is completely deployed so that it is a normal bracket which stays in a fixed state without any relative motion. The whole frame can be simplified as orthogonal beam system shown in Fig. 10.10.

The structure model could be solved out by displacement method step by step. Firstly, it comes to the boundary value matrix equation. Figure 10.11 shows a general beam system, which bears a series of concentrated forces $p_i$ and concentrated couples $m_i$ ($i = 0, 1, 2, 3, \ldots, n$). Suppose the distance between each two neighboring points $i-1$ and $i$ is $l_i$ ($i = 0, 1, 2, 3, \ldots, n$), the end reactions of the beam are separately $Y_0$, $Y_n$, and the support moments are $M_0$, $M_n$. The deflection and intersection angle of vertex $i$ are defined as $v_i$ and $\theta_i$ ($i = 0, 1, 2, 3, \ldots, n$), respectively.

Express the parameters in matrix form:

$$\begin{cases} R_0 = [\,Y_0\ M_0\,]^{\mathrm{T}} \\ R_n = [\,Y_n\ M_n\,]^{\mathrm{T}} \\ P_i = [\,p_i\ m_i\,]^{\mathrm{T}} \\ \Delta_i = [\,v_i\ \theta_i\,]^{\mathrm{T}} \end{cases} \quad (i = 0, 1, 2, \ldots, n) \tag{10.20}$$

**Fig. 10.10** An equivalent structure model

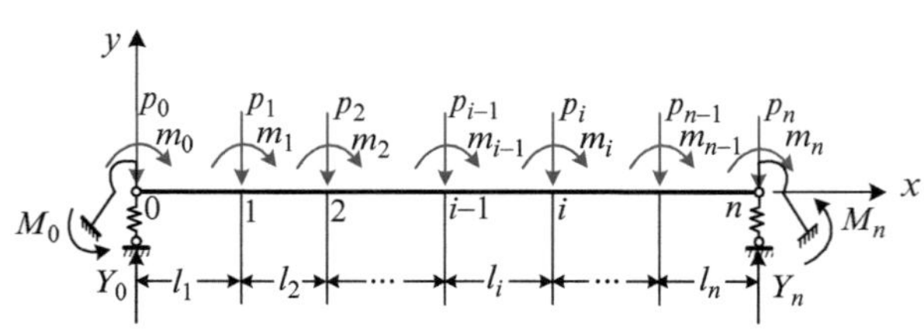


**Fig. 10.11** A general beam system bearing loads

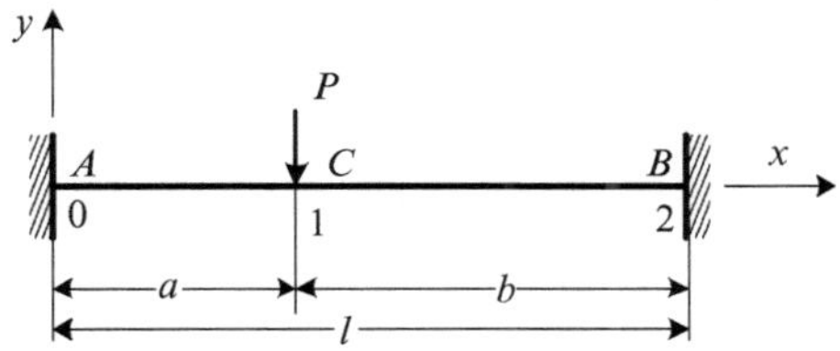


**Fig. 10.12** A general beam

$$A_{0n} = \begin{bmatrix} 1 & -L_n \\ 0 & 1 \end{bmatrix}, \quad C_{0n} = \begin{bmatrix} 1 & 0 \\ L_n & 1 \end{bmatrix}, \quad B_{0n} = \begin{bmatrix} \frac{L_n^3}{6EI} & \frac{L_n^2}{2EI} \\ -\frac{L_n^2}{2EI} & -\frac{L_n}{EI} \end{bmatrix}, \quad L_n = \sum_{i=1}^{n} l_i \tag{10.21}$$

where $E$ is the elastic modulus of beam and $I$ is the moment of inertia.

From Fig. 10.11 we obtain the boundary values of the bracket that should satisfy

$$\begin{bmatrix} I_0 & 0 & -A_{0n} & B_{0n} \\ 0 & I_0 & 0 & C_{0n} \end{bmatrix} \begin{bmatrix} \Delta_0 \\ R_0 \\ \Delta_2 \\ R_2 \end{bmatrix} = \sum_{i=0}^{n} \begin{bmatrix} B_{0i} \\ C_{0i} \end{bmatrix} P_i \tag{10.22}$$

where $I_0$ is a unit matrix.

Then we can utilize the boundary constraint Eq. (10.22) to analyze a general beam with both ends fixed under the action of the bearing force $P$ shown in Fig. 10.12. The length of $AC$, $AB$, and $BC$ are denoted as $a$, $l$, and $b$.

Using the boundary constraint Eq. (10.22), we gain that

$$\begin{bmatrix} I_0 & 0 & -A_{02} & B_{02} \\ 0 & I_0 & 0 & C_{02} \end{bmatrix} \begin{bmatrix} \Delta_0 \\ R_0 \\ \Delta_2 \\ R_2 \end{bmatrix} = \begin{bmatrix} B_{00} \\ C_{00} \end{bmatrix} P_0 + \begin{bmatrix} B_{01} \\ C_{01} \end{bmatrix} P_1 + \begin{bmatrix} B_{02} \\ C_{02} \end{bmatrix} P_2 \tag{10.23}$$

From Eq. (10.20), we obtain

$$\begin{cases} R_0 = [Y_0 \; M_0]^{\mathrm{T}} \\ R_2 = [Y_2 \; M_2]^{\mathrm{T}} \\ \Delta_0 = \Delta_2 = 0 \\ P_0 = P_2 = [0 \; 0]^{\mathrm{T}} \\ P_1 = [P \; 0]^{\mathrm{T}} \end{cases} \tag{10.24}$$

Therefore, there is

$$B_{02} R_2 = B_{01} P_1 \tag{10.25}$$

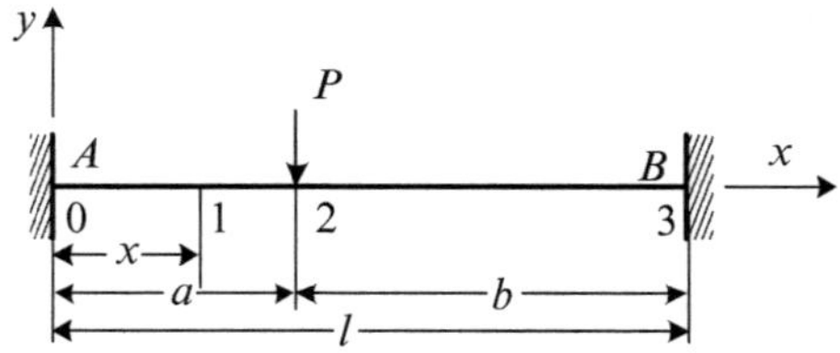


**Fig. 10.13** Arbitrary point between point 0 and point 2

Based on Eq. (10.21), there is

$$B_{01} = \begin{bmatrix} \frac{a^3}{6EI} & \frac{a^2}{2EI} \\ -\frac{a^2}{2EI} & -\frac{a}{EI} \end{bmatrix}, \quad B_{02} = \begin{bmatrix} \frac{l^3}{6EI} & \frac{l^2}{2EI} \\ -\frac{l^2}{2EI} & -\frac{l}{EI} \end{bmatrix} \tag{10.26}$$

Substituting the parameters into Eq. (10.25) yields

$$\begin{bmatrix} \frac{l^3}{6EI} & \frac{l^2}{2EI} \\ -\frac{l^2}{2EI} & -\frac{l}{EI} \end{bmatrix} \begin{bmatrix} Y_2 \\ M_2 \end{bmatrix} = \begin{bmatrix} \frac{a^3}{6EI} & \frac{a^2}{2EI} \\ -\frac{a^2}{2EI} & -\frac{a}{EI} \end{bmatrix} \begin{bmatrix} P \\ 0 \end{bmatrix} \tag{10.27}$$

Therefore, we obtain

$$\begin{bmatrix} Y_2 \\ M_2 \end{bmatrix} = \begin{bmatrix} \frac{Pa^2}{l^2}\left(1 + \frac{2b}{l}\right) \\ -\frac{Pa^2 b}{l^2} \end{bmatrix} \tag{10.28}$$

For force balance requirement, there is

$$\begin{bmatrix} Y_0 \\ M_0 \end{bmatrix} = \begin{bmatrix} \frac{Pb^2}{l^2}\left(1 + \frac{2a}{l}\right) \\ \frac{Pab^2}{l^2} \end{bmatrix} \tag{10.29}$$

Next, we can calculate the deflection of arbitrary point on the beam shown in Fig. 10.13. The deflections of point 0 and point 3 are both zeros. The force of point 0 is zero and the force of point 2 is $P$. Supposing that the distance between point 0 and point 1 is $x$, there is

$$\Delta_0 = 0, \Delta_3 = 0, \quad P_1 = [0\ 0]^{\mathrm{T}}, \quad P_2 = [P\ 0]^{\mathrm{T}} \tag{10.30}$$

Based on the displacement-method equation, there is

$$\begin{cases} \Delta_0 = A_{0,1}\Delta_1 + B_{0,1}F_{1,0} \\ F_{0,1} = -C_{0,1}F_{1,0} \\ F_{0,1} = P_0 - R_0 \end{cases} \tag{10.31}$$

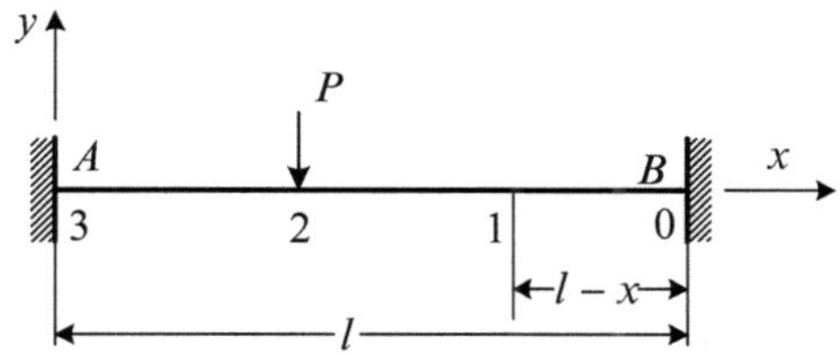


**Fig. 10.14** An arbitrary point between point 2 and point 0

where

$$A_{0,1} = \begin{bmatrix} 1 & -x \\ 0 & 1 \end{bmatrix}, \quad B_{0,1} = \begin{bmatrix} \frac{x^3}{6EI} & \frac{x^2}{2EI} \\ -\frac{x^2}{2EI} & -\frac{x}{EI} \end{bmatrix}, \quad C_{0,1} = \begin{bmatrix} 1 & 0 \\ x & 1 \end{bmatrix},$$

$$R_0 = \begin{bmatrix} Y_0 \\ M_0 \end{bmatrix}, \quad P_0 = \begin{bmatrix} 0 \\ 0 \end{bmatrix} \tag{10.32}$$

Substituting Eq. (10.32) into Eq. (10.31) yields

$$\begin{cases} \begin{bmatrix} 1 & -x \\ 0 & 1 \end{bmatrix} \begin{bmatrix} v_1 \\ \theta_1 \end{bmatrix} + \begin{bmatrix} \frac{x^3}{6EI} & \frac{x^2}{2EI} \\ -\frac{x^2}{2EI} & -\frac{x}{EI} \end{bmatrix} \begin{bmatrix} Y_{1,0} \\ M_{1,0} \end{bmatrix} = 0 \\ -\begin{bmatrix} 1 & 0 \\ x & 1 \end{bmatrix} \begin{bmatrix} Y_{1,0} \\ M_{1,0} \end{bmatrix} = -\begin{bmatrix} Y_0 \\ M_0 \end{bmatrix} \end{cases} \tag{10.33}$$

Simplifying Eq. (10.33) yields

$$\begin{cases} v_1 = \frac{Pb^2x^2}{6EIl^2}\left[3a - \left(1 + \frac{2a}{l}\right)x\right] \\ \theta_1 = \frac{Pb^2x}{2EIl^2}\left[2a - \left(1 + \frac{2a}{l}\right)x\right] \end{cases} \quad (0 \le x \le a) \tag{10.34}$$

Using the same method, we can get the similar conclusion under the conditions shown in Fig. 10.14.

$$\begin{cases} v_1 = \frac{Pa^2(l-x)^2}{6EIl^2}\left[3b - \left(1 + \frac{2b}{l}\right)(l-x)\right] \\ \theta_1 = -\frac{Pa^2(l-x)}{2EIl^2}\left[2b - \left(1 + \frac{2b}{l}\right)(l-x)\right] \end{cases} \quad (a \le x \le l) \tag{10.35}$$

For the situation shown in Fig. 10.10, latitudinal beam 1 and latitudinal beam 2 have the same stress condition because of symmetrical characteristic. Therefore, we

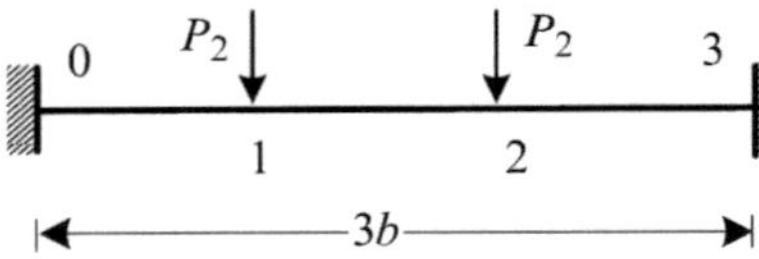


**Fig. 10.15** Stress condition of the longitudinal beam

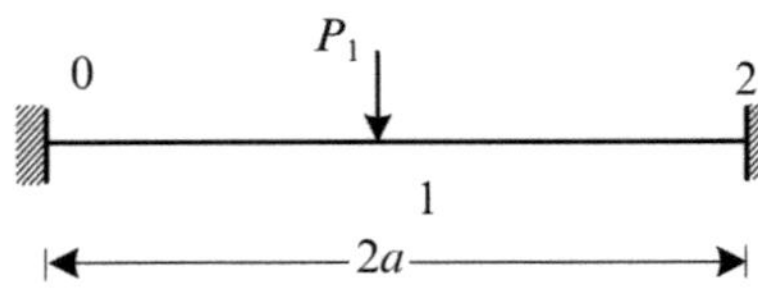


**Fig. 10.16** Stress condition of the latitudinal beam

can just analyze the stress conditions of one latitudinal beam and one longitudinal beam. Supposing the force of latitudinal beam 1 is $P_1$ and the force of longitudinal beam is $P_2$ shown in Figs. 10.15 and 10.16, there should be

$$P_1 + P_2 = P \tag{10.36}$$

Firstly, it comes to the situation of the longitudinal beam in Fig. 10.15. Considering force $P_2$ of point 1 without the influence caused by force of point 2 in Fig. 10.15, we can obtain the vertical displacement of point 1 based on Eq. (10.34)

$$v_{11} = \frac{Pb^2x^2}{6EIl^2}\left[3a - \left(1 + \frac{2a}{l}\right)x\right] = \frac{P_2(2a)^2a^2}{6EI(3a)^2}\left[3a - \left(1 + \frac{2a}{3a}\right)a\right] = \frac{8P_2a^3}{81EI} \tag{10.37}$$

Then let us consider the vertical displacement of point 1 caused by force point 2 based on Eq. (10.35)

$$\begin{aligned} v_{21} &= \frac{Pa^2(l-x)^2}{6EIl^2}\left[3b - \left(1 + \frac{2b}{l}\right)(l-x)\right] \\ &= \frac{Pa^2(l-b)^2}{6EIl^2}\left[3b - \left(1 + \frac{2b}{l}\right)(l-b)\right] \\ &= \frac{11P_2a^3}{162EI} \end{aligned} \tag{10.38}$$

where $l$ equals $3a$ and $b$ equals $2a$.

Therefore, the whole vertical displacement of point 1 is

$$v_1 = v_{11} + v_{21} = \frac{8P_2a^3}{81EI} + \frac{11P_2a^3}{162EI} = \frac{P_2a^3}{6EI} \tag{10.39}$$

Secondly, let us consider the situation of latitudinal beam in Fig. 10.16. Based on Eq. (10.34), the vertical displacement of point 1 is

$$v = \frac{Pb^2x^2}{6EIl^2}\left[3a - \left(1 + \frac{2a}{l}\right)x\right] = \frac{P_1a^2a^2}{6EI(2a)^2}\left[3a - \left(1 + \frac{2a}{2a}\right)a\right] = \frac{P_1a^3}{24EI} \tag{10.40}$$

Based on the principle of deflections consistency, we immediately obtain

$$\frac{P_1{a_1}^3}{24EI_1} = \frac{P_2{a_2}^3}{6EI_2} \tag{10.41}$$

Substituting $a_1 = 1.58$ m, $a_2 = 0.808$ m, $I_1 = I_2$, $E = 210$ GPa into Eq. (10.40) yields

$$P_1 = 0.535P_2 \tag{10.42}$$

Associating Eqs. (10.36) and (10.42) presents

$$P_1 = 0.349P,\ P_2 = 0.651P \tag{10.43}$$

For the longitudinal beam, there is

$$\begin{cases} \Delta_0 - \begin{bmatrix} 1 & -3b \\ 0 & 1 \end{bmatrix}\Delta_3 + \begin{bmatrix} \dfrac{27b^3}{6EI_b} & \dfrac{9b^2}{2EI_b} \\ -\dfrac{9b^2}{2EI_b} & -\dfrac{3b}{EI_b} \end{bmatrix} R_3 = \begin{bmatrix} \dfrac{b^3}{6EI_b} & \dfrac{b^2}{2EI_b} \\ -\dfrac{b^2}{2EI_b} & -\dfrac{b}{EI_b} \end{bmatrix}\begin{bmatrix} P_2 \\ 0 \end{bmatrix} \\ \quad + \begin{bmatrix} \dfrac{8b^3}{2EI_b} & \dfrac{4b^2}{2EI_b} \\ -\dfrac{4b^2}{2EI_b} & -\dfrac{2b}{EI_b} \end{bmatrix}\begin{bmatrix} P_2 \\ 0 \end{bmatrix} \\ R_0 + \begin{bmatrix} 1 & 0 \\ 3b & 1 \end{bmatrix} R_3 = \begin{bmatrix} 1 & 0 \\ b & 1 \end{bmatrix}\begin{bmatrix} P_2 \\ 0 \end{bmatrix} + \begin{bmatrix} 1 & 0 \\ 2b & 1 \end{bmatrix}\begin{bmatrix} P_2 \\ 0 \end{bmatrix} \end{cases} \tag{10.44}$$

Substituting the boundary condition values into Eq. (10.44), we obtain

$$R_0 = \begin{bmatrix} Y_0 \\ M_0 \end{bmatrix} = \begin{bmatrix} P_2 \\ \dfrac{2}{3}bP_2 \end{bmatrix},\quad R_3 = \begin{bmatrix} Y_3 \\ M_3 \end{bmatrix} = \begin{bmatrix} P_2 \\ -\dfrac{2}{3}bP_2 \end{bmatrix} \tag{10.45}$$

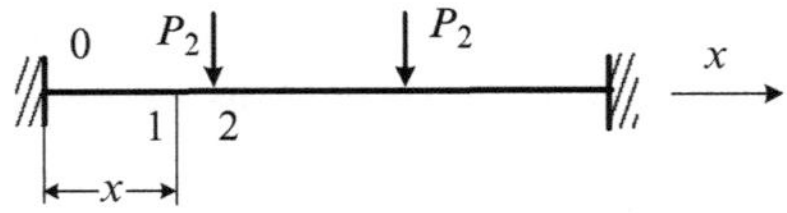


**Fig. 10.17** The longitudinal beam

where $Y_0$ and $Y_3$ denote the forces and $M_0$ and $M_3$ denote bending moments. The positive direction of the bending moment is clockwise, so the minus sign in Eq. (10.45) is that the direction of bending moment is the anticlockwise.

There is

$$\begin{cases} \Delta_0 = A_{01}\Delta_1 + B_{01}F_{10} \\ F_{01} = -C_{01}F_{10} \\ F_{01} = P_0 - R_0 \end{cases} \tag{10.46}$$

Then, it is easy to obtain

$$\begin{cases} v_1 = \dfrac{b^3}{6EI_b}P_2 \\ \theta_1 = \dfrac{b^2}{6EI_b}P_2 \end{cases} \tag{10.47}$$

We can calculate the displacement of other points with the same method

$$\Delta_0 = \begin{bmatrix} 0 \\ 0 \end{bmatrix}, \quad \Delta_1 = \begin{bmatrix} \dfrac{b^3}{6EI_b}P_2 \\ \dfrac{b^2}{6EI_b}P_2 \end{bmatrix}, \quad \Delta_2 = \begin{bmatrix} \dfrac{b^3}{6EI_b}P_2 \\ -\dfrac{b^2}{6EI_b}P_2 \end{bmatrix}, \quad \Delta_3 = \begin{bmatrix} 0 \\ 0 \end{bmatrix} \tag{10.48}$$

Next, rename the point of the beam shown in Fig. 10.17.

Based on the analysis above, we get

$$\begin{cases} \Delta_0 = A_{01}\Delta_1 + B_{01}F_{10} \\ F_{01} = -C_{01}F_{10} \\ F_{01} = P_0 - R_0 \end{cases} \tag{10.49}$$

and

$$A_{01} = \begin{bmatrix} 1 & -x \\ 0 & 1 \end{bmatrix}, \quad C_{01} = \begin{bmatrix} 1 & 0 \\ x & 1 \end{bmatrix}, \quad B_{01} = \begin{bmatrix} \dfrac{x^3}{6EI_b} & \dfrac{x^2}{2EI_b} \\ -\dfrac{x^2}{2EI_b} & -\dfrac{x}{EI_b} \end{bmatrix},$$

$$\Delta_0 = \begin{bmatrix} 0 \\ 0 \end{bmatrix}, \quad \Delta_2 = \begin{bmatrix} \dfrac{b^3}{6EI_b} P_2 \\ \dfrac{b^2}{6EI_b} P_2 \end{bmatrix} \tag{10.50}$$

Substituting Eq. (10.50) into Eq. (10.49) yields

$$\begin{cases} v = \dfrac{2bx^2 - x^3}{6EI_b} P_2 \\ \theta = \dfrac{4bx - 3x^2}{6EI_b} P_2 \end{cases} \quad (0 < x \le b) \tag{10.51}$$

Using the same method, we obtain

$$\begin{cases} v = \dfrac{3b^2x - bx^2 - b^3}{6EI_b} P_2 \\ \theta = \dfrac{3b^2 - 2bx}{6EI_b} P_2 \end{cases} \quad (b < x \le 2b)$$

$$\begin{cases} v = \dfrac{(x-3b)^2(x-b)}{6EI_b} P_2 \\ \theta = \dfrac{3x^2 - 14bx + 15b^2}{6EI_b} P_2 \end{cases} \quad (2b < x \le 3b) \tag{10.52}$$

Next, the situation of latitudinal beam is shown below

$$\begin{cases} v = \dfrac{3ax^2 - 2x^3}{24EI_a} P_1 \\ \theta = \dfrac{ax - x^2}{4EI_a} P_1 \end{cases} \quad (0 \le x \le a)$$

$$\begin{cases} v = \dfrac{(2x-a)(2a-x)^2}{24EI_a} P_1 \\ \theta = \dfrac{(x-a)(x-2a)}{4EI_a} P_1 \end{cases} \quad (a < x \le 2a) \tag{10.53}$$

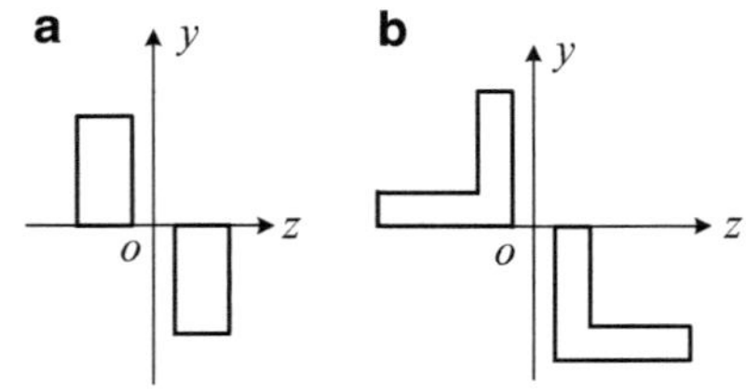


**Fig. 10.18** Two alternative designs of section. (**a**) Rectangle section, and (**b**) L section

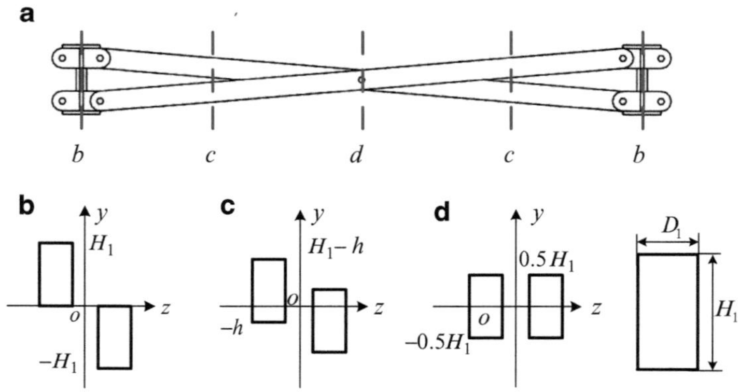


**Fig. 10.19** Rectangular cross section. (**a**) An SLE, (**b**) Equivalent section of the frame at *b*, (**c**) Equivalent section of the frame at *c*, and (**d**) Equivalent section of the frame at *d*

## *10.4.2 Effect of Links with Different Cross Sections*

In this section, we will discuss the sectional geometry of the linkages. Considering the requirement of the structure, there are two alternative designs of section of the linkages. One is the rectangular cross section; the other is a section of angle steel shown in Fig. 10.18.

### 10.4.2.1 Rectangular Cross Section

When the solar photovoltaic mounting bracket is in the fully extended position, the upper and lower mounting sets are connected by bolts, clamping the solar battery between them. In this regard, we can consider the whole bracket as a rigid structure. The equivalent bending stiffness factor is *EI*. The rectangular cross section is shown in Fig. 10.19.

The structure shown in Fig. 10.19a is considered as a rigid beam which has different cross-section shape in different position. Figure 10.19b–d shows the cross-section shapes corresponding to the different positions which are noted by the dashed lines in Fig. 10.19a. We next take Fig. 10.19c, for example, to analyze its equivalent stiffness.

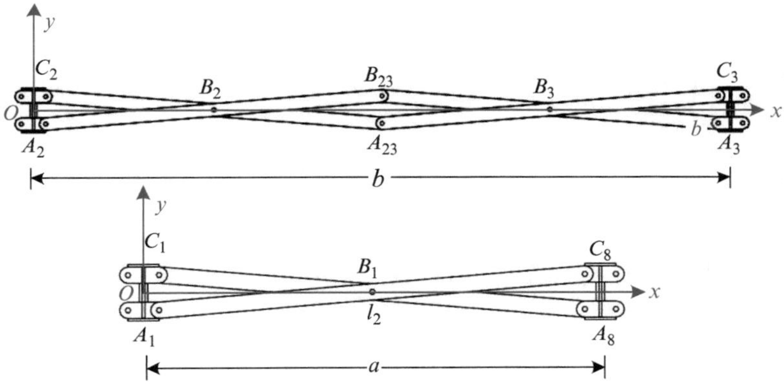


**Fig. 10.20** A frame with the coordinate system. (**a**) Coordinate system in the long side, and (**b**) Coordinate system in the short side

The moment of inertia of the $z$-axis is

$$I_z = \int y^2\,\mathrm{d}A = \int y^2 z\,\mathrm{d}y = 2D_1 \int_{-h}^{H_1-h} y^2\,\mathrm{d}y = \frac{2D_1}{3}\left[(H_1-h)^3 + h^3\right] \tag{10.54}$$

where $D_1$ and $H_1$ are the dimensions of rectangular and $-h$ is the position of the lower width. The size of $h$ is related to the length of scissorlike element. Establish a coordinate system by setting the origin in the middle of joint pairs, $A_2C_2$ and $A_1C_1$; the $x$-axis is along the rightward horizontal direction and the y-axis is upward and passes through the origin as shown in Fig. 10.20. We can immediately obtain the equation.

According to the geometry, we can immediately obtain the relation for the structure in (a) of Fig. 10.20.

$$h(x) = \begin{cases} \dfrac{0.5H_1}{0.5pb}x = \dfrac{H_1}{pb}x & (0 < x \le 0.5pb) \\ 0.5H_1 - \dfrac{0.5H_1}{0.5(1-p)b}(x - 0.5pb) = \dfrac{0.5H_1}{1-p} - \dfrac{H_1}{(1-p)b}x & (0.5pb < x \le 0.5b) \\ \dfrac{0.5H_1}{0.5(1-p)b}(x - 0.5b) = \dfrac{H_1}{(1-p)b}(x - 0.5b) & (0.5b < x \le a - 0.5pb) \\ 0.5H_1 - \dfrac{0.5H_1}{0.5pb}\left[x - 0.5b - 0.5(1-p)b\right] = \dfrac{H_1}{p} - \dfrac{H_1}{pb}x & (b - 0.5pb < x \le b) \end{cases} \tag{10.55}$$

Substituting Eq. (10.55) into Eq. (10.54) yields

$$I_1(x) = \begin{cases} \frac{2D_1}{3}\left[\left(H_1 - \frac{H_1}{pb}x\right)^3 + \left(\frac{H_1}{pb}x\right)^3\right] & (0 < x \leq 0.5pb) \\ \frac{2D_1}{3}\left[\left(H_1 - \frac{H_1}{2-2p} + \frac{H_1}{b-pb}x\right)^3 + \left(\frac{H_1}{2-2p} - \frac{H_1}{b-pb}x\right)^3\right] & (0.5pb < x \leq 0.5b) \\ \frac{2D_1}{3}\left\{\left[H_1 - \frac{H_1}{(1-p)b}\left(x - \frac{b}{2}\right)\right]^3 + \left[\frac{H_1}{(1-p)b}\left(x - \frac{b}{2}\right)\right]^3\right\} & (0.5b < x \leq a - 0.5pb) \\ \frac{2D_1}{3}\left[\left(H_1 - \frac{H_1}{p} + \frac{H_1}{pb}x\right)^3 + \left(\frac{H_1}{p} - \frac{H_1}{pb}x\right)^3\right] & (b - 0.5pb < x \leq b) \end{cases} \tag{10.56}$$

Using the same method, we get the similar conclusion of the structure in (b) of Fig. 10.20.

$$h(x) = \begin{cases} \frac{0.5H_1}{0.5a}x = \frac{H_1}{a}x & (0 \leq x \leq 0.5a) \\ 0.5H_1 - \frac{0.5H_1}{0.5a}(x - 0.5a) = H_1 - \frac{H_1}{a}x & (0.5a \leq x \leq a) \end{cases} \tag{10.57}$$

$$I_2(x) = \begin{cases} \frac{2D_1}{3}\left[\left(H_1 - \frac{H_1}{a}x\right)^3 + \left(\frac{H_1}{a}x\right)^3\right] & (0 \leq x \leq 0.5a) \\ \frac{2D_1}{3}\left[\left(\frac{H_1}{a}x\right)^3 + \left(H_1 - \frac{H_1}{a}x\right)^3\right] & (0.5a \leq x \leq a) \end{cases} \tag{10.58}$$

#### 10.4.2.2 Section of Angle Steel

The moment of inertia of the section will be changed when choosing angle steel as the linkages. The shape is shown in Fig. 10.21.

The structure shown in subfigure (a) of Fig. 10.21 is considered as a rigid beam which has different cross-section shapes at different positions. Subfigures (b), (c), and (d) in Fig. 10.21 show the cross-section shapes corresponding to the different positions which are noted by the dashed lines in Fig. 10.21a. Take Fig. 10.21c for example, we next analyze its equivalent stiffness.

Based on the parallel axis theory, we can calculate the moment of inertia about the $z$-axis.

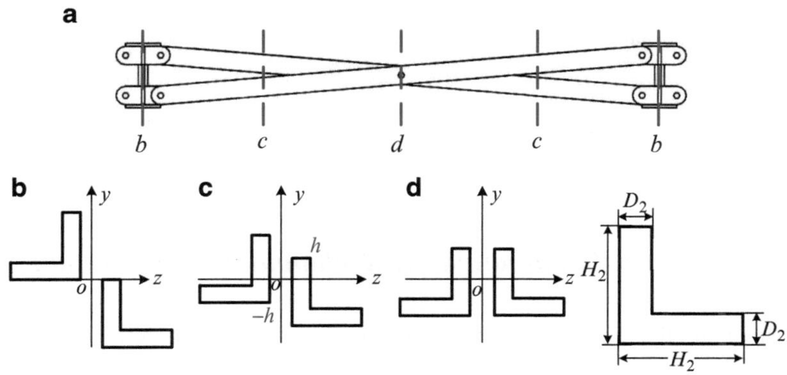


**Fig. 10.21** Section of angle steel. (**a**) An SLE, (**b**) Equivalent section of the frame at $b$, (**c**) Equivalent section of the frame at $c$, and (**d**) Equivalent section of the frame at $d$

$$\begin{aligned} I_z =& \frac{1}{12} D_2 H_2{}^3 + \frac{1}{12}(H_2 - D_2) D_2{}^3 + D_2 H_2 (0.5 H_2 - h)^2 \\ &+ D_2 (H_2 - D_2)(H_2 - 0.5 D_2 - h)^2 + \frac{1}{12} D_2 H_2{}^3 + \frac{1}{12}(H_2 - D_2) D_2{}^3 \\ &+ D_2 H_2 (0.5 H_2 - h)^2 + D_2 (H_2 - D_2)(0.5 D_2 - h)^2 \\ =& \frac{1}{6}\left(D_2 H_2{}^3 + H_2 D_2{}^3 - D_2{}^4\right) + 2 D_2 H_2 (0.5 H_2 - h)^2 \\ &+ D_2 (H_2 - D_2)\left(H_2{}^2 + 0.5 D_2{}^2 + 2h^2 - H_2 D_2 - 2h H_2\right) \end{aligned} \tag{10.59}$$

where $D_2$ and $H_2$ are the dimensions of angle steel.

Using the same analytic method, we can obtain the same relationships between the size of $h$ and length of scissorlike element as Eqs. (10.55) and (10.57) represent.

Substituting Eq. (10.55) or (10.57) into (10.59) yields the equivalent moment of inertia.

## 10.5 Numerical Simulations on Photovoltaic Mounting Bracket

Based on the analysis of Sect. 10.4, we execute computer simulations. We program at MATLAB and gain some numerical simulations. The primary structural parameters are discussed in Sect. 10.4. There are two different results by choosing different sections of linkages.

**Table 10.1** Primary parameters for rectangular cross section

| | |
|---|---|
| $a$ (mm) | 1,580 |
| $b$ (mm) | 808 |
| $E$ (GPa) | 210 |
| $I_1$ (mm$^4$) | 15,552 |
| $I_2$ (mm$^4$) | 15,552 |
| $P$ (N) | 916 |

## 10.5.1 Numerical Simulations of Rectangular Cross Section

Suppose that the original parameters of the rectangular cross section is 18 mm × 8 mm. Substitute the parameters into Eqs. (10.56) and (10.58). Based on the mathematical integral, we calculate the equivalent moment of inertia of every side by MATLAB.

$$\begin{cases} \bar{I}_1 = \frac{1}{b}\int I_1(x)\mathrm{d}x = 15,552\,\mathrm{mm}^4 \\ \bar{I}_2 = \frac{1}{a}\int I_2(x)\mathrm{d}x = 15,552\,\mathrm{mm}^4 \end{cases} \tag{10.60}$$

where $\bar{I}_1$ is the equivalent moment of inertia of the long side of basal frame and $\bar{I}_2$ is the equivalent moment of inertia of the short side.

The primary mechanical parameters are listed in Table 10.1.

Substituting the parameters into Eqs. (10.51), (10.52), and (10.53), we get the curve of the stiffness of the solar photovoltaic mounting bracket with respect to deflection and intersection angle.

Figures 10.22 and 10.23 separately show that deflection and intersection angle of the longitudinal beam change with different positions of beam. The deflection curve is almost symmetrical on line $x = 1.212$ and the maximum deflection is 20.06 mm, which is located at point (1.2, −0.02006). Meanwhile, the intersection angle curve has symmetry, whose maximum intersection angle is 1.581° at points (0.545, 1.581) and (1.88, -1.581).

Likewise, Figs. 10.24 and 10.25 show the change of deflection and intersection angle of the latitudinal beam. The deflection curve is almost symmetrical about $x = 1.57$ and the maximum deflection is 16.09 mm, which is located at point (1.57, −0.01609). Meanwhile, the intersection angle curve has symmetry, whose maximum intersection angle is 0.875° at point (0.785, 0.875) and (2.37, −0.875).

## 10.5.2 Numerical Simulations of Section of the Angle Steel

Supposing the original parameters of section of angle steel is 20 mm×20 mm×4 mm. Substitute the parameters into Eqs. (10.55) and (10.57). Based on the mathematical integral, we can calculate out the equivalent moment of inertia by MATLAB according to Eq. (10.59).

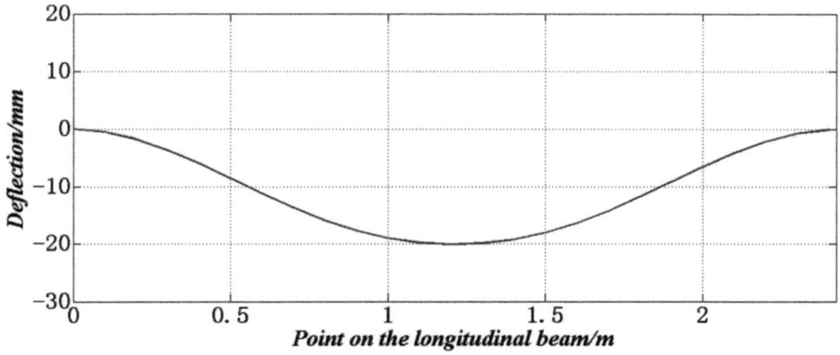


**Fig. 10.22** Deflection curve of the longitudinal beam

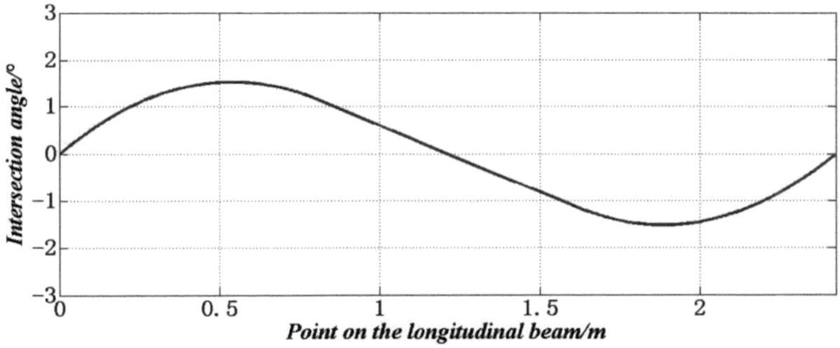


**Fig. 10.23** Intersection angle curve of the longitudinal beam

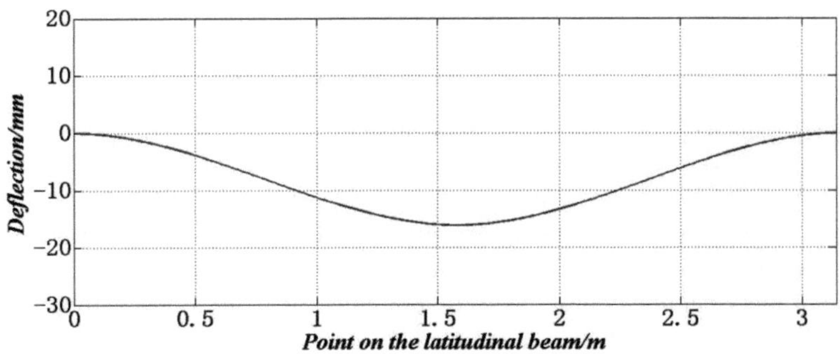


**Fig. 10.24** Deflection curve of the latitudinal beam

$$\begin{cases} \bar{I}_1 = \dfrac{1}{b}\displaystyle\int I_1(x)\mathrm{d}x = 23{,}296\,\mathrm{mm}^4 \\ \bar{I}_2 = \dfrac{1}{a}\displaystyle\int I_2(x)\mathrm{d}x = 23{,}296\,\mathrm{mm}^4 \end{cases} \tag{10.61}$$

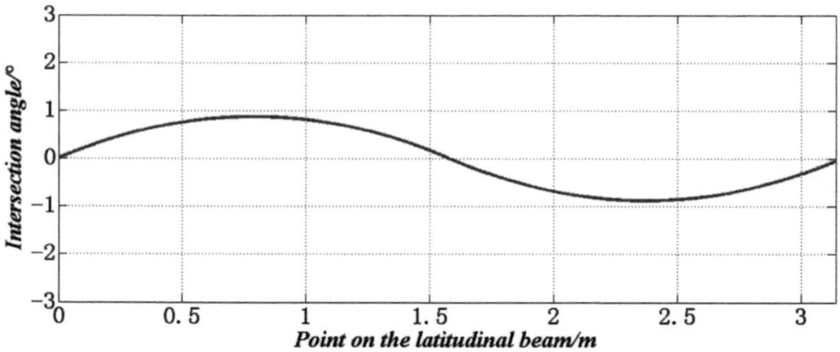


**Fig. 10.25** Intersection angle curve of the latitudinal beam

**Table 10.2** Primary parameters for the section of angle steel

| | |
|---|---|
| $a$ (mm) | 1,580 |
| $b$ (mm) | 808 |
| $E$ (GPa) | 210 |
| $I_1$ (mm$^4$) | 23,296 |
| $I_2$ (mm$^4$) | 23,296 |
| $P$ (N) | 916 |

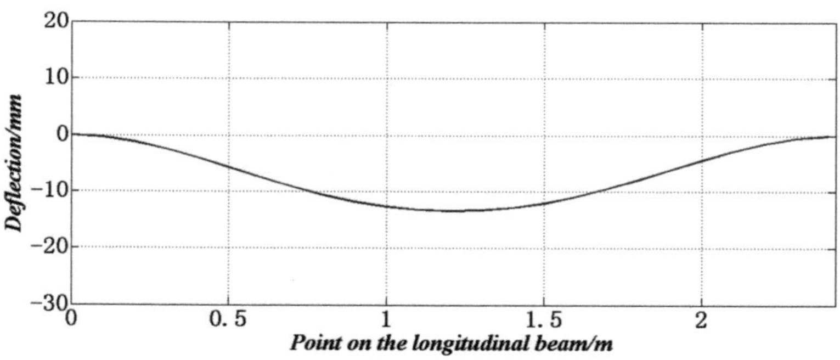


**Fig. 10.26** Deflection curve of the longitudinal side

where $\bar{I}_1$ is the equivalent moment of inertia of the long side of basal frame and $\bar{I}_2$ is the equivalent moment of inertia of the short side.

The primary parameters for the section of angle steel are listed in Table 10.2.

Substituting the parameters into Eqs. (10.51), (10.52), and (10.53), we get the curve of the stiffness of the solar photovoltaic mounting bracket with respect to deflection and intersection angle. Figures 10.26 and 10.27 separately show that deflection and intersection angle of the longitudinal beam change at different positions of beam. The deflection curve is almost symmetrical about $x = 1.212$ and the maximum deflection is 13.39 mm, which is located at point (1.2, −0.01339). Meanwhile, the intersection angle curve has symmetry, whose maximum intersection angle is 1.031° at points (0.545, 1.031) and (1.895, −1.013).

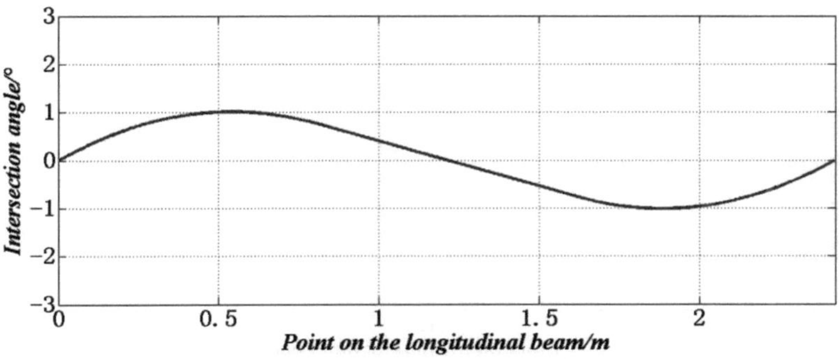


**Fig. 10.27** Intersection angle curve of the longitudinal side

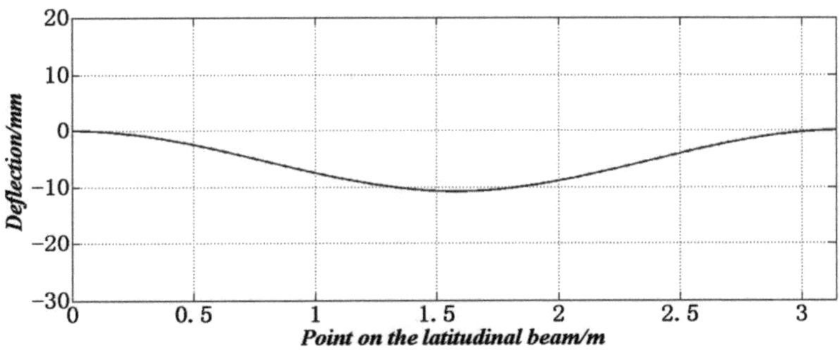


**Fig. 10.28** Deflection curve of the latitudinal side

Likewise, Figs. 10.28 and 10.29 show the change of deflection and intersection angle of the latitudinal beam. The deflection curve is almost symmetrical about $x = 1.6$ and the maximum deflection is 10.73 mm, which is located at point (1.6, −0.01073). Meanwhile, the intersection angle curve has symmetry, whose maximum intersection angle is 0.5841° at points (0.78, 0.5841) and (2.365, −0.5841).

Next, comparisons between the numerical simulations of rectangular cross section and that of the section of angle steel will be discussed. We can easily find the sectional area of each section:

$$\begin{cases} S_1 = 18 \times 8 = 144\,(\mathrm{mm}^2) \\ S_2 = 2 \times 16 \times 4 + 4 \times 4 = 144\,(\mathrm{mm}^2) \end{cases} \tag{10.62}$$

where $S_1$ and $S_2$ denote the sectional area of the rectangular cross section and that of the angle steel, both areas of which are equal to 144 $\mathrm{mm}^2$. In other words, the amounts of material are identical when the bracket is designed for the same solar panel. Therefore, the weights of the whole structure are identical.

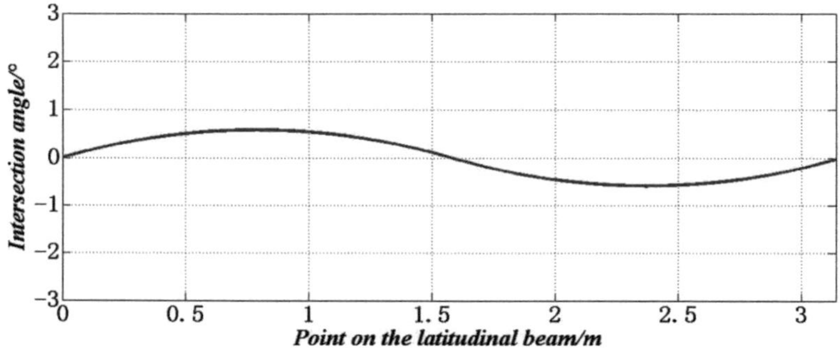


**Fig. 10.29** Intersection slope curve of the latitudinal side

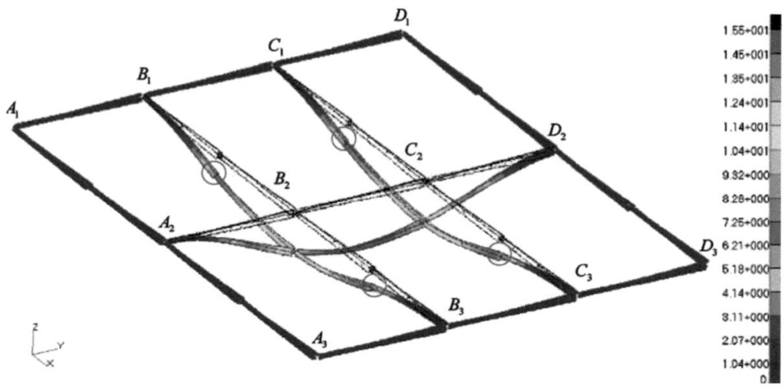


**Fig. 10.30** Simulation via ANSYS

From the viewpoint of theoretical calculation of the equivalent moment of inertia, angle steel has larger equivalent stiffness than that of the steel with rectangular section. The consequence of numerical simulations illustrates that the deformation of angle steel is less than that of steel with rectangular section, wherever the deflection or intersection angle of the longitudinal side (the latitudinal side) is. Here, we choose the second model shown in Fig. 10.21 as the example, which has rich resources and cheaper price compared with the rectangular section. To sum up, the section of angle steel is a better choice for the solar photovoltaic mounting equipment.

### *10.5.3 Simulation Analysis*

Simulations of the real load on the solar photovoltaic mounting equipment by ANSYS software and the consequences are shown in Fig. 10.30.

As the surrounding points $A_1, B_1, C_1, D_1, A_2, D_2, A_3, B_3, C_3$, and $D_3$ are fastened by connecting elements, the deformations of them equal zero. The deformation is symmetrical because the whole bracket is a symmetrical structure, which is in accordance with the actual condition. From Fig. 10.30, we can obtain that the maximum deflection is about 15.5 mm. The deflections of point $B_2$ and $C_2$ are both equal to 11.2 mm. From the statics analysis of Sect. 10.3, we can figure out the theoretical deflections of point $B_2$ (or point $C_2$) and the maximum value is

$$\begin{aligned} v_x(0.808) &= \frac{3b^2x - bx^2 - b^3}{6EI_\text{b}} P_2 \\ &= \frac{3 \times 0.808^3 - 0.808^3 - 0.808^3}{6 \times 210 \times 10^9 \times 23{,}296 \times 10^{-12}} \times 0.651 \times 916 = 0.0107\,\text{m} \end{aligned} \tag{10.63}$$

Compared with the simulation value, the relative error is

$$\delta = \frac{|10.7 - 11.2|}{11.2} \times 100\,\% = 4.46\,\% \tag{10.64}$$

where δ denotes the relative error. Likewise, the maximum deflection equals

$$\begin{aligned} v_\text{max} = v_x(1.424) &= \frac{3b^2x - bx^2 - b^3}{6EI_\text{b}} P_2 \\ &= \frac{3\times0.808^2\times1.424 - 0.808\times1.424^2 - 0.808^3}{6\times210\times10^9\times23,296 \times 10^{-12}} \times 0.651 \times 916\,\text{m} = 0.014\,\text{m} \end{aligned} \tag{10.65}$$

Comparing it with the simulation value, we can find that the relative error is

$$\delta = \frac{|14.0 - 15.5|}{15.5} \times 100\,\% = 9.7\,\% \tag{10.66}$$

The relative errors are within a reasonable range for engineering requirement. The theoretical results can be used to pre-adjust the deployable structure for real applications.

## 10.6 Conclusions

This chapter focuses on the solar photovoltaic mounting equipment which should be used as the support of solar panels. It investigates the geometry, statics of the structure, and simulation experiments to verify the feasibility. The advantage of the flexibility to fold and unfold can be utilized for long-distance transportation with convenient assembly and disassembly. The stiffness of the foldable bracket is especially high when it is deployed completely. Meanwhile, it also allows installing the arc-shaped or spherical-shaped solar panels by changing the position of the rod hinge hole and the assembly of scissor units.

## References

1. Zhou LY (2001) World energy crisis and China's energy security. Energy of China, No. 1:12–13
2. Qin W, Li DQ (2003) Solar energy-energy star in 21th century. New Economy Weekly, 01:44–46
3. Wang S-C (2009) Four key problems of photovoltaic power generation. China Pow Enterp Manag 7:24–25
4. Gantes CJ, Konitopoulou E (2004) Geometric design of arbitrarily curved bi-stable deployable arches with discrete joint size. Int J Solids Struct 41:5517–5540
5. Escrig F, Valcarcel P (1996) Geometry of expandable space structures. Int J Space Struct 11(1):257–274
6. Kaveh A, Davaran A (1996) Analysis of pantograph foldable structures. Comput Struct 59(1):131–140
7. Rosenfeld Y, Logcher RD (1988) New concepts for deployable collapsible structures. Int J Space Struct 3(1):20–32
8. Escrig F, Valcarcel JP, Sanchez J (1996) Deployable cover on a swimming pool in seville. J Int Assoc Shell Spat Struct 37(1):39–70
9. Chen W-J, Luo L-Y et al (2001) A study on space masts based on octahedral truss family. Int J Space Struct 16(1):75–82
10. Langbecker T (1999) Kinematic analysis of deployable scissor structures. Int J Space Struct 14(1):1–15

MIX
Papier aus verantwortungsvollen Quellen
Paper from responsible sources
FSC® C105338

If you have any concerns about our products,
you can contact us on
**ProductSafety@springernature.com**

In case Publisher is established outside the EU,
the EU authorized representative is:
**Springer Nature Customer Service Center GmbH**
**Europaplatz 3, 69115 Heidelberg, Germany**

Printed by Libri Plureos GmbH
in Hamburg, Germany